Intelligent Optimization

Changhe Li • Shoufei Han • Sanyou Zeng •
Shengxiang Yang

Intelligent Optimization

Principles, Algorithms and Applications

Changhe Li
School of Artificial Intelligence
Anhui University of Science
and Technology
Hefei, Anhui, China

Shoufei Han (iD)
School of Artificial Intelligence
Anhui University of Sciences
& Technology
Hefei, Anhui, China

Sanyou Zeng
School of Mechanical Engineering
and Electronic Information
China University of Geosciences
Wuhan, Hubei, China

Shengxiang Yang (iD)
School of Computer Science
and Informatics
De Montfort University
Leicester, UK

ISBN 978-981-97-3285-2   ISBN 978-981-97-3286-9   (eBook)
https://doi.org/10.1007/978-981-97-3286-9

Jointly published with China University of Geosciences Press, Hubei Province, P.R. China
The print edition is not for sale in Mainland China. Customers from Mainland China please order the
print book from: China University of Geosciences Press.

This Springer imprint is published by the registered company Springer Nature Singapore Pte Ltd.
The registered company address is: 152 Beach Road, #21-01/04 Gateway East, Singapore 189721,
Singapore

If disposing of this product, please recycle the paper.

To honor the memory of Lishan Kang for his significant contribution to parallel computing and evolutionary computation

Preface

Optimization problems widely exist in our world, manifesting themselves in various domains, from the realms of science and engineering to our everyday routines. For example, determining the ideal combination of breakfast foods, finding the most efficient route to the office, organizing the optimal sequence for grocery shopping, or managing room temperature to maximize comfort while minimizing energy consumption, these scenarios all pose optimization questions that demand optimal solutions. However, obtaining these optimal solutions can often be challenging, and decision-making may not always come easily to us.

Generally speaking, optimization means the search for the best option among different options. Broadly, the intelligence of intelligent optimization refers to the usage of a priori and/or a posteriori knowledge about the problem, which is expressed by implicit/explicit rules/models, to guide the search. More specifically, it refers to the mechanism of probabilistic search based on these rules or models, which are introduced in this book.

This book offers teaching and learning materials of intelligent optimization regarding fundamental introduction, applications, and research topics for the following readers:

(1) Lecturers, graduate and undergraduate students. This book is devoted to providing materials for beginning courses. It gives a thorough description of intelligent optimization, including the fundamentals of optimization problems and common intelligent optimization algorithms, especially evolutionary computation algorithms.
(2) Users who wish to apply intelligent optimization methods to a particular problem. This book also introduces applications to variant types of optimization problems, such as the constrained optimization problem, the multi-objective optimization problem, the dynamic optimization problem, the large-scale optimization problem, the robust optimization problem, and the expensive optimization problem. Particularly, we end this book with four real-world applications, i.e., antenna design, vehicle routing, contamination source identification, and games.

(3) Researchers in the domain of intelligent optimization. For each topic of this book, we present the recent development and propose several challenging issues that should be addressed in this domain. Potential solutions to these challenging issues are also discussed.

The overall structure of this book contains four parts. Part I presents the fundamentals of optimization from Chaps. 1 to 3. Part II presents evolutionary computation algorithms from Chaps. 4 to 5. Part III presents optimization techniques from Chaps. 6 to 8. Part IV presents advanced topics and applications. The four parts are followed by the references with nearly 400 entries. We may have inevitably missed some and please feel free to send us an email in case a really important one is forgotten.

This book is distributed as part of the OFEC project, which is an open-source project that aims to develop a common platform for students, users, and researchers to quickly design and visualize their algorithms (see Appendix B). OFEC has collected a plenty of algorithms and problems and is constantly maintained and updated. The source code of the project and other materials of the book can be obtained by following the link https://gitlab.com/ofec-release.

Hefei, China   Changhe Li
May 2024

Acknowledgments

This book could not have been possible without the support of many people. The authors thank the Intelligent Optimization and Learning team for their hard work on the research. Furthermore, the authors thank Yong He, Min Wu, Chuanke Zhang, Michalis Mavrovouniotis, Trung Thanh Nguyen, Carlos A. Coello Coello, Ming Yang, Xin Yao, Kay Chen Tan, Boyang Qu, Jing Liang, Juan Zou, Qingfu Zhang, Yew-Soon Ong, Jinliang Ding, Ponnuthurai Nagaratnam Suganthan, Miqing Li, Juergen Branke, Xiaodong Li, Hui Wang, Aimin Zhou, Xinye Cai, Qunfeng Liu, Ke Li, Yong Liu, Bob McKay, and Yuping Chen for their sincere support.

I would like to thank my wife, Fengxia Wang, for her endless love and care for the family and children. I also would like to thank my parents and my parents-in-law for their support and encouragement.

The genesis of this book is deeply rooted in the desire to honor the memory of Lishan Kang. Lishan Kang made significant contributions to the fields of parallel computing and evolutionary computation. His enduring scientific spirit serves as a perpetual source of inspiration, motivating us to continually explore the frontiers of knowledge.

The publication of this book was supported by the National Natural Science Foundation of China under Grant 62076226, in part by the Hubei Provincial Natural Science Foundation of China under Grant 2023AFA049.

Acknowledgment

Contents

Acronyms

ABC	artificial bee colony
ACO	ant colony optimization
ACS	ant colony system
AI	artificial intelligence
ANN	artificial neural network
APGA	adaptive population size for genetic algorithm
AR	axial ratio
AST	abstract syntax tree
BOA	Bayesian optimization algorithm
BFS	breadth-first search
BWAS	best-worst ant system
CBCC	contribution-based cooperative co-evolution
CC	cooperative co-evolution
cGA	compact genetic algorithm
CEI	constrained expected improvement
CEP	classic evolutionary programming
CGP	Cartesian genetic programming
COMIT	combining optimizer with mutual information tree
CMA-ES	covariance matrix adaptation evolution strategy
CMODE	constrained multi-objective differential evolution
CMOP	constrained multi-objective optimization problem
COP	constrained optimization problem
CVRP	capacitated vehicle routing problem
DCBG	dynamic composition benchmark generator
DCMOEA	dynamic constraint multi-objective genetic algorithm framework
DCMOP	dynamic constrained multi-objective problem
DE	differential evolution
DFS	depth-first search
DI	dependency identification
DOP	dynamic optimization problem
DPDE	co-evolutionary dual-population differential evolution

DRPBG	dynamic rotation peak benchmark generator
DTLZ	Deb-Thiele-Laumanns-Zitzler benchmark
EA	evolutionary algorithm
EAS	elitism ant system
EC	evolutionary computation
ECGA	extended compact genetic algorithm
ECHT	ensemble of constraint handling techniques
EDA	estimation of distribution algorithm
EGNA	estimation of Gaussian networks algorithm
EI	expected improvement
EMNA	estimation of multivariate normal algorithm
EOP	expensive optimization problems
EP	evolutionary programming
ES	evolution strategy
FEP	faster evolutionary programming
FER-PSO	fitness Euclidean-distance ratio based particle swarm optimization
FIFO	first-in-first-out
FSM	finite-state-machine
GA	genetic algorithm
GD	generational distance
GDBG	generalized dynamic benchmark generator
GEP	gene expression programming
GL	genetic learning
GLT	Gu-Liu-Tan benchmark
GOP	global optimization problem
GP	Gaussian processes
GP	genetic programming
GPM	genotype-phenotype mapping
HV	hypervolume
IBEA	indicator-based evolutionary algorithm
IDEA	iterated density evolutionary algorithm
IFEP	improved faster evolutionary programming
IGD	inverted generational distance
IPSOLS	incremental particle swarm optimizer with local search
LCB	lower confidence bound
LGP	linear genetic programming
LIPS	locally informed particle swarm model
LSGO	large-scale global optimization
MaOP	many-objective optimization problem
MIMIC	mutual-information-maximization for input clustering
MLCC	multilevel cooperative co-evolution
MLP	multi-layer perception
MMAS	max-min ant system
MMOP	multimodal optimization problem
MOEA	multi-objective optimization evolutionary algorithm

MOEA/D	multi-objective optimization algorithm based on decomposition
MOP	multi-objective optimization problem
MOVRPRTC	multi-objective vehicle routing problem with real-time traffic condition
MPB	moving peak benchmark
MS	maximum spread
MSE	mean square error
NAS	neural architecture search
NEA	niching based evolutionary algorithm
NP	non-deterministic polynomial time
NSGA-II	non-dominated sorting genetic algorithm
PAES	Pareto archived evolution strategy
PBI	penalty-based boundary intersection
PBIL	population-based incremental learning
PESA	Pareto envelope-based selection algorithm
PF	Pareto optimal front
PoI	probability of improvement
PR	polynomial regression
PRAM	probabilistic rule-driven adaptive model
PS	Pareto optimal set
PSO	particle swarm optimization
RBF	radial basis function model
RL	reinforcement learning
RSM	response surface method
RTS	restricted tournament selection
SaDE	self-adaptive differential evolution
SA	simulated annealing
SAEA	surrogate-assisted evolutionary algorithm
SGA	standard genetic algorithm
SI	swarm intelligence
SLL	side lobe level
SLPSO	self-learning particle swarm optimization
SPEA	strength Pareto evolutionary algorithm
SPSO	standard particle swarm optimization
SVM	support vector machine
TCH	Tchebycheff
TSP	travelling salesman problem
UMDA	univariate marginal distribution algorithm
VNS	variable neighborhood search
VRP	vehicle routing problem
VRPDP	vehicle routing problem with simultaneous delivery and pickup
VRPTW	vehicle routing problem with time window
WDS	water distribution system
WFG	walking fish group
WS	weighted sum
ZDT	Zitzler-Deb-Thiele benchmark

Part I
Introduction and Fundamentals

This part introduces the basics of optimization, including the relationship between optimization and machine learning, categories of optimization problems and optimization algorithms, the fitness landscape, and computational complexity. Several typical canonical optimization algorithms, i.e., numerical optimization algorithms, state space search algorithms, and single-solution-based random search algorithms, are introduced.

Chapter 1
Introduction

Abstract This chapter briefly introduces the background of optimization. Firstly, the relationship between optimization and machine learning is discussed, and an example of machine learning task is given. Secondly, the mathematical formulation of an optimization problem is defined. Optimization problems are categorized into continuous optimization problems and discrete optimization problems. Finally, optimization algorithms are introduced and categorized into deterministic algorithms and probabilistic algorithms, where several terms regarding intelligent optimization are introduced.

1.1 Optimization and Machine Learning

Artificial intelligence (AI), which was coined in 1956, is a wide-ranging branch of technology that mimics human intelligence to execute tasks. AI is an interdisciplinary science with multiple approaches in mathematics, computer science, biology, neuroscience, psychology, sociology, linguistics, philosophy, and more. Recent advances in machine learning, especially deep learning, has been successfully applied in games, transportation, medical treatment, industry, etc.

Maybe you have seen applications of machine learning somewhere in your daily routine. Actually, it has been around for a long time in so many scenarios. For example, when you find what you want on the Internet, it helps the search engine judge which search results (advertisements at the same time) are suitable for you; you do not need to label all spammers in your mailbox by hand since it has helped you do most of the work (if you have never checked your spam box, then check it, and you will find something you have never noticed); your smartphone automatically identifies and ignores spam calling from auto-dialers; and if you go to YouTube or some other video-sharing platforms, you can find that they suggest something you are interested in and even automatically generates caption for some videos. In addition, whenever you use facial-recognition payment and speech recognition input, you are enjoying the benefits of machine learning.

© China University of Geosciences Press 2024
C. Li et al., *Intelligent Optimization*,
https://doi.org/10.1007/978-981-97-3286-9_1

Fig. 1.1 Chirps per minute as temperature changes

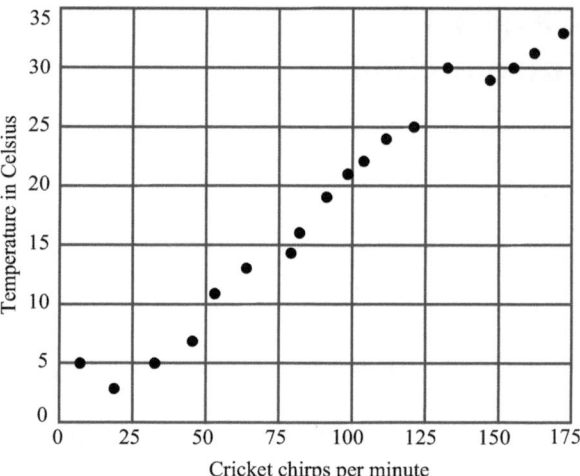

Generally speaking, the target of application of a machine learning method is to create a model to reflect the relationship between input features (weather for tomorrow, all the pixels of your photo) and desired output (the rate of students being late for the first class, who you are); then we can use this model to predict the result according to new input features.

However, have you ever thought about how machine learning methods can do such things? To answer this question, we have to get deeper into the inner world of machine learning. Do not worry, we just need to take maybe the simplest machine learning problem, a linear regression problem, for example.

Figure 1.1 shows the number of chirps of crickets per minute at different temperatures in Celsius;[1] one can say that the relationship is linear.[2] Of course, we can use an equation to represent this relationship

$$y' = wx + b \tag{1.1}$$

where y' means temperature in Celsius, the label we predict; x means the number of chirps of crickets per minute, the input feature; w is the slope of the line or the weight of the feature; and b is y-intercept or the bias.

This equation is the model we need to build. w and b are its model parameters, and now, our job is to find the right values for them. In supervised learning, we build our model by examining a set of examples and trying to find a model that minimizes loss. So, how to evaluate the loss? A simple and popular way is to use mean square error (MSE). MSE is the average squared loss per example over the entire dataset.

---

[1] https://developers.google.com/machine-learning/crash-course/descending-into-ml/linear-regression

[2] https://en.wikipedia.org/wiki/Dolbear%27s_law

To calculate MSE, we sum up all squared losses for each example and then divide by the number of examples.

$$\text{MSE} = \frac{1}{|\mathbb{D}|} \sum_{(x,y)\in\mathbb{D}} (y - prediction(x))^2 \qquad (1.2)$$

where $\mathbb{D}$ is a dataset containing labeled examples and $prediction(x)$ is a function of the weights and bias with the input of x.

If we replace $prediciton(x)$ with our linear equation of Eq. (1.1), we get

$$\text{MSE} = \frac{1}{|\mathbb{D}|} \sum_{(x,y)\in\mathbb{D}} (y - wx - b)^2 \qquad (1.3)$$

Now we can examine this equation, all the x and y are known for us, and now, the problem is very explicit to find the right values of w and b, which make MSE minimal. The final problem here is a pure optimization problem. Once we solve this problem, we can get the parameters needed in our model in Eq. (1.1), i.e., we have built the model.

Generally speaking, as we have seen in the previous example, optimization lies at the heart of most machine learning algorithms [1]. No matter how complex the problems look, most machine learning problems can be reduced to optimization problems. The modeler formulates the problem by selecting an appropriate family of models. Then the model is typically trained by solving an optimization problem that optimizes the parameters of the model given a selected loss function.

1.2 Optimization Problems

Optimization is very important for decision-makers to make good decisions before taking actions. It exists widely in science and engineering as long as there are multiple choices of solutions. To solve a real-world problem, the first step is to establish a mathematical model to describe the problem.

1.2.1 Mathematical Formulation

In general, to find the best solution of an optimization problem, we first need to construct a mathematical model to describe the problem. Normally, modelling an optimization problem needs to identify three components, which are objectives, variables, and constraints. They can be formulated as

$$\min_{x} \text{ or } \max_{x} \ f_i(\boldsymbol{x}), \quad i = 1, \cdots, M \qquad (1.4)$$

subject to

$$g_j(x) \leq 0, \quad j = 1, \cdots, N$$
$$h_k(x) = 0, \quad k = 1, \cdots, P$$

where $f_i(x)$ are objective functions to be minimized or maximized; they are measures to quantify the performance of a solution.

In manufacturing, we want to maximize profit and minimize production cost, while in data fitting, we want to minimize the difference error between observed data and predicted results like Eq. (1.3). x are variables represented for a solution. The aim is to find solutions that have the best performance in terms of the objective functions. In manufacturing, they may be the amount of each resource consumed in production process, whereas in data fitting, they may be the parameters of a model like the weights and bias defined in Eq. (1.3). $g_j(x) \leq 0$ and $h_k(x) = 0$ are called inequality constraints and equality constraints, respectively. They describe the relationships among variables and define the possible range of values for variables. A solution is feasible only if it satisfies all constraints; otherwise, the solution is infeasible.

The term minimization problem refers to minimizing the objective function, whereas the maximization problem refers to maximizing the objective function, which can be defined as negating the minimized objective function. If both N and P are zero, the problem is an unconstrained optimization problem.

If M is larger than one, which means that there are more than one objectives that need to be optimized synchronously, then the problem is called multi-objective optimization problem. A simple example of a multi-objective optimization problem is building a computer: we need to use limited money to balance performance, appearance, comfort, and other conflicting indicators; the variables are configurations of hardware components; the constraints may be the limited budget, or we only want a 32-inch display.

1.2.2 Continuous Optimization Versus Discrete Optimization

The optimization problem can be divided into two types according to the type of input: continuous optimization and discrete optimization. Here is an example to help us understand continuous optimization and discrete optimization.

A chemical company has two factories F_1 and F_2 and three retailers R_1, R_2, R_3. Each factory F_i produces a_i tons of a certain chemical product each week; each retailer R_j demands b_j tons of the product weekly. c_{ij} is the cost of shipping one ton of the product from factory F_i to retailer R_j. Figure 1.2 shows the problem.

The problem is to determine how much of the product to ship from each factory to all retailers to satisfy all the requirements and minimize the total cost. The variables

Fig. 1.2 A transportation
problem

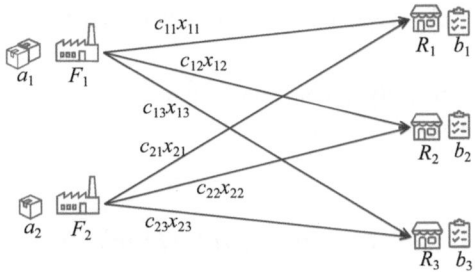

of the problem are x_{ij}, $i = 1, 2$, $j = 1, 2, 3$, where x_{ij} is the amount of the product
shipped from F_i to R_j. The problem can be formulated as

$$\min \sum_{ij} c_{ij} x_{ij} \tag{1.5}$$

subject to

$$\sum_{j=1}^{3} x_{ij} \leq a_i, \quad i = 1, 2$$

$$\sum_{i=1}^{2} x_{ij} \geq b_j, \quad j = 1, 2, 3$$

$$x_{ij} \geq 0, \quad i = 1, 2, \quad j = 1, 2, 3$$

This is a continuous optimization problem, since the variables used in the
objective function are continuous variables, i.e., x_{ij} to be chosen from real values
normally within a range.

But in some optimization problems, the variables make sense only if they
take integer values or binary values. Suppose that in the transportation problem
mentioned above, the factories produce cars rather than chemicals. In this case, x_{ij}
would represent integers (i.e., the number of cars shipped) rather than real numbers.
We call this type of problems as discrete optimization problems; particularly, in this
example, the problem is also known as integer programming problems. For some
discrete optimization problems, discrete variables mean a set of objects to be chosen
or combinatorial structures, such as combinations, assignments, schedules, and
sequences. These problems are also known as combinatorial optimization problems.

Some problems contain both continuous and discrete variables. These problems
are referred to mixed-integer programming problems. For example, those factories
not only produce chemical materials weighed by real values but also produce
chemical equipment, which must be counted by integers. The problem turns into
a mixed-integer programming problem.

1.3   Optimization Algorithms

Generally, optimization algorithms can be categorized into deterministic algorithms and probabilistic algorithms in terms of whether random sources are used.

1.3.1   Deterministic Algorithms and Probabilistic Algorithms

A deterministic algorithm refers to an algorithm that will always produce the same output given a particular input. It is often used in a problem where the objective function is clear and the solution can be efficiently searched with a direction-guided strategy. For example, the gradient information is normally used as the search direction for continuous optimization problems that are smooth and particularly differentiable. However, when the relationship between a solution and its objective value is not obvious or calculating its objective value is very time-consuming or the problem has multiple objectives or all possible solutions cannot be examined by each due to the large number, particularly for problems where the number of possible solutions increases exponentially as the number of variables increases, then probabilistic algorithms come into play.

A probabilistic optimization algorithm refers to an algorithm that takes random inputs during the search process, which are typically uniformly and independently distributed. A typical characteristic of the probabilistic optimization algorithm is that the algorithm may produce different outputs even when given the same input due to randomness. However, we can get a stable output as long as the randomization process is pseudo-random, which is usually driven by a random seed generator. The randomness is typically used to help the algorithm discover new solutions and makes the algorithm less sensitive to modeling errors. The algorithm normally produces a satisfactory solution within an affordable amount of time. Furthermore, the injected randomness can help the algorithm avoid trapping into local optima and eventually find a global optimum.

Indeed, this randomization principle is known to be a simple yet effective way to achieve good performance uniformly across many sorts of problems, particularly for complex problems mentioned above. The probabilistic algorithm belongs to the family of Monte Carlo method devised in the 1930s. Now, it has become one of the most important research branches of optimization since it was proposed. Due to the successful application of these intelligent optimization algorithms for solving complicated optimization problems where deterministic algorithms fail, a variety of probabilistic algorithms have been implemented in some commercial software, such as TOMLAB,[3] MIDACO,[4] and IOSO,[5] Most of the methods we will introduce in the rest of this book are probabilistic optimization algorithms.

1.3.2 Intelligent Optimization Techniques

Intelligent optimization techniques have developed from the simple low-level heuristic to the complex high-level metaheuristic and to the recent autonomous search of hyper-heuristics.

A heuristic [4, 5] is a part of an optimization algorithm that uses the information gathered by the algorithm so far to help decide which search direction to take next. This is normally achieved by designing a heuristic function, which is usually problem dependent. A heuristic function takes the goal into account and hence provides a search direction for an algorithm. A heuristic can accelerate the search process, which is very important for complex optimization problems, especially for NP-hard problems that make it the only viable option in real-world scenarios.

For instance, if you are trying to solve the travelling salesman problem (TSP), which is "Given a list of cities and the distances between each pair of cities, what is the shortest route that visits each city and returns to the origin city?" Suppose now that you are in a city, then your heuristic could be "next take the closest unvisited city to the current city where you are." Usually, this provides a very quick solution; however, it can easily get stuck at the local optima since it does not examine all possible choices from a global point of view.

As mentioned already, ① heuristics use domain-specific knowledge and therefore cannot be easily used to solve problems that belong to other domains; ② heuristics can easily get stuck in the local optima. Meta-heuristics aim to address both of these two drawbacks. A metaheuristic is an algorithmic framework with guidelines for designing heuristic optimization algorithms for solving very general classes of problems. Generally, it combines heuristic optimization techniques with randomness within its framework in an abstract and, hopefully, efficient way, usually without utilizing deeper insight into their structures, i.e., by treating them as black-box procedures [3]. Metaheuristic optimization algorithms can produce good-enough solutions within a small enough computational time. They can often take a good trade-off between computing time and solution quality. There are a wide variety of metaheuristics developed for optimization problems of different domains, which can be divided into single-solution-based methods and population-based methods. Single-solution-based methods include the hill climbing, simulated annealing, iterated local search, and variable neighborhood search. Population-

---

[3] https://tomopt.com/tomlab/products/midaco/

[4] http://www.midaco-solver.com/

[5] http://iosotech.com/index.htm

based metaheuristics include genetic algorithms, differential evolution, particle swarm optimization, ant colony optimization, etc.

Experimental studies have shown that different metaheuristics perform differently on different types of problems. In addition, for the same type of problem, different metaheuristics perform differently on different instances. Even different metaheuristics perform differently at different stages of the search process on the same instance. Unlike metaheuristics, a hyper-heuristic [2] framework aims to search for a right heuristic or a sequence of heuristics within a search space of heuristics. The idea is to automatically devise algorithms on low-level heuristics or heuristic components in a way of selecting or generating heuristics with the purpose of combining the strength and compensating for the weakness of known heuristics (see Sect. 6.2.2 for more details).

References

1. Bennett, K.P., Parrado-Hernández, E.: The interplay of optimization and machine learning research. J. Mach. Learn. Res. **7**(46), 1265–1281 (2006)
2. Cowling, P., Kendall, G., Soubeiga, E.: A hyperheuristic approach to scheduling a sales summit. In: International Conference on the Practice and Theory of Automated Timetabling, pp. 176–190. Springer, Berlin (2000)
3. Glover, F.W., Kochenberger, G.A.: Handbook of Metaheuristics. Kluwer, Dordrecht (2003)
4. Michalewicz, Z., Fogel, D.B.: How to Solve It: Modern Heuristics. Springer Science & Business Media, Berlin (2000)
5. Pearl, J.: Heuristics: Intelligent Search Strategies for Computer Problem Solving. Addison Wesley, Reading (1984)

Chapter 2
Fundamentals

Abstract This chapter introduces the fundamentals of optimization. These fundamentals include the definition of fitness landscapes, properties of fitness landscapes, and computational complexity. They help us more profoundly understand optimization, especially in understanding the difficulties in solving optimization problems. They will lay the foundation for learning and designing optimization methods.

2.1 Fitness Landscapes

The concept of fitness landscapes, originating from theoretical biology more than 90 years ago [15], has played an important role in optimization. It is a valuable tool for analyzing the optimization process, especially for studying the relationship between the performance of optimization algorithms and the characteristics of optimization problems [12]. In practice, it attempts to draw an analogy with real landscapes to gain a better understanding of how and where algorithms operate [7].

A fitness landscape is a mapping from every possible solution, which represents a candidate solution of a problem to be solved, into a fitness value. More formally, a fitness landscape consists of three main components: a set of solutions, a fitness function that evaluates these solutions, and a neighborhood relationship between the solutions (i.e., the definition of distance between solutions). The fitness function determines how good a solution is in terms of the objective value. It is the driving force behind the search. Neighborhood relations are crucial in the analysis of the fitness landscape. Typically, the shape of a fitness landscape depends on the representation/encoding (see Sect. 4.2) of solutions and the chosen distance measures between solutions for a given problem. That is, a particular problem can have different fitness landscapes. For example, the landscape of real-valued and binary encoding for a continuous optimization problem varies significantly.

More specially, a fitness landscape describes a surface of fitness values over the solution space where optimization methods are searched. When an optimization problem is a minimization problem, searching for an optimal solution to the problem can be interpreted as walking on the surface of its fitness landscape toward the lowest valleys, with the ability to overcome other valleys [3]. Therefore, the

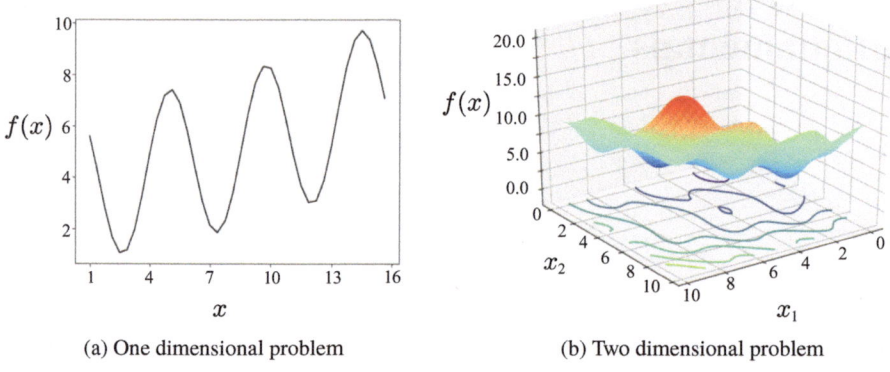

(a) One dimensional problem   (b) Two dimensional problem

Fig. 2.1 Fitness landscapes of one-dimensional (a) and two-dimensional (b) continuous optimization problems

fitness landscape has been found to be effective for analyzing the behavior of the optimization method in an optimization process and can help design an effective optimization method. Figure 2.1 shows the fitness landscapes of two continuous optimization problems in one dimension and two dimensions, respectively.

2.1.1 Solution Space

The first question when considering using an optimization method to solve a problem is: What space does the optimization method to search in? The solution space, also called decision space or search space, consists of all possible solutions that connect with each other by a user-defined distance measure. For a practical example, if we want to minimize the cost of manufacturing a car, all possible solutions might represent all alternative ways the manufacturing process could be configured. In this case, these solutions comprise the solution space.

We call the union of all possible solutions x of an optimization problem as its solution space $\mathbb{X}$. The representation of x could be a vector of bit strings, real values, integers, or even more complex structures depending on the problem and the algorithm to be used. The solution space can be represented as

$$\mathbb{X} = \{x_1, x_2, \cdots\} \tag{2.1}$$

Normally, the solution space is uncountably infinite for continuous optimization problems and countably finite for discrete optimization problems. Note that a countably finite solution space does not mean that it is easier for an algorithm to search than an uncountably infinite solution space. For example, we consider the TSP as a hard problem since there does not exist such an algorithm so far that

Fig. 2.2 Infeasible and
feasible regions of a
constrained optimization
problem

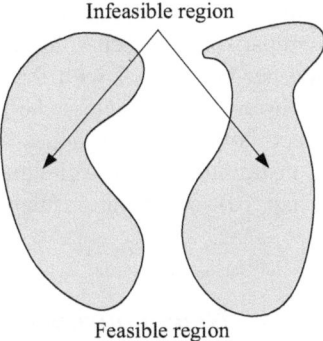

Infeasible region

Feasible region

can solve it in polynomial time. However, the continuous sphere function with
an uncountably infinite solution space can be easily solved by a gradient descent
method since it is convex and unimodal.

For a constrained optimization problem, only parts of the solution space are
feasible (see the feasible region in Fig. 2.2). A solution is feasible only if it satisfies
all constraints, i.e., it is within the feasible region of the solution space; otherwise,
the solution is infeasible, i.e., it is within the infeasible region in the solution space.
Infeasible solutions cannot be deployed. We need to find the feasible solution with
the best fitness value (see Chap. 10 for methods for finding feasible solutions).

2.1.2   Objective Space

As we know, the main objective of optimization methods is to find a promising
solution (or a set of promising solutions; see the multimodal optimization in
Chap. 8 and multi-objective optimization in Chap. 9) as possible from the solution
space. Generally, we are interested in the solution with the best fitness value. This
evaluation is carried out by an objective function (or a set of objective functions in
the multi-objective optimization scenario).

The objective values corresponding to all possible solutions comprise the
objective space. The objective space F can be represented as

$$F = \{f(x_1), f(x_2), \cdots\} \tag{2.2}$$

where f is the objective functions.

For unconstrained optimization problems, the objective function often is the only
guide for an algorithm to search in the search space. A solution's fitness value is
normally positively correlated to its objective value for maximization problems.
Note that a surrogate model may be needed to describe the relationship between
the input x and output f (see Chap. 14) in a situation where the relationship is
vague or the objective evaluation is time-consuming, especially for problems in the

real world. For optimization problems with more than one objective function, the comparison between solutions becomes complicated. Therefore, special methods are needed to deal with the comparison difficulty (see Chap. 9). For constrained optimization problems, the comparison between solutions is also complicated since both objective values and constraint values need to be taken into account. Constraint-handling techniques are normally needed to deal with constraints (see Chap. 10) since solutions that violate any constraint are infeasible.

2.1.3 Neighbourhood

As a basic concept in topological space, a neighborhood of a solution x is a set of solutions containing that point where one can move any direction away from that solution without leaving the set. Neighborhood solutions of x are normally close to x in some sense because they share a certain amount of structure and can be easily obtained by a neighborhood generating function with input of x. The neighborhood of a solution x is often written as

$$\mathbb{N}_\tau(x) = \{p | p \in \mathbb{X}, p \neq x, d(x, p) \leq \tau\} \tag{2.3}$$

where d is a distance function and $\tau > 0$ is the threshold of a neighborhood.

The neighborhood topology is used to define or describe the arrangement of the elements of a communication network for a set of solutions. Some typical topologies are ring network, global network, random network, and star network (see the examples in Sect. 5.5). The neighborhood relations among solutions are often defined on the basis of the distance function. Many distance measurement methods can be used, such as Manhattan distance, Euclidean distance, Chebyshev distance, Hamming distance, etc.

From the perspective of a search algorithm, the neighborhood topology depends not only on the distance function with a given solution representation but also on the way an algorithm searches in the solution space. Given a starting solution in the search space, if the search of the algorithm follows a particular direction, the algorithm cannot reach the space that is not in the search direction even if it is very close to the given solution in terms of the distance function.

2.1.4 Global Optimum

For an optimization problem, a global optimum is an optimum of the whole solution space. A global optimum is a solution that has the best objective value in the search space, as shown in Fig. 2.3.

$$x^* \text{ is a global optimal solution, if } \forall x \in \mathbb{X}, f(x^*) \leq f(x) \tag{2.4}$$

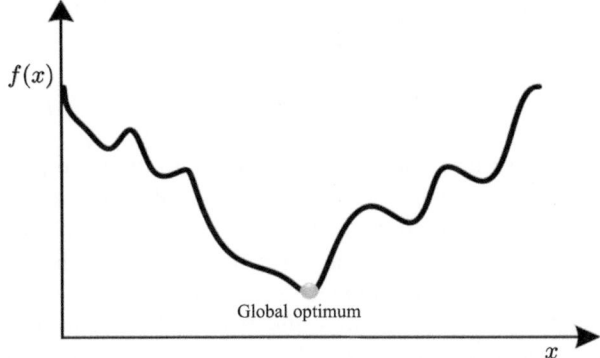

Fig. 2.3 The global optimum for a single objective minimization problem

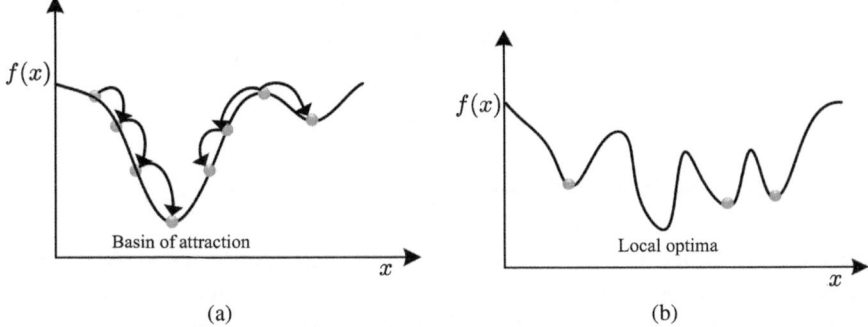

(a)   (b)

Fig. 2.4 Illustration for basin of attraction (**a**) and local optima (**b**)

Normally, the distribution and the number of optimal solutions of a problem are strongly dependent on the type of the problem. In single-objective optimization problems, there is one global optimum in general or multiple global optima in the scenario of multimodal optimization (see Chap. 8). In multi-objective optimization problems, there is a set of optimal solutions (widely known as Pareto-optimal solutions; see Chap. 10), instead of a single optimal solution.

2.1.5 Local Optimum

Before introducing the definition of the local optimum, we introduce the concept of the basin of attraction. The basin of attraction of an optimum is the region of all its neighborhood solutions whose objective values change monotonically as they move closer to the optimum, as shown in Fig. 2.4a.

A local optimum is a solution that has the best fitness value within its neighborhood in the search space, as shown in Fig. 2.4b.

$$x \text{ is a local optimal solution, if } \forall p \in \mathbb{N}_\tau(x), \, f(x) \leq f(p) \qquad (2.5)$$

In some interesting cases, the local optimum is not a single point but a set of points on ridges or plateaus with the same objective value. The shape of ridges and plateaus can be linearly or non-linearly distributed in the fitness landscape.

To find the global optimum of a problem is not an easy task due to many factors, such as a huge number of local optima, comparison difficulty between solutions in constrained or multi-objective situations, dynamic environments, etc. To design an efficient optimization algorithm, we should ensure that the algorithm can converge to the region of global optimum, i.e., carry out exploitation for local optima. To achieve the objective, the algorithm should be able to avoid trapping into local optima and continue to explore promising areas that are not exploited, i.e., carry out exploration for global optima (see the balance between exploration and exploitation in Chap. 7).

2.2   Properties of Fitness Landscape

The fitness landscape has played an important role in the theory of complex systems. It has been found to be effective in optimization theory, shown to be powerful for understanding the behavior of optimization algorithms, and to help improve their performance. Therefore, by analyzing the structure of the underlying fitness landscape, the behavior of optimization methods can be understood, and the design of an effective algorithm becomes easier.

The structure of the fitness landscape significantly impacts the performance of an optimization algorithm. Several properties associated with the structure of fitness landscapes [8] will be discussed in detail as follows.

2.2.1   Modality

During landscape analysis in evolutionary theory, there has always been a tendency to focus on the relationship between the problem difficulty and the landscape modality. Generally speaking, more modality of the fitness landscape means more difficulty of the problem. This is because modality biases search algorithms, as more modality implies more candidate solutions with the same objective value, making it more challenging for algorithms to determine the direction to be taken.

A landscape that has only one optimum is commonly called unimodal, as shown in Fig. 2.5. On the contrary, a landscape with more than one optimum is known as multimodal, as shown in Fig. 2.6. To some extent, the number of local optima in a

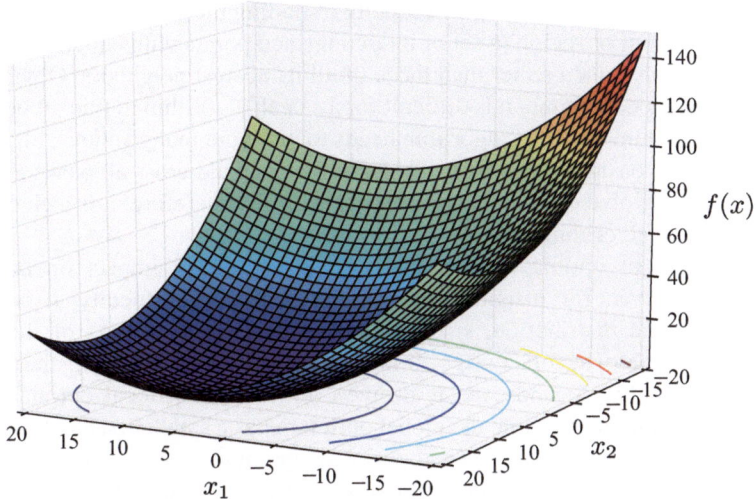

Fig. 2.5 An example for a 2-D unimodal shifted sphere function

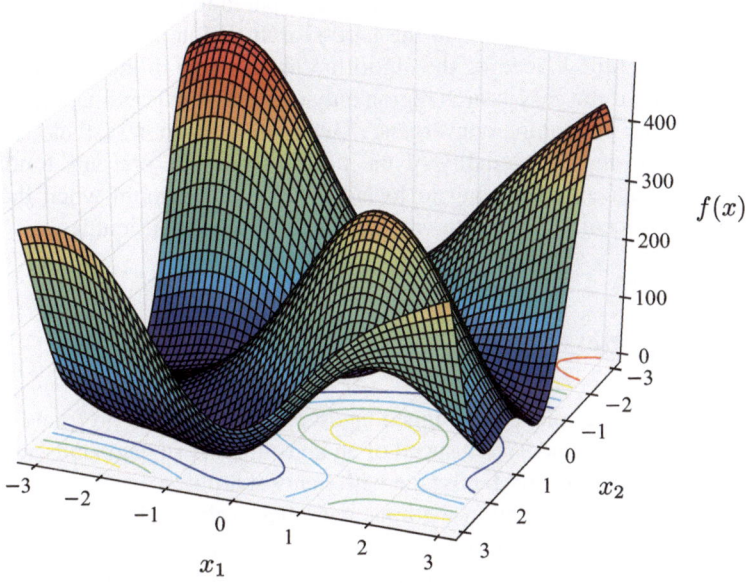

Fig. 2.6 An example for a 2-D multimodal Schwefel's function

landscape significantly impacts the difficulty in finding the global optimum. A local optimum is a point or region (a set of interconnected points with equal fitness) with a fitness function value greater than those of all its nearest neighbors. Obviously, if there are many local optima, it is difficult for a search algorithm to find all optima or the global optimum since the algorithm needs to jump out many valleys, i.e., points, which are in the basin of attraction of local optima, should cross all peaks along the direction to the global optimum as shown in Fig. 2.4b. Therefore, it is a challenging issue for search algorithms to address.

The number of optima does indeed have a significant impact on landscape difficulty. However, for insightful analysis of landscape difficulty, it is crucial to investigate the distribution, size, shape, and depth (or height) of the basins of attraction. These factors may be more important in determining the level of landscape difficulty. A landscape is deemed difficult when local optima possess large basins of attraction, while the global optimum has a smaller basin of attraction. An extreme scenario occurs when the global optimum is an isolated spike in a plateau. Optima with basins of attraction of asymmetrical shapes are generally more difficult to navigate than those with symmetrical shapes.

An objective-driven search algorithm inherently favors exploring local regions that are easy to search. Solutions in such regions typically exhibit faster and larger improvements in terms of the objective value compared to those in harder-to-search regions. Consequently, solutions in the hard-search region are gradually removed due to their poor fitness, leading the algorithm to converge in the easiest-to-search region. However, if the easy-search region only contains local optima, the algorithm is susceptible to premature convergence, wherein it becomes stuck in a local optimum. Therefore, the modality of the fitness landscape presents fundamental challenges for algorithms aiming to locate the global optimum when the search space comprises local regions with varying levels of search difficulties.

2.2.2 Ruggedness

The search often takes place in a rugged landscape, which has a profound effect on the problem difficulty and search speed of an algorithm. A rugged landscape can be regarded as a multi-modality landscape with many sharp descents and ascends.

In simple cases, optimization methods can effectively track optimum solutions using gradient information from fitness landscapes, particularly for objective functions with continuous and low total variation. However, when objective functions exhibit instability or frequent oscillations, as depicted in Fig. 2.7, it becomes more challenging to determine the correct directions to proceed during the optimization process. The ruggedness of these functions directly correlates with their difficulty to optimize, with more rugged functions presenting greater challenges.

In short, ruggedness has to do with the level of variation in fitness values in a fitness landscape, where the fitness values of neighborhood solutions have very different fitness values. An extreme case is the needle-in-a-haystack problem, where

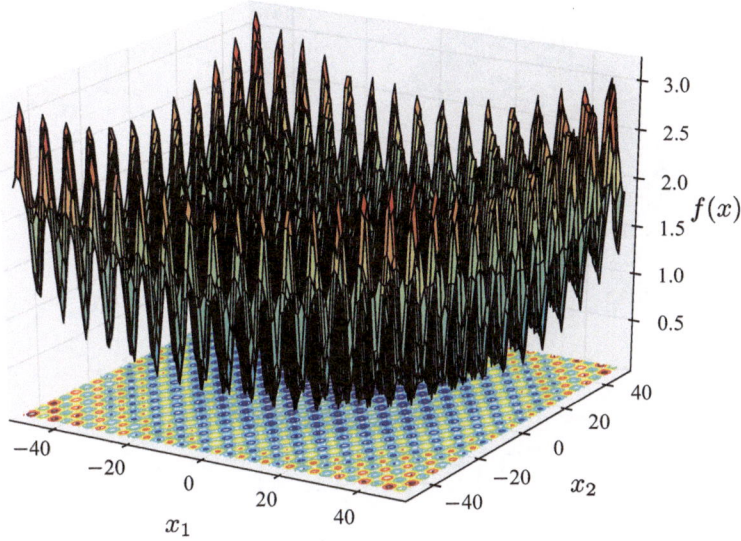

Fig. 2.7 An example for ruggedness for a 2-D shifted Griewank function

each peak is like a needle without basin of attraction. The opposite of a very rugged landscape would be a landscape with a single large basin/peak of attraction or a flat landscape with no features. For the rugged landscape, a search algorithm is hardly to get knowledge of where to go, causing the algorithm to be difficult to converge.

2.2.3 Deceptiveness

A typical fitness landscape should not only provide sufficient information to guide the search for an algorithm, but the information should also guide the search in the right direction, i.e., the direction toward the global optimum. In contrast, a fitness landscape guides the optimization process far away from the optimum (and thus also in the wrong direction), which is sometimes known as fitness deceptiveness as shown in Fig. 2.8. The gradient of deceptive fitness landscapes usually leads the search away from the global optimum.

Deceptiveness is intricately tied to the structure of optima within a landscape. The relative positions of local optima to the global optimum and the presence of isolation significantly influence the level of deceptiveness. For example, when an algorithm discovers a region with superior fitness compared to others, it tends to exploit this area under the assumption that it likely contains the global optimum. However, in reality, the optimal solution may be distant from the currently searched area [6]. Interestingly, certain optimization algorithms demonstrate an advantage in solving problems with high levels of deceptiveness [2], whereas increased

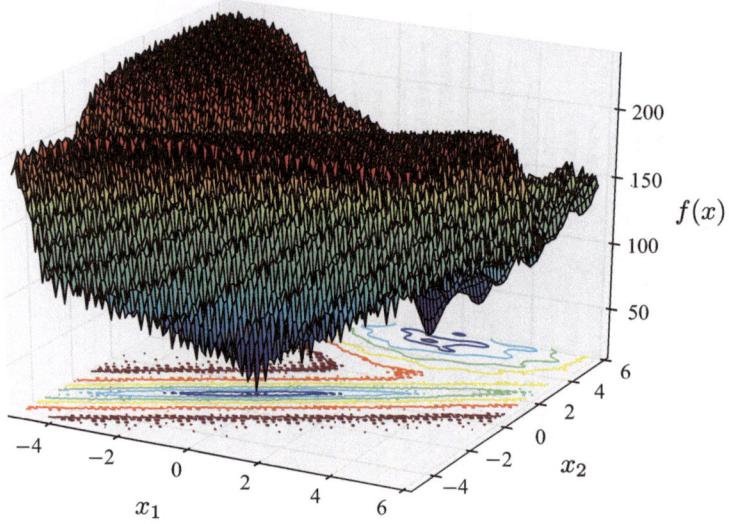

Fig. 2.8 An example for deceptiveness for a 2-D rotated hybrid composition function

deceptiveness generally escalates problem complexity [14]. It is important to note that deceptiveness holds meaning only in relation to a specific search algorithm, not universally across all algorithms.

2.2.4   Neutrality

In a landscape where the fitness values of the neighborhood solutions of a solution remain unchanged, the region surrounding that solution can be considered neutral [9]. The size of the neutral area around the solution is called the neutral degree of the solution. A neutral landscape does not imply a flat landscape across the whole search space (i.e., the objective function is constant) but rather the presence of neutrality in local areas, such as plateaus and ridges in the landscape mentioned above.

Neutrality has a profound effect on the success of a search algorithm [12]. When a population moves to a neutral area of a fitness landscape, where the fitness values of all solutions are the same, as shown in Fig. 2.9, then this could be misunderstood as stagnation in the search space or convergence in the objective space on a local optimum. But in real case, the population is moving just across the neutral area. In the other case, when the current best solution is located in a flat area of fitness landscapes, and all adjacent solutions have the same fitness values, there is no gradient information for algorithms to guide the search with a direction, as shown in Fig. 2.10. For the two situations, they make the algorithm cannot move any more.

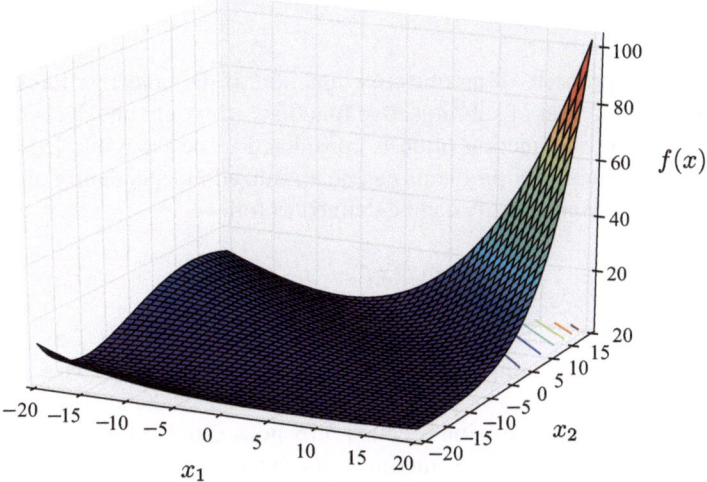

Fig. 2.9 An example for neutrality, 3-D map for 2-D shifted and rotated Rosenbrock function

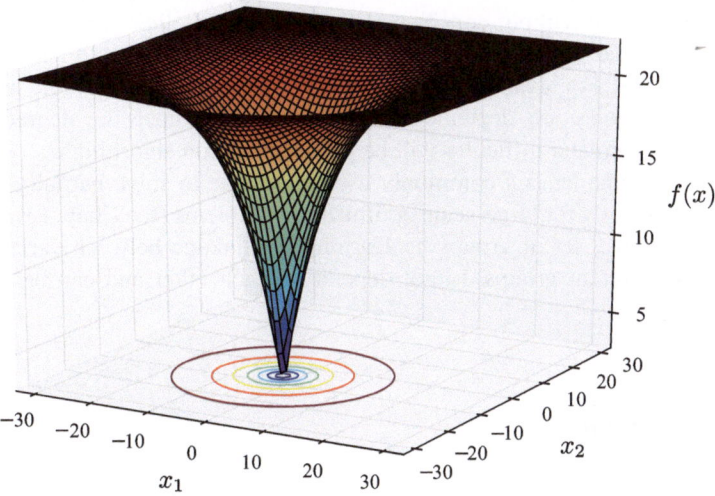

Fig. 2.10 An example for neutrality, 3-D map for 2-D shifted Ackley function

Interestingly, a study [1] of neutral landscapes found that neutrality has a smoothing effect on problem difficulty where adding neutrality to a deceptive landscape makes the problem easier and adding neutrality to an easy landscape makes it harder on a binary landscape.

2.2.5 *Separability*

Separability is a property of an objective function. If the objective function can be written as a sum of a set of sub-objective functions with only one decision variable, this type of objective function often is considered to be separable [5]. Therefore, these decision variables of problems can be optimized independently of each other. The concept of the separability can be defined as follows:

$$\min f(x_1, \cdots, x_n) = (\min f(x_1, \cdots), \cdots, \min f(\cdots, x_n)) \tag{2.6}$$

A fitness function is deemed as separable if and only if Eq. (2.6) holds. Otherwise, it is called nonseparable. If a fitness function $f(x)$ is separable, the decision variables $x_1, \cdots, x_n$ are called independent, and the fitness function is decomposable. A fitness function is called fully nonseparable if there is dependence among all decision variables. Sometimes, the interaction among variables is also referred to as epistasis, which is the degree of inter-dependency between variables (also known as nonlinear separability). When variables in an optimization problem are dependent on each other, it is impossible to find the optimum solutions by tuning one variable without considering the changes of other variables. There are partially separable fitness functions between separable and fully non-separable fitness functions. For example, a fitness function is k-non-separable if at most k decision variables are dependent. In general, the higher the degree of non-separability, the greater difficulty will be for optimization algorithms.

Divide and conquer is a commonly used technique to solve partially separable problems, especially for large-scale optimization problems (see Chap. 13). It divides all variables into a set of groups by learning the linkage between variables [16]. Variables in different groups do not depend on each other and can be optimized separably.

2.2.6 *Scalability*

In the dimension of the objective space, an increasing number of objective functions can greatly impact the performance of an algorithm due to the contradictory gradient for the same variables with different objectives and the exponential decrease in selection pressure among solutions. In general, this is termed multi-objective optimization problems (see Chap. 9). Likewise, in the dimension of the solution space, the increase in decision variables also has an impact on the performance of an algorithm due to the exponentially increased search space, which is often called the "curse of dimensionality". Generally, scalability refers to the solution space. Some of the problems have the property of scalability. These problems are scalable if they can scale dimensions without affecting the inherent difficulty of the problems.

The difference between small-scale and large-scale problems in discrete or continuous search spaces focuses on the number of decision variables. For large-scale problems, the main difficulty is the slow search and the expensive runtime requirement. Parallel and distributed computing can be used to solve the expensive problem to some extent. The other way is to divide the large-scale problem into a set of small-scale problems by the divide-and-conquer method mentioned in Sect. 2.2.5 using linkage learning methods between variables.

Note that increasing the number of decision variables normally exponentially increases the number of optima for most problems. For some particular problems, increasing the number of variables may change the problem property. For example, the Rosenbrock function is unimodal in two dimensions but becomes multimodal when the number of variables increases [11].

2.2.7 Domino Convergence

Domino convergence has begun to be considered since the emergence of literature [10]. In general, domino convergence usually describes the characteristic of solutions that contribute to significantly different degrees to the total objective values of optimization problems [13]. When these solutions are encoded as separate variables, due to these variables contributing to significantly different degrees to the total fitness, thus, they will be considered with different priorities in optimization methods.

In a solution, some of the variables have a very important influence on the fitness. Therefore, these variables play an important role in the optimization process and usually converge fast. However, for other variables with a smaller contribution, they will be ignored or have a small impact on the process. In the worst case, variables with less contribution may look like noise and not be optimized at all. In this situation, the search for optimization methods easily traps into local optima, resulting in the global optimum being difficult to find. Usually, enhancing the exploration of optimization methods will reduce the phenomenon of premature convergence in a certain degree.

2.2.8 Property Control, Analysis, and Visualization

Some properties, such as modality, ruggedness, deceptiveness, and neutrality, can be physically viewed in a fitness landscape, while separability, scalability, and domino convergence cannot be easily viewed in the landscape. In addition, these properties often interact with each other: changing one will have an effect on others. This makes it difficult for algorithm designers to analyze factors related to algorithm performance. To address this issue, an open framework was proposed to construct continuous optimization problems by partitioning the solution space with the k-

d tree [4]. In each subspace, the properties can be fine-tuned separately, such as dimensionality, modality, the number of objectives, optima positions, shape and size of basin of attraction, landscape structure (distribution of optima), inter-relationship between decision variables, and domino convergence.

How to quantify the properties mentioned above is very important in fitness landscape analysis. Normally, special techniques are needed to understand the structure of the fitness landscape. In addition, how to visualize the fitness landscape for discrete optimization problems or high-dimensional continuous optimization problems while keeping the visual effect on these properties is also important yet a challenging issue.

2.3 Computational Complexity

A problem that can be solved by a computer is called a computational problem. To solve a computational problem by an algorithm, a certain amount of resources is needed. Providing the bound of resource usage is an important role of computational complexity theory. Another task of computational complexity theory is to analyze the complexity classes of problems by taking account of the type of the computational problem (e.g., decision problems, function problems, optimization problems, etc.), the model of computation (e.g., deterministic Turing machines, Boolean circuits, quantum Turing machines, etc.), and the bound of resource usage.

2.3.1 Complexity Measures

Efficient algorithms are preferred in terms of time and storage usage. Time and storage are the two most commonly used measures to quantify the amount of resources used. The efficiency of an algorithm depends on its time complexity and space complexity.

2.3.1.1 Time Complexity

The time complexity of an algorithm refers to the computational effort required to execute this algorithm. The number of statements executed in an algorithm is known as the statement frequency or time frequency, notes for $T(n)$, where n is the scale of the problem. In general, the number of repeats of the basic operation of an algorithm is a function $f(n)$, so the time complexity of the algorithm is denoted as $T(n) = O(f(n))$. When calculating the time complexity of an algorithm, one should first find out the basic operation of this algorithm and then determine its execution times in terms of the corresponding statements. Finally, find the highest order of magnitude of $T(n)$ as the time complexity, i.e., the "Big O" notation. Note that sorted by increasing order of magnitude, the commonly used time complexity is $1, \lg n, n, n \lg n, n^2, n^3, 2^n$, and $n!$.

When calculating the time complexity of an algorithm, one way is to look at how many "for" loops there are, i.e., only one is $O(n)$, two is $O(n^2)$, and so on. If the upper bound of an algorithm is independent of the input size, it has a constant time, denoted by $O(1)$. If the algorithm's $T(n) = O(\lg n)$, it has a logarithmic time, i.e., binary tree operations and binary search.

2.3.1.2 Space Complexity

An algorithm's space complexity means the amount of memory space required by the algorithm in its life cycle. The space required by the algorithm is made up of two parts: ① fixed part, which is the space needed to store some data and variables (i.e., simple variables and constants, program size, etc.) that are independent of the size of the problem, and ② auxiliary part, which is the temporary or extra space required by variables when the algorithm is running, and its size depends on the size of the problem, such as recursive stack space and dynamic memory allocation.

Similar to the time complexity, the space complexity of an algorithm is defined as $S(n)$, where n denotes the scale of the problem. In order to obtain the space complexity, we have to know how much memory is needed to store values of different data types (depending on the compiler). If an algorithm requires that all input values have a fixed amount of space, then this space complexity is called constant space complexity, denoted $S(n) = O(1)$. If the amount of space needed by an algorithm increases with the input value, then the space complexity is called linear space complexity.

2.3.1.3 Ways of Measures

The best, worst, and average case complexity are the three common ways of measuring the time complexity (also applied to other complexity measures) regarding different inputs with the same size. Take the quick sort algorithm as an example. The best-case complexity of quit sort is $O(n \lg(n))$ when each pivot divides the set into half. The worst-case complexity occurs when there is no division at all during the whole sorting process, which takes $O(n^2)$. The average-case complexity is also $O(n \lg(n))$ by taking all possible permutations of n elements with equal probability.

2.3.1.4 Time Versus Space

Sometimes, we can optimize an algorithm for reducing both the time complexity and the space complexity. However, in practice, it is not always possible to achieve the two objectives. A different algorithm may take less time but at the expense of requiring more memory space and vice versa.

Therefore, we should balance the trade-off between space and time according to our constraints. That is to say, if we have memory constraints, then we can take algorithms requiring less memory at the cost of more running time. On the contrary, if we have running-time constraints, then we can take algorithms using less time at the cost of more memory. For example, our task needs to implement a lookup table. We can implement it by storing all entries in the memory, and then a result can be quickly obtained for each enquiry. Other than that, we can recalculate each entry as needed for each enquiry to save memory but at the cost of increasing computing time.

2.3.2   P Versus NP Problem

The P versus NP problem is one of the seven Millennium Prize Problems selected by the Clay Mathematics Institute in 2000. It is also one of the hardest problems in the field of computer science concerning the speed of a computer to finish a task. The P and NP problem is the question of whether all so-called NP problems are equal to the P problems. Although this question is still an open question, the latest investigation shows that 99% of experts believe that $P \neq NP$.[1]

A problem in class P represents the problem that could be addressed in polynomial time with respect to the size of the problem. The problem is also called deterministic polynomial problem, the P problem, for short. To be specific, a P problem means that a solution can be obtained by an algorithm running in an amount of time that is polynomial to its problem scale.

NP problems denote the non-deterministic polynomial problems. If the solution of a problem can be verified in polynomial time, then it could be regarded as NP. Verifying a solution in polynomial time does not mean finding a solution in polynomial time. For example, finding priming factors of a large number will try many different candidates, but verifying these candidates can be achieved by simply multiplying these factors.

NP-hard problems refer to problems that are at least as hard as NP problems, i.e., all NP problems can be reduced to NP-hard problems. Nevertheless, it is worth noting that not all NP-hard problems can be classified as NP problems, which means that verifying a solution may not take polynomial time.

The NP-complete problem is a problem that is both NP and NP-hard. Therefore, finding an effective algorithm for any NP-complete problem suggests that an efficient algorithm could be found for all NP problems as a solution for any problem belonging to this type could be transformed into a solution for any other member of the type. The TSP is a classical NP-complete problem, which widely exists in logistic scheduling. It is unknown whether there are polynomial-time algorithms for NP-complete problems, since it is still one of the most important problems in

---

[1] Guest Column: The Third P =? NP Poll, 2020.

theoretical computer science to determine whether these problems can be handled or not.

References

1. Beaudoin, W., Verel, S., Collard, P., Escazut, C.: Deceptiveness and neutrality the ND family of fitness landscapes. In: Proceedings of the 8th Annual Conference on Genetic and Evolutionary Computation, pp. 507–514. ACM SIGEVO, Association for Computing Machinery, New York (2006)
2. Goldberg, D.E.: Genetic algorithms and Walsh functions: Part II, deception and its analysis. Complex Syst. **3**(2), 153–171 (1989)
3. Jones, T.: Evolutionary algorithms, fitness landscapes and search. Ph.D. thesis, The University of New Mexico, Albuquerque (1995)
4. Li, C., Nguyen, T.T., Zeng, S., Yang, M., Wu, M.: An open framework for constructing continuous optimization problems. IEEE Trans. Cybern. **49**(6), 2316–2330 (2018)
5. Li, X., Tang, K., Omidvar, M.N., Yang, Z., Qin, K.: Benchmark functions for the CEC'2013 special session and competition on large-scale global optimization. Technical report, RMIT University, Cancun (2013)
6. Liepins, G.E., Vose, M.D.: Deceptiveness and genetic algorithm dynamics. In: Rawlins, G.J. (ed.) Foundations of Genetic Algorithms, pp. 36–50. Elsevier, Amsterdam (1991)
7. Pitzer, E., Affenzeller, M.: A comprehensive survey on fitness landscape analysis. In: Fodor, J., Klempous, R., Araujo, C.P.S. (eds.) Recent Advances in Intelligent Engineering Systems, pp. 161–191. Springer, Berlin (2012)
8. Reeves, C., Rowe, J.E.: Genetic Algorithms: Principles and Perspectives: A Guide to GA Theory. Springer Science & Business Media, Boston (2002)
9. Reidys, C.M., Stadler, P.F.: Neutrality in fitness landscapes. Appl. Math. Comput. **117**(2–3), 321–350 (2001)
10. Rudnick, W.M.: Genetic algorithms and fitness variance with an application to the automated design of artificial neural networks. Ph.D. thesis, Oregon Graduate Institute of Science & Technology, Hillsboro (1992)
11. Shang, Y.W., Qiu, Y.H.: A note on the extended Rosenbrock function. Evol. Comput. **14**(1), 119–126 (2006)
12. Smith, T., Husbands, P., O'Shea, M.: Fitness landscapes and evolvability. Evol. Comput. **10**(1), 1–34 (2002)
13. Thierens, D., Goldberg, D.E., Pereira, A.G.: Domino convergence, drift, and the temporal-salience structure of problems. In: 1998 IEEE International Conference on Evolutionary Computation Proceedings. IEEE World Congress on Computational Intelligence, pp. 535–540. IEEE, Anchorage (1998)
14. Weise, T., Niemczyk, S., Skubch, H., Reichle, R., Geihs, K.: A tunable model for multi-objective, epistatic, rugged, and neutral fitness landscapes. In: Proceedings of the 10th Annual Conference on Genetic and Evolutionary Computation, pp. 795–802. ACM SIGEVO, New York (2008)
15. Wright, S.: The roles of mutation, inbreeding, crossbreeding, and selection in evolution. In: Proceedings of the VI International Congress of Genetics, pp. 356–366. The International Genetics Federation, The VI International Congress of Genetics, New York (1932)
16. Yang, M., Zhou, A., Li, C., Yao, X.: An efficient recursive differential grouping for large-scale continuous problems. IEEE Trans. Evol. Comput. **25**(1), 159–171 (2020)

Chapter 3
Canonical Optimization Algorithms

Abstract This chapter introduces canonical optimization algorithms, including numerical optimization methods for continuous optimization problems and state space search methods for discrete optimization problems. Several popular numerical optimization methods based on line search are presented, e.g., steepest descent method, Newton method, and conjugate gradient method. State space search methods are categorized into uninformed search methods and informed search methods. Uninformed search methods include breadth-first search, depth-first search, and depth-limited search. Informed search methods include greedy search, A* search, and Monte-Carlo tree search. Several single-solution-based metaheuristic search algorithms are also introduced in this chapter, e.g., hill climbing, simulated annealing, iterated local search, and variable neighborhood search.

3.1 Numerical Optimization Algorithms

This section introduces several simple numerical optimization algorithms to solve unconstrained continuous optimization problems. These algorithms assume that the fitness landscape is like a valley. To reach the bottom of the valley from a certain position, one firstly finds a direction to the bottom according to the gradient and walks toward the direction for a step with a certain length, then recalculates the direction, and walks forward one step again until reaches the bottom.

3.1.1 Line Search

Numerical optimization algorithms use the above idea to solve the problem in a recursive way. Suppose that x_k is the point obtained at the k-th iteration and d_k is the direction along which the fitness function $f(x)$ declines at point x_k. $\alpha_k > 0$ is the step size along d_k. Then the next point x_{k+1} is obtained by

$$x_{k+1} = x_k + \alpha_k d_k \tag{3.1}$$

where $f(\mathbf{x}_{k+1}) < f(\mathbf{x}_k)$ must hold. Such an algorithm is also called a line search.

Line search starts from an initial solution $\mathbf{x}_0$. It finds a direction along which the fitness value will decline and decides the step size to calculate the next solution. The procedure is repeated until the termination is met. Algorithm 3.1 shows the procedures of the algorithm.

Algorithm 3.1: Line search

Input: Initial solution $\mathbf{x}_0$
Output: Final solution $\mathbf{x}_n$
1 Set initial counter $k = 0$ **while** *termination is not met* **do**
2 $\quad$ Compute a descent direction $\mathbf{d}_k$;
3 $\quad$ Choose α_k to 'loosely' minimize $f(\mathbf{x}_{k+1}) = f(\mathbf{x}_k + \alpha_k \mathbf{d}_k)$ over $\alpha \in \mathbb{R}_+$;
4 $\quad$ Update $\mathbf{x}_{k+1} = \mathbf{x}_k + \alpha_k \mathbf{d}_k$ and $k = k + 1$;

An important aspect in the algorithm is the termination criterion, which can be defined as one of the following:

(1) The procedure is stopped when the computing resources (i.e., time) are exhausted, that is, the maximum number of iterations is reached.
(2) The procedure is stopped when the gradient at $\mathbf{x}_k$ is smaller than a gradient norm tolerance ϵ_g, i.e., the algorithm has converged at a critical point.

$$\|\nabla f(\mathbf{x}_k)\| < \epsilon_g \tag{3.2}$$

where ϵ_g is a user-defined parameter.
(3) The procedure is stopped when the difference between $\mathbf{x}_{k+1}$ and $\mathbf{x}_k$ is smaller than a small constant value of ϵ_x, i.e., the algorithm converges at an x point, and the difference between $f(\mathbf{x}_{k+1})$ and $f(\mathbf{x}_k)$ is also smaller than a small constant value of ϵ_f

$$\|\mathbf{x}_{k+1} - \mathbf{x}_k\| < \epsilon_x$$
$$\|f(\mathbf{x}_{k+1}) - f(\mathbf{x}_k)\| < \epsilon_f \tag{3.3}$$

where ϵ_x and ϵ_f are both user-defined parameters.
(4) The procedure is stopped when $f(\mathbf{x}_{k+1})$ is larger than $f(\mathbf{x}_k)$, i.e., the algorithm diverges.

Another important aspect of Algorithm 3.1 is how to choose the step size α. Obviously, the optimum choice of α is to make the function $\phi(\alpha) = f(\mathbf{x}_k + \alpha \mathbf{d}_k)$ smallest. The difficulty to solve this problem is the same as that to solve the current minimization problem. Thus, we can only find an approximate α that makes $\phi(\alpha)$ smaller. It seems that α just needs to satisfy the condition $f(\mathbf{x}_{k+1}) < f(\mathbf{x}_k)$. But in practice, we should take the other factor into account. For example, Fig. 3.1 shows that although the selected α satisfies the condition, the algorithm still traps at two

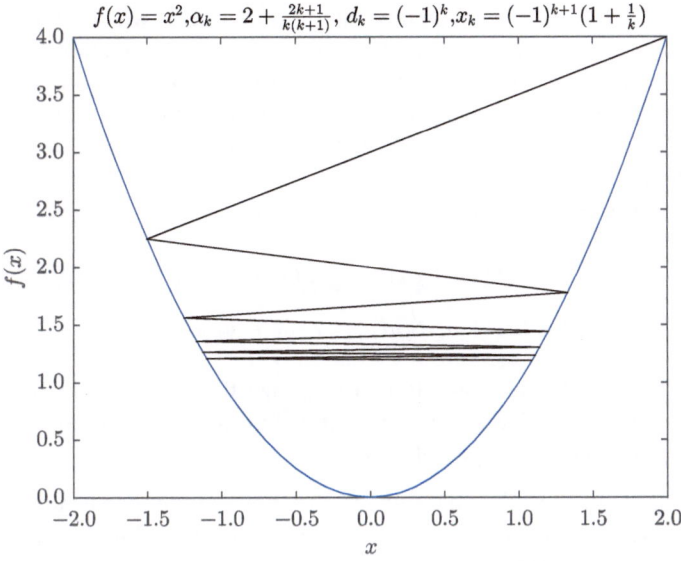

Fig. 3.1 Insufficient reduction in f

points, $x = -1$ and $x = 1$, where $x_1 = 2$. The key issue is that there is no sufficient reduction in f.

There are some famous conditions to find such α to obtain a sufficient reduction in f:

(1) The Armijo condition

$$f(\boldsymbol{x}_k + \alpha_k \boldsymbol{d}_k) \leq f(\boldsymbol{x}_k) + c_1 \alpha_k \boldsymbol{d}_k^{\mathrm{T}} \nabla f(\boldsymbol{x}_k) \tag{3.4}$$

where constant c_1 belonging to $(0, 1)$ generally is set to $c_1 = 10^{-4}$ in practice. $\boldsymbol{d}_k$ is the descent direction, so $\boldsymbol{d}_k^{\mathrm{T}} \nabla f(\boldsymbol{x}_k) < 0$ and the right-hand side of the inequality is the subtractive function of α. This condition guarantees that there is some reduction in f as long as α is not too small. There is an issue with the Armijo condition. Any small α that satisfies the Armijo condition cannot make sufficient progress. To address this issue, we add another constraint to the Armijo condition as follows:

(2) The Wolfe condition

$$f(\boldsymbol{x}_k + \alpha_k \boldsymbol{d}_k) \leq f(\boldsymbol{x}_k) + c_1 \alpha_k \boldsymbol{d}_k^{\mathrm{T}} \nabla f(\boldsymbol{x}_k)$$
$$\nabla f(\boldsymbol{x}_k + \alpha_k \boldsymbol{d}_k) \boldsymbol{d}_k^{\mathrm{T}} \geq c_2 \boldsymbol{d}_k^{\mathrm{T}} \nabla f(\boldsymbol{x}_k) \tag{3.5}$$

where $0 < c_1 < c_2 < 1$, c_2 is a relatively large number compared with c_1 where $c_2 = 0.9$ for Newton or quasi-Newton methods and $c_2 = 0.1$ for the nonlinear conjugate gradient method. The second inequality guarantees that the slope of

α in $f(x_k + \alpha_k d_k)$ is not smaller than $c_2 d_k^T \nabla f(x_k)$, which indicates that we can further reduce f significantly taking the chosen direction. It also means that we cannot get much reduction of f in the chosen direction if the slope is slightly negative or even positive, so we can stop the line search. Therefore, the second condition in Eq. (3.5) makes the algorithm reasonable progress by filtering unacceptable short steps.

(3) The strong Wolfe condition

$$f(x_k + \alpha_k d_k) \le f(x_k) + c_1 \alpha_k d_k^T \nabla f(x_k)$$
$$|\nabla f(x_k + \alpha_k d_k) d_k^T| \le c_2 |\nabla f(x_k) d_k^T|$$

(3.6)

The Wolfe conditions may allow a step size that is not particularly close to a minimizer of ϕ. The second inequality forces α to locate in the optimal or local-optimal neighborhood of $\phi(\alpha)$.

In fact, finding a step size that satisfies the Wolfe conditions is not an easy task. In practice, we can use the backtracking line search method to find a proper value of α as shown in Algorithm 3.2. The initial step length α' is often set to 1 in Newton and quasi-Newton methods but can be set different values for other methods. After a finite number of trials, an acceptable step length can be found since α_k will become small enough to satisfy the sufficient reduction condition in Eq. (3.4).

Algorithm 3.2: Backtracking step length search

Input: Initial value $\alpha' > 0$, contraction factor $\rho \in (0, 1)$
Output: Final value of α_k
1 **while** $f(x_k + \alpha d_k) > f(x_k) + c_1 \alpha d_k^T \nabla f(x_k)$ **do**
2    $\lfloor$ $\alpha = \rho \alpha$;

3.1.2   Steepest Descent Method

To obtain faster convergence, a proper descent direction should be configured with a fine-tuned step size. A simple idea is to take the direction of the negative gradient $-\nabla f$ at x_k. The negative gradient is the locally steepest descent direction to reduce the objective value. This method is called the gradient descent method or the steepest descent method.

Suppose $f(x)$ is differentiable at x_k, the first-order Taylor polynomial of $f(x)$ about x_k is

$$f(x) = f(x_k) + \nabla f(x_k) \cdot (x - x_k) + O(\|x - x_k\|^2)$$

(3.7)

Given a sufficiently small step size α_k, we want to find a unit vector $\boldsymbol{d}$ that minimizes $f(\boldsymbol{x}_{k+1})$. Let $\boldsymbol{x} = \boldsymbol{x}_{k+1} = \boldsymbol{x}_k + \alpha_k \boldsymbol{d}_k$, we have

$$f(\boldsymbol{x}_k + \alpha_k \boldsymbol{d}_k) - f(\boldsymbol{x}_k) = \alpha_k \nabla f(\boldsymbol{x}_k) \cdot \boldsymbol{d}_k + \alpha_k^2 O(\|1\|) \tag{3.8}$$

where the insignificant term α_k^2 can be ignored as α_k is small enough. To make $f(\boldsymbol{x}_k + \alpha_k \boldsymbol{d}_k) - f(\boldsymbol{x}_k)$ smallest, we must minimize $\nabla f(\boldsymbol{x}_k) \cdot \boldsymbol{d}_k$. Let θ be the angle between $\boldsymbol{d}_k$ and $\nabla f(\boldsymbol{x}_k)$; then we have

$$\nabla f(\boldsymbol{x}_k) \cdot \boldsymbol{d}_k = \|\nabla f(\boldsymbol{x}_k)\| \|\boldsymbol{d}_k\| \cos(\theta) \tag{3.9}$$

Therefore, we have $\theta = \pi$ that makes $\nabla f(\boldsymbol{x}_k) \cdot \boldsymbol{d}_k$ smallest, i.e., $\boldsymbol{d}$ is negative of $\nabla f(\boldsymbol{x}_k)$, so we have

$$\boldsymbol{d}_k = -\frac{\nabla f(\boldsymbol{x}_k)}{\|\nabla f(\boldsymbol{x}_k)\|} \tag{3.10}$$

Ideally, the gradient descent method can achieve a linear convergence rate. However, the convergence rate can be very slow in cases where the direction to the local optimum does not comply with the gradient at the current point. Take a simple quadratic function $f(\boldsymbol{x}) = ax_1^2 + bx_2^2$ with two positive coefficients in two dimensions as an example. The optimum is at the origin, and the gradient of f is

$$\nabla f(\boldsymbol{x}) = \begin{bmatrix} 2ax_1 \\ 2bx_2 \end{bmatrix} \tag{3.11}$$

If $a = b$, then the value of f increases isotropically from the origin in any direction. That is, for any point $\boldsymbol{x}$ in the search space, the negative gradient $-\nabla f(\boldsymbol{x})$ points exactly to the origin, and then the algorithm works well by following the negative gradient direction.

If $a \neq b$, say $b = 3$, $a = 1$, then at a point $(x_1, x_2)^{\mathrm{T}}$, the negative descent direction $(-2x_1, -6x_2)^{\mathrm{T}}$ is much steeper along the x_2 axis than that along the x_1 axis. That is to say, the moving direction does not direct to the origin, resulting in the moving path in a zigzag mode as shown in Fig. 3.2. Therefore, the convergence toward optimum will be very slow, especially in functions with ill conditioning.

To speed up the convergence rate of the steepest descent method, a two-point step size gradient method was proposed by Barzilai and Borwein [1], namely, the Barzilai-Borwein gradient method. In the method, the step size is derived from a two-point approximation to the secant equation underlying quasi-Newton methods.

$$\alpha_k = \frac{(\boldsymbol{x}_k - \boldsymbol{x}_{k-1})^{\mathrm{T}} [\nabla f(\boldsymbol{x}_k) - \nabla f(\boldsymbol{x}_{k-1})]}{\|\nabla f(\boldsymbol{x}_k) - \nabla f(\boldsymbol{x}_{k-1})\|} \tag{3.12}$$

Fig. 3.2 Zigzag steps in the steepest descent method with an initial point $(-1, -1)^{\mathrm{T}}$, where an optimal step size is used in each step

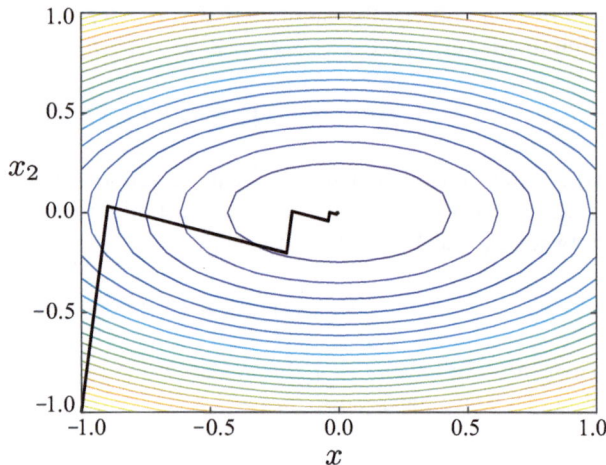

3.1.3   Newton Method

Although the gradient descent is quite simple, it has its own disadvantages of slow speed in the slope near the optimum. The Newton method was proposed to overcome this shortcoming.

Suppose that $f(x)$ has a continuous second derivative at x_k. The second-order Taylor polynomial of $f(x)$ about x_k is

$$f(x) \approx f(x_k) + f'(x_k)(x - x_k) + \frac{1}{2}f''(x_k)(x - x_k)^2 \qquad (3.13)$$

Finding a minimum of $f(x)$ is equivalent to find the solution whose derivative is zero in the equation below

$$\min f(x) \iff f'(x) = 0 \qquad (3.14)$$

so we have

$$f'(x_k) + f''(x_k)(x - x_k) = 0 \qquad (3.15)$$

Suppose that the solution of Eq. (3.15) is x_{k+1}; then,

$$x_{k+1} = x_k - \frac{f'(x_k)}{f''(x_k)} \qquad (3.16)$$

Similarly, in a high-dimensional problem, we have

$$f(\boldsymbol{x}) \approx f(\boldsymbol{x}_k) + \nabla f(\boldsymbol{x}_k)^{\mathrm{T}}(\boldsymbol{x} - \boldsymbol{x}_k) + \frac{1}{2}(\boldsymbol{x} - \boldsymbol{x}_k)^{\mathrm{T}}\nabla^2 f(\boldsymbol{x}_k)(\boldsymbol{x} - \boldsymbol{x}_k) \qquad (3.17)$$

To find out the stationary point of $f(\boldsymbol{x})$, let the derivative of $f(\boldsymbol{x})$ be zero.

$$\nabla f(\boldsymbol{x}_k) + \nabla^2 f(\boldsymbol{x}_k)(\boldsymbol{x} - \boldsymbol{x}_k) = 0 \qquad (3.18)$$

Suppose $\nabla^2 f(\boldsymbol{x}_k)$ is invertible and $\boldsymbol{x}_{k+1}$ is the solution to Eq. (3.18). We have

$$\boldsymbol{x}_{k+1} = \boldsymbol{x}_k - H^{-1}(\boldsymbol{x}_k)\nabla f(\boldsymbol{x}_k) \qquad (3.19)$$

where $H(\boldsymbol{x}_k)$ is the Hessian matrix $\nabla^2 f(\boldsymbol{x}_k)$ of function $f(\boldsymbol{x})$

$$H(\boldsymbol{x}_k) = \left[\frac{\partial^2 f(\boldsymbol{x}_k)}{\partial x_i \partial x_j}\right]_{n \times n} \qquad (3.20)$$

We now try to start from different initial point to solve the following problem using Newton method:

$$\min f(x) = 3x_1^2 + 3x_2^2 - x_1^2 x_2 \qquad (3.21)$$

We used the basic Newton method to solve this problem from three initial points $x_0 = (1.5, 1.5)^{\mathrm{T}}$, $x_0 = (-2, 4)^{\mathrm{T}}$, and $x_0 = (0, 3)^{\mathrm{T}}$. The two graphs in Fig. 3.3 show that when the method starts from $(1.5, 1.5)^{\mathrm{T}}$, it can find the global optimum quickly and successfully. But when it starts from $(-2, 4)^{\mathrm{T}}$, it finally sticks to the saddle point $(-3\sqrt{2}, 3)^{\mathrm{T}}$. Worse still is that when it starts from $(0, 3)^{\mathrm{T}}$, the Hessian matrix is singular, and the Newton method fails to work. The Newton method is quite sensitive to the initial point.

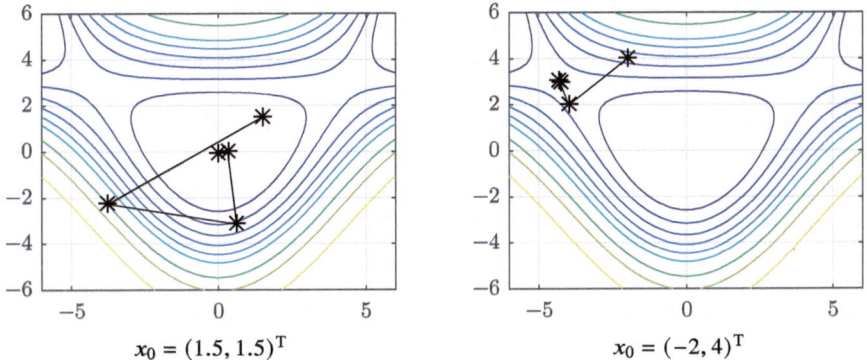

Fig. 3.3 Newton method with different initial points

The Newton method has quadratic convergence rate, while the gradient descent method converges with first-order rate, so the Newton method has a faster convergence speed. However, the Newton method works only if certain conditions are satisfied, i.e., the function must have a continuous second derivative, and the Hessian matrix must be invertible. Apart from the strict conditions, it takes a lot of calculation to find the inverse of the Hessian. In addition, if the Hessian is not positive definite, it's inverse matrix does not exist. Therefore, the quasi-Newton method was proposed to overcome this disadvantage. In the quasi-Newton method, a positive definite symmetric matrix that approximates the Hessian matrix (or the inverse of the Hessian matrix) is constructed without the use of the second derivative.

3.1.4 Conjugate Gradient Method

Both the steepest descent method and the Newton method have their own disadvantages. Steepest descent is slow, while Newton method is quick, but the calculation of the inverse of the Hessian matrix is time-consuming. Conjugate gradient method can seem as a trade-off method between the steepest descent method and Newton method, which accelerates the convergence rate of steepest descent while avoiding the high computational cost of Newton method.

Conjugate gradient method was originally developed for solving the convex quadratic problem

$$\min_{x \in \mathbb{R}^n} f(x) = \frac{1}{2} x^{\mathrm{T}} Q x - b^{\mathrm{T}} x \tag{3.22}$$

Assume that matrix $Q \in \mathbb{R}_{n \times n}$ is symmetric and positive definite. This problem is equivalent to $Qx = b, x \in \mathbb{R}^n$, which means that the conjugate direction method can solve this problem in most n iterations. The basic idea here is to find a search direction d_k that is orthogonal to all previous search directions, at each iteration, and we find the exact value of one variable x_k of the global optimum in each iteration. Unfortunately, we do not know the global optimum beforehand. To solve this problem, we introduce the Q-conjugate (orthogonal) directions.

Let Q be a symmetric matrix. Vectors $\{d_1, d_2, \cdots, d_k\}$ ($d_i \in \mathbb{R}^n, d_i \neq 0$) are Q-conjugate, if

$$d_i^{\mathrm{T}} Q d_j = 0, \forall i \neq j \tag{3.23}$$

Let Q be a positive definite matrix. If vectors $\{d_1, d_2, \cdots, d_k\}$ are Q-conjugate, then they are linearly independent. Here is the proof by contradiction.

Assume that vectors $\{d_1, d_2, \cdots, d_k\}$ are linearly dependent with each other; then,

$$d_k = \alpha_1 d_1 + \cdots + \alpha_{k-1} d_{k-1} \tag{3.24}$$

multiply $d_k^{\mathrm{T}} Q$; we have

$$
\begin{aligned}
d_k^{\mathrm{T}} Q d_k &= d_k^{\mathrm{T}} Q (\alpha_1 d_1 + \cdots + \alpha_{k-1} d_{k-1}) \\
&= \alpha_1 d_k^{\mathrm{T}} Q d_1 + \cdots + \alpha_{k-1} d_k^{\mathrm{T}} Q d_{k-1} \\
&= 0
\end{aligned}
\tag{3.25}
$$

$d_k^{\mathrm{T}} Q d_k > 0$ as $d_k \neq 0$. There is a contradiction in the assumption; thus, $\{d_1, d_2, \cdots, d_k\}$ are linearly independent.

Let x^* denote the optimal solution to the quadratic problem defined in Eq. (3.22) (it is also the unique solution to $Qx = b$) and vectors $\{d_1, d_2, \cdots, d_k\}$ be Q-conjugate. Since $\{d_1, d_2, \cdots, d_n\}$ vectors are independent, then

$$x^* = \alpha_1 d_1 + \cdots + \alpha_n d_n \tag{3.26}$$

Therefore

$$d_i^{\mathrm{T}} Q x^* = d_i^{\mathrm{T}} Q (\alpha_1 d_1 + \cdots + \alpha_n d_n) = d_i^{\mathrm{T}} \alpha_i Q d_i \tag{3.27}$$

then

$$\alpha_i = \frac{d_i^{\mathrm{T}} Q x^*}{d_i^{\mathrm{T}} Q d_i} = \frac{d_i^{\mathrm{T}} b}{d_i^{\mathrm{T}} Q d_i} \tag{3.28}$$

Now α_i is calculated without the need for x^*. Then the optimal solution x^* can be calculated

$$x^* = \sum_{i=1}^{n} \alpha_i d_i = \sum_{i=1}^{n} \frac{d_i^{\mathrm{T}} b}{d_i^{\mathrm{T}} Q d_i} d_i \tag{3.29}$$

The optimal solution can be calculated without the need for inverse of any matrixes. It can be regarded as the process of n iterations where $\alpha_i d_i$ is added at each iteration.

Let vectors $\{d_1, d_2, \cdots, d_{n-1}\}$ be Q-conjugate and $x_1 \in \mathbb{R}^n$ be an arbitrary starting point. By the update rule $x_{k+1} = x_k + \alpha_k d_k$ in Eq. (3.1), we have

$$\alpha_k = -\frac{g_k^{\mathrm{T}} d_k}{d_k^{\mathrm{T}} Q d_k} \tag{3.30}$$

where $g_k = \nabla f(x_k) = Qx_k - b$, and then after n steps, $x_n = x^*$. This is also the update rules for conjugate direction method.

Firstly we prove that $x_{n+1} = x^*$, since vectors $\{d_1, d_2, \cdots, d_n\}$ are independent, and there exist some $\alpha_1, \cdots, \alpha_n$ that satisfy the following equation:

$$x^* - x_1 = \alpha_1 d_1 + \cdots + \alpha_n d_n \tag{3.31}$$

using the update rule $x_{k+1} = x_k + \alpha_k d_k$

$$\begin{aligned}
x_2 &= x_1 + \alpha_1 d_1 \\
x_3 &= x_1 + \alpha_1 d_1 + \alpha_2 d_2 \\
x_k &= x_1 + \alpha_1 d_1 + \cdots + \alpha_{k-1} d_{k-1} \\
x_{n+1} &= x_1 + \alpha_1 d_1 + \cdots + \alpha_n d_n = x^*
\end{aligned} \tag{3.32}$$

Next, we prove that $\alpha_k = -\frac{g_k^T d_k}{d_k^T Q d_k}$. We already know

$$\begin{cases}
x^* - x_1 = \alpha_1 d_1 + \cdots + \alpha_n d_n \\
x_k - x_1 = \alpha_1 d_1 + \cdots + \alpha_{k-1} d_{k-1}
\end{cases} \tag{3.33}$$

Therefore

$$\begin{cases}
d_k^T Q(x^* - x_1) = d_k^T Q(\alpha_1 d_1 + \cdots + \alpha_n d_n) = \alpha_k d_k^T Q d_k \\
d_k^T Q(x_k - x_1) = d_k^T Q(\alpha_1 d_1 + \cdots + \alpha_{k-1} d_{k-1}) = 0
\end{cases} \tag{3.34}$$

where $d_k^T Q(x^* - x_1)$ can be re-written as

$$\begin{aligned}
d_k^T Q(x^* - x_1) &= d_k^T Q(x^* - x_k + x_k - x_1) \\
&= d_k^T Q(x^* - x_k) + d_k^T Q(x_k - x_1) \\
&= d_k^T Q(x^* - x_k) = d_k^T(Qx^* - Qx_k) \\
&= -d_k^T(Qx_k - b) = -d_k^T g_k = \alpha_k d_k^T Q d_k
\end{aligned} \tag{3.35}$$

Thus, we have

$$\alpha_k = -\frac{d_k^T g_k}{d_k^T Q d_k} \tag{3.36}$$

We can use the Gram-Schmidt process to construct the Q-conjugate vectors $\{d_1, d_2, \cdots, d_n\}$, where each conjugate direction d_{k+1} can be obtained by

$$d_{k+1} = -g_{k+1} + \frac{g_{k+1}^T Q d_k}{d_k^T Q d_k} d_k \tag{3.37}$$

where $d_1 = -g_1 = b - Qx_1$.

In comparison with the Newton method, the conjugate gradient method does not involve matrix inversion. Although the conjugate gradient method was first proposed to solve Quadratic programming and linear equation problems, later it was expanded to nonlinear functions. Two well-known methods are Fletcher-Reeves method [3] and the Hestenes-Stiefel method [4].

The conjugate gradient method is a method between the steepest descent and the Newton method. It needs the first derivative information, but it overcomes the shortcoming of slow convergence of the steepest descent method and avoids the shortcoming of the Newton method, which needs to store the Hessian matrix and calculate its inverse. It is one of the most useful methods to solve a large-scale system of linear equations. But it is still sensitive to initial points and cannot avoid being trapped into local optima.

This section introduces three famous line search methods. The steepest descent method is the most simple one, but it is relatively slow in convergence. While the Newton method has fast convergence speed, it requires strict conditions and needs to calculate the inverse of the Hessian matrix, which needs a large amount of calculation. The conjugate descent method is the method between the two. It has relatively fast convergence speed and avoids the calculation of the inverse of any matrix. However, these three methods still cannot avoid being trapped into local optima. As shown in Fig. 3.4, line search methods are shortsighted only with gradient information and may converge to a local optimum with the wrong choice of step size. Since all the immediate neighboring points around a local optimum are worse than it in the performance value, line search cannot proceed once trapped in a locally optimal point.

Fig. 3.4 Trapped into the local optimum

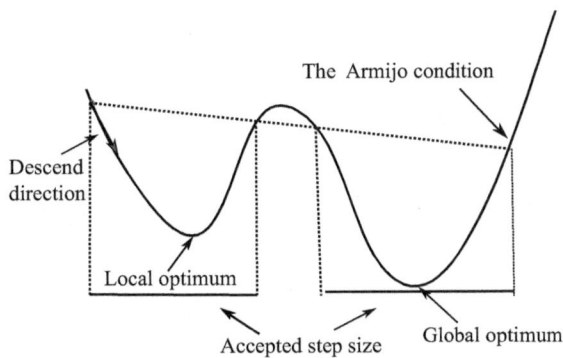

3.2   State Space Search

In continuous optimization, decision variables are real numbers typically within an interval. Contrary to continuous optimization, the set of decision variables in discrete optimization is a countably finite set, which typically can be mapped to a range of integral numbers. This section will introduce several well-known algorithms for discrete optimization problems.

In state space search, the discrete problem with countably finite decision variables is modelled as $\mathbf{S} = (\mathbb{S}, \mathbb{A}, action(s), result(s, a), cost(s, a))$, where $\mathbb{S}$ is the set of all possible states, $\mathbb{A}$ is the set of all possible actions, $action(s)$ is the set of all feasible actions of the state s, $result(s, a)$ is the next state after performing action a at state s, and $cost(a, s)$ returns the cost by taking action a at state s. Note that all the cost in the state space should be positive to guarantee the optimality of the algorithms discussed later. The solution is the goal state or the path from the initial state to the goal state, depending on the model of the problem.

The state space search starts from an initial state and records the current state s. Then it selects a feasible unvisited action a and records the states to be expanded from s. If there is no feasible unvisited action connected to s, then it will go back to get a new state. It will repeat this procedure until all states are visited or reach the goal state. Note that all the visited states are labeled as visited to avoid the algorithm being stuck into an infinite loop. Figure 3.5 shows the search process of state space search with a tree structure, where each state is denoted by a node and each action

Fig. 3.5 State space search

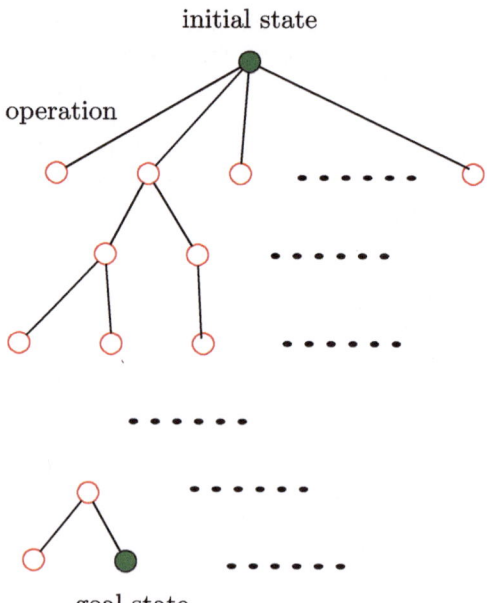

is denoted by an edge. The algorithm starts from an initial state and tries to find a path to the goal state with a set of actions.

The state space search can be divided into uninformed search and informed search. Uninformed search algorithms (also known as blind search) generate the search tree without any domain knowledge, while informed search tries to use some heuristic information to reach the goal state faster. There are several measures regarding the performance of a space state search algorithm:

(1) Completeness. Whether the algorithm can find a goal state when there is at least one goal state.
(2) Optimality. Whether the algorithm guarantees to find the optimal solution. The optimality of an algorithm means that the algorithm is able to find the shortest path from the initial state to the goal state in the search tree.
(3) Space complexity. The maximum space it takes to solve the problem, which indicates whether it can be implemented with the limited memory.
(4) Time complexity. The time it takes to solve the problem in the worst case, which indicates the run time of the algorithms.

Space complexity and time complexity are generally related to the number of states $\|\mathbb{S}\|$ and the number of actions $\|\mathbb{A}\|$ of the search tree.

3.2.1   State Space

In the state space search, the problem needs to be defined in the form of the above five components. This subsection will introduce two cases of problem formulation.

3.2.1.1   The Shortest Path Problem

Figure 3.6 shows the map of the part of China. It is a simple problem to find the shortest path from the capital city of Jilin province to the capital city of Hubei province. Bold lines show the optimal path, and numbers along the edges are the distance between the two connected cities. The formulation of the problem is defined as follows:

(1) Solution x. The solution to this problem is any path from Jilin to Hubei, e.g., the shortest path shown below.

$$x = \langle \text{Jilin, Inner Mongolia, Shanxi, Shaanxi, Henan, Hubei} \rangle$$

(2) State s. Each city is a state and all the cities form the whole state space. The number of total states in this problem is the number of total cities. The initial state $s_0 = $ Jilin, and the goal state $s_g = $ Hubei. The search starts from Jilin and tries to find a path to Hubei.

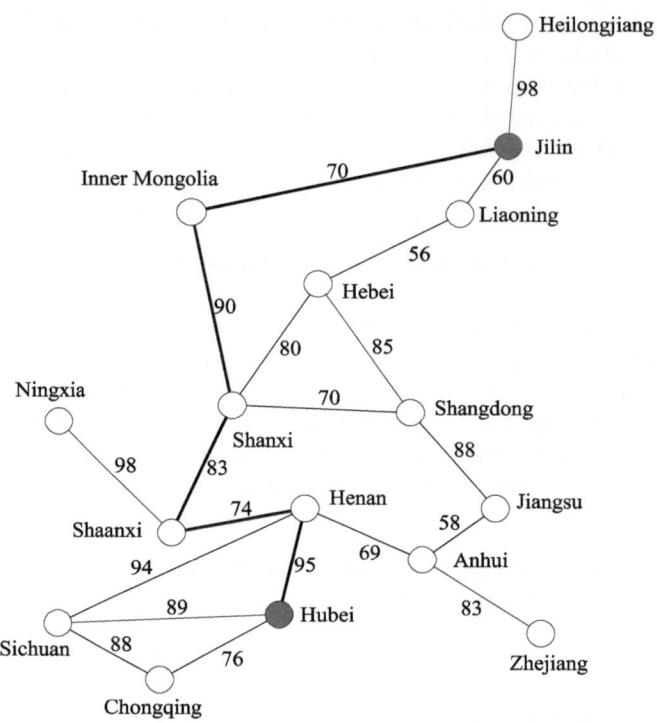

Fig. 3.6 The shortest path problem

(3) Action a. All roads that connect two cities consist of the whole action space. An action, which is also a road that connects two cities s_i and s_j, is written as $a_{s_i}^{s_j}$, e.g., $a_{Hebei}^{Shangdong}$.

(4) $action(s)$. All roads that connect the city s consist of the action set of $action(s)$, e.g., there are three roads that connect the city $Hebei$, then $action(Hebei) = \{a_{Hebei}^{Liaoning}, a_{Hebei}^{Shangdong}, a_{Hebei}^{Shanxi}\}$.

(5) $result(s, a)$. The search starts from a city, selects a road, and then moves to another city, e.g., $result(Hebei, a_{Hebei}^{Shangdong}) = Shangdong$.

(6) $cost(s, a)$. The cost of an action can be the distance of the road represented by the action, e.g., $cost(Hebei, a_{Hebei}^{Shangdong}) = 85$.

3.2.1.2 The Travelling Salesman Problem

The TSP is a famous problem for discrete optimization algorithms, which aims to find the shortest path loop around all cities. In the loop, all the cities are visited, and each city is visited only once. Figure 3.7 shows an example of a two-dimensional

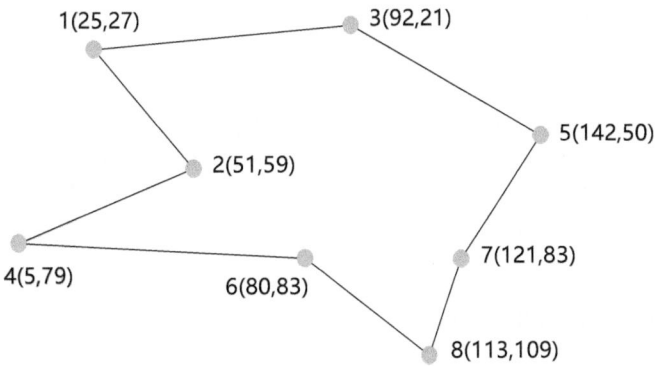

Fig. 3.7 The travelling salesman problem

TSP. Each node on the figure denotes a city, and the numbers besides denote its serial number and coordinates.

The state space of TSP can be modeled as follows:

(1) Solution **x**. A feasible solution in TSP is a loop that visits all cities only once, which is also one of the goal state. The solution in the figure can be coded **x** = 135786421, which represents the visiting order of each city.

(2) State s. Each state in TSP is a partial solution, say "135" is a state that indicates that cities 1, 3, and 5 have been visited in order. The initial state $s_0 = 1$ means that an algorithm starts from city 1 to construct a loop. The algorithm can start from any city, and here we take city 1 as the departing city. The goal state is a loop, and solutions with different initial cities but the same visiting order belong to one goal state. Starting from an initial city for a TSP with n cities, we have $n - 1$ choices for the first stop and $n - 2$ choices for the next stop, so there are totally $1 + n - 1 + (n - 1) * (n - 2) + \cdots + (n - 1)!$ states in total.

(3) Action a. All connections between cities consist of the whole action space. An action refers to a road that connects two cities, for example, action a_1^2 represents a road from city 1 to city 2.

(4) $action(s)$. The actions connected to the state s are all connections between the last visited city in state s and the remaining cities, e.g., $action(135) = \{a_5^2, a_5^4, a_5^6, a_5^7, a_5^8\}$.

(5) $result(s, a)$. The result of taking an action a in state s, that is, to append a city to s by taking action a, e.g., $result(135, a_5^7) = 1357$.

(6) $cost(s, a)$. The cost of an action can be the distance (w) of the road represented by the action, e.g., $cost(135, a_5^7) = w_5^7$.

3.2.2   Uninformed Search

Uninformed search does not use any information about the goal state, and it is also known as blind search. It only decides the rules of when to generate new states or when to reverse along the search tree. This subsection will introduce three classical uninformed search methods: breadth-first search, depth-first search, and depth limited search.

3.2.2.1   Breadth-First Search

Breadth-first search (BFS) always expands the states with the lowest depth, and it usually uses a queue (typically first in first out queue) to record states to be expanded. As Algorithm 3.3 shows, it always expands the state in the front of the queue and search all the unvisited states connected to s.

Figure 3.8 shows the search order of BFS in a search tree, where the letter inside each node is the name of each state and the number beside the edge is the visit order of each action. The initial state is A, and the goat states are P and V. BFS always finds the goal state with the lowest depth, i.e., P in Fig. 3.8, and stops when it finds the goal state.

Algorithm 3.3: Breadth-first search

Input: a problem $\mathbf{S} = (\mathbb{S}, \mathbb{A}, action(s), result(s, a), cost(s, a))$
Output: the found solution
1 Put an initial state s_0 into a set $\mathbb{S}_v$ (the set of all states to be expanded);
2 Set the initial state s_0 visited;
3 **while** $\mathbb{S}_v$ *is not empty* **do**
4   Select and remove the front state s from $\mathbb{S}_v$;
5   **if** s *is the goal state* **then** return the found solution
6   **for** *all feasible action a connected to s* **do**
7          $s' = action(s, a)$;
8   **if** s' *is not visited* **then**
9   Insert s' into $\mathbb{S}_v$;
10   Set s' visited;

Here we will discuss the performance of BFS from the following four respects:

(1) Completeness. BFS can always find a feasible solution as long as there is at least one goal state in the search tree.
(2) Optimality. BFS does not guarantee to find the optimal solution. BFS always finds the solution with the minimum level to the root of the search tree. But if the cost of all actions in the problem is the same, it guarantees to find the optimal solution.

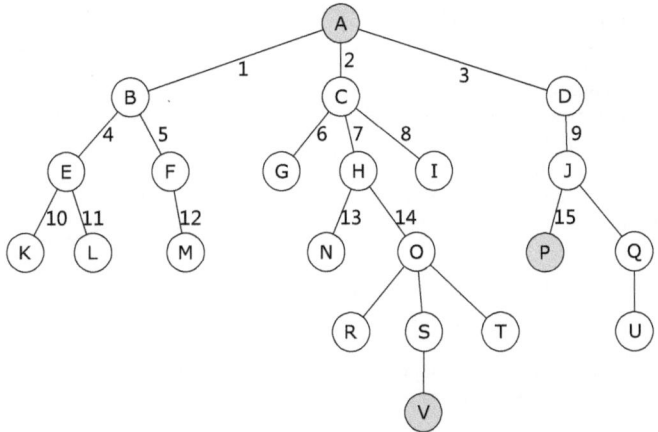

Fig. 3.8 The search order of breadth-first search in a search tree

(3) Space complexity. In the worst case, that is, the goal state is situated at the right-bottom of the search tree. BFS should store the whole tree nodes in the queue before finding the goal state. So the space complexity of BFS is $O(\|\mathbb{S}\|)$ for explicit search tree.

(4) Time complexity. BFS should search the whole search tree in the worst case, so the time complexity of BFS is $O(\|\mathbb{S}\| + \|\mathbb{A}\|)$ for exploring every node and each edge.

3.2.2.2   Depth-First Search

The goal state is usually situated at the bottom of the search tree, while BFS usually expands the states with the lowest depth, so it may take a long time to reach a goal state for most cases. To gain the goal state faster, depth-first search (DFS) can be used.

Unlike BFS, DFS always expands the states with the deepest depth and uses a stack (first in last out queue) in the algorithm. As shown in Algorithm 3.4, it always expands the first state in the stack, which is one of the states with the deepest depth explored so far.

Figure 3.9 shows the search order of DFS in the same search tree as shown in Fig. 3.8. From the figure, we can see that DFS always finds the goal state on the far left of the search tree, that is, V, where DFS takes 14 actions to reach V.

Similarly, we will discuss the performance of DFS from the following four aspects:

(1) Completeness. DFS guarantees to find a feasible solution if there is at least one goal state in the problem.
(2) Optimality. DFS does not guarantee to find the optimal solution.

Algorithm 3.4: Depth-first search

Input: a problem $\mathbf{S} = \{\mathbb{S}, \mathbb{A}, action(s), result(s, a), cost(s, a)\}$
Output: the found solution
1 Insert the initial state s_0 into the set $\mathbb{S}_v$ (the set of all states to be expanded);
2 **while** $\mathbb{S}_v$ *is not empty* **do**
3 $\quad$ Select the front state s from $\mathbb{S}_v$;
4 $\quad$ **if** *No feasible unvisited action connected to s* **then**
5 $\quad\quad$ Remove s from $\mathbb{S}_v$ and go to step 3;
6 $\quad$ **else**
7 $\quad\quad$ Get a feasible unvisited action a of s;
8 $\quad$ Set the selected action a visited;
9 $\quad$ $s' = action(s, a)$;
10 $\quad$ **if** s' *is the goal* **then** return the found solution
11 $\quad$ **if** s' *is not visited* **then** Insert s' into $\mathbb{S}_v$

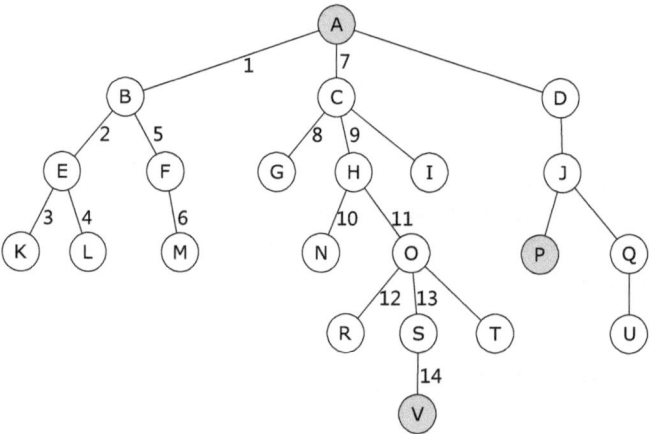

Fig. 3.9 The search order of depth-first search in a search tree

(3) Space complexity. In the worst case, that is, the goal state is situated at the right-bottom of the search tree, DFS should search the whole tree to find the goal state. It should record all states and actions, so the space complexity of DFS is $O(\|\mathbb{S}\|)$ for the explicit search tree.

(4) Time complexity. DFS should search the whole search tree in the worst case, so the time complexity of DFS is $O(\|\mathbb{S}\| + \|\mathbb{A}\|)$ for expanding the whole search tree.

3.2.2.3   Depth-Limited Search

Sometimes, the search tree of a problem may be very deep, and it takes a long time to reach a goal state even by DFS. For example, in the Go game, it takes almost 200 actions to calculate the outcome. To reach a goal state faster, the idea of limiting the depth of the search can be applied to DFS. This idea may also help find better solutions, because if the solution is at a deeper node, it contains more edges, and the total cost is relatively larger.

Algorithm 3.5 is the framework of depth-limited search. It is quite similar to DFS. The only difference between DFS and depth-limited search is that depth-limited search introduces the depth of the current state, and if the depth is larger than the limited depth L, it does not expand the current state and goes back along the search tree.

Algorithm 3.5: Depth-limited search

Input: a problem $S = \{S, A, action(s), result(s, a), cost(s, a)\}$
Output: the found solution
1 Insert the initial state s_0 into the set S_v (set of all states to be expanded);
2 $d_{s_0} = 0$ (d_{s_0} is the depth of the state s_0);
3 **while** S_v *is not empty* **do**
4 | Select the front state s from S_v;
5 | **if** *No feasible unvisited action connected to s* **then**
6 | | Remove s from S_v and go to step 4;
7 | **else**
8 | | Get a feasible unvisited action a of s;
9 | Set the selected action a visited;
10 | $s' = action(s, a)$;
11 | $d_{s'} = d_s + 1$;
12 | **if** s' *is a goal state* **then** return the found solution
13 | **if** s' *is not visited and* $d_{s'} < L$ **then** Insert s' into S_v

---

Figure 3.10 shows the search order of the depth-first search in a search tree. It will never visit the state with depth greater than L, which is 4 in the figure; thus, it can reach the goal state V. Note that if the limited depth is equal to or larger than the depth of the search tree, the depth-limited search is the same as DFS.

The four measurements of depth-limited search are discussed as follows:

(1) Completeness. The depth-limited search cannot find a feasible solution when the limited depth L is smaller than the depth of any feasible state. To overcome this shortcoming, iterative deepening depth-first search can be used with an automatic adjustment depth.
(2) Optimality. The depth-limited search also does not guarantee to find the optimal solution.
(3) Space complexity. The space complexity of depth-limited search is similar to that of DFS, but the states at depths deeper than the limited depth are never visited. So the space complexity of the depth-limited search is $O(\|S\|)$.
(4) Time complexity. The time complexity of the depth-limited search is also similar to that of DFS. It only visited states with a depth lower than the limited depth. So, the time complexity is $O(\|S\| + \|A\|)$.

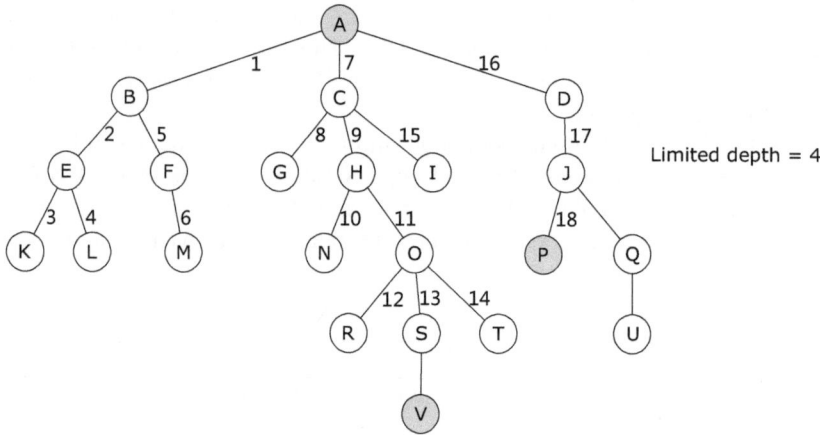

Fig. 3.10 The search order of depth-limited search in a search tree

3.2.3 Informed Search

Unlike uninformed search, informed search makes use of problem-domain knowledge to reach the goal state faster and find a better solution. This subsection will introduce three informed search methods: greedy search, A* search, and Monte-Carlo tree search.

3.2.3.1 Greedy Search

Greedy search is a well-known informed search. The idea of greedy search is quite simple using any greedy idea. Algorithm 3.6 shows the framework of greedy search, which is quite similar to that of DFS where $\mathbb{S}_v$ is a stack. Greedy search always selects the feasible unvisited action with the minimum cost to fulfill its greedy idea. This way is reasonable with a positive cost in the state space.

Figure 3.11 shows the search order of each state of greedy search to solve the shortest path problem introduced in Sect. 3.2.1. The number inside the node is the search order of each state, and the bold line is the path found by greedy search. From the figure, we can see that greedy search always selects the unvisited action with the minimum cost.

The four measurements of greedy search are discussed as follows:

(1) Completeness. Similar to DFS, greedy search guarantees to find the goal state with the search strategy.
(2) Optimality. Although the greedy idea is reasonable, it may misguide the search to select a bad action and then get a bad solution. Greedy search can quickly obtain a solution, but the solution is normally a local optimum. However, if the optimal choice can be made at each step, then it can obtain the global optimum.

Algorithm 3.6: Greedy search

Input: a problem $\mathbf{S} = \{\mathbb{S}, \mathbb{A}, action(s), result(s, a), cost(s, a)\}$
Output: the found solution

1 Insert the initial state s_0 into the set $\mathbb{S}_v$;
2 **while** $\mathbb{S}_v$ *is not empty* **do**
3   Select the front state s from $\mathbb{S}_v$;
4   **if** *No feasible unvisited action connected to s* **then**
5   Pop s from S_v and go to step 3;
6   Set current cost $c = inf$ and the selected action $a_s = null$;
7   **for** *all feasible unvisited action a connected to s* **do**
8   **if** $c > cost(s, a)$ **then** $c = cost(s, a); a_s = a$
9   Label the selected action a_s as visited;
10     $s' = action(s, a_s)$;
11   **if** s' *is the goal state* **then** Return the found solution
12   **if** s' *is not visited* **then** Insert s' into $\mathbb{S}_v$

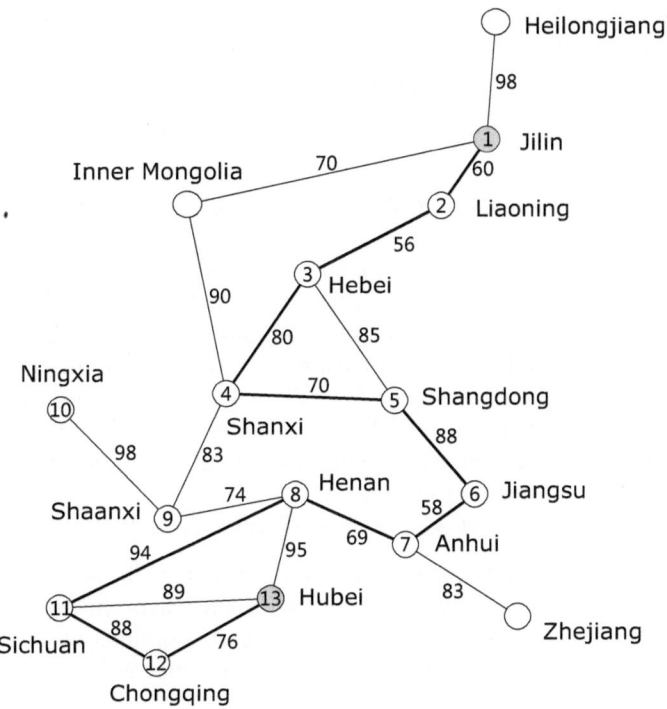

Fig. 3.11 Search order of greedy search in a shortest path problem

(3) Space complexity. It is possible that the greedy search visits the whole search tree until it reaches the goal state. So in the worst case, greedy search stores the whole search tree, and its space complexity is $O(\|\mathbb{S}\|)$.

(4) Time complexity. The time complexity of greedy search is similar to that of DFS. In the worst case, it visits the whole search space until it reaches the goal state. So the time complexity of greedy search is $O(\|\mathbb{S}\| + \|\mathbb{A}\|)$.

The greedy idea is quite simple and is widely used to combine with evolutionary algorithms, which will be discussed later to solve complex optimization problems.

3.2.3.2 A* Search

Different from the previous search algorithms, A* search uses an intelligent way of using heuristic information to guide the search. It guarantees to find the global optimal. The idea of A* search is to select the next state along which the path has the minimum expected cost from the initial state to the goal state.

To achieve the above idea, it introduces a heuristic function $h(s)$ to estimate the cost of the shortest path from the current state s to the goal state. It uses the estimated total cost $f(s)$ to decide which state to select:

$$f(s) = g(s) + h(s) \tag{3.38}$$

where $g(s)$ is the true cost of the shortest path from the initial state to the current state s.

The framework of A* search is similar to that of BFS, which is shown in Algorithm 3.3. The only difference is that A* search uses the priority queue for $\mathbb{S}_v$, which is a heap with the state with the minimum estimated total cost $f(*)$ at the top. It should be noted that all states must be expanded before they are goal-tested in A* search to find the shortest path from the initial state to the goal state, that is, to find the global optimal. It is modeled as the shortest path from the initial state to the goal state or the goal state with the minimum total cost to the initial state, while there are no such requirements in BFS.

The four measurements of A* search are discussed as follows:

(1) Completeness. The framework of A* search is almost the same as that of BFS, so it guarantees to find a feasible solution if there exists one as BFS does.

(2) Optimality. A* search always can find the global optimal when the designed heuristic function satisfies the admissibility constraint as follows:

$$0 \le h(s) \le h^*(s) \tag{3.39}$$

where $h^*(s)$ is the true optimal cost from the current state s to the goal state.

Suppose that there are at least two goal states in the problem: s_g is the goal state with the global shortest path from the initial state, and s_s is the goal state with the suboptimal shortest path from the initial state. It can be concluded that

$$g(s_g) < g(s_s) \tag{3.40}$$

the cost from the initial state to s_g must be smaller than that of s_s. Since s_g and s_s are both goal state, thus their heuristic functions satisfy

$$0 \le h(s_g) \le h^*(s_g) = 0 \Rightarrow h(s_g) = 0$$
$$0 \le h(s_s) \le h^*(s_s) = 0 \Rightarrow h(s_s) = 0 \tag{3.41}$$

With Eq. (3.40) and Eq. (3.41)

$$f(s_g) = g(s_g) + h(s_g) = g(s_g) < g(s_s) = g(s_s) + h(s_s) = f(s_s) \tag{3.42}$$

thus the expected total cost of s_g is smaller than that of s_s. Supposed that s_a is any state through the global shortest path from the initial state to the goal state s_g and its heuristic function also satisfies the admissibility constraint

$$h(s_a) < h^*(s_a) \tag{3.43}$$

since s_a is a state on the optimal shortest path, $h^*(s_a)$ is actually the shortest distance of the path from s_a to s_g, and it can be concluded that

$$f(s_a) = g(s_a) + h(s_a) < g(s_a) + h^*(s_a) = g(s_g)$$
$$= g(s_g) + h(s_g) = f(s_g) \tag{3.44}$$

With Eq. (3.42) and Eq. (3.44), it can be concluded that

$$f(s_a) < f(s_g) < f(s_s) \tag{3.45}$$

the expected total cost of any state through the shortest path is smaller than that of s_s, and thus they are expanded before s_s. The algorithm can then finally find the global shortest path from the initial state to s_g.

(3) Space complexity. Similar to BFS, if the heuristic function is poorly designed like $h(*) = 0$, the information is not useful, and A* search still needs to visit the whole search tree until it reaches the goal state. In the worst case, A* search should record all states in the search tree, and its space complexity is $O(\|\mathbb{S}\|)$.

(4) Time complexity. The time complexity of A* search is similar to that of BFS. In the worst case, it visits the whole search space until it reaches the goal state. So the space complexity of A* search is $O(\|\mathbb{S}\| + \|\mathbb{A}\|)$.

Another key problem to A* search is the design of the heuristic function $h(*)$. It should be very close to the real optimal cost function $h^*(*)$ to provide enough information to the search; otherwise, the performance of A* search will degenerate to that of BFS. For example, in TSP, the largest distance between the end city c_e of the current sub-path s and the unvisited cities left can be a heuristic function

$$0 \le h_1(s) = \max_{c \in \mathbb{S}_u} \|c_e, c\| \le f^*(s) \tag{3.46}$$

where $\mathbb{S}_u$ is the set of unvisited cities.

The heuristic information can also be defined as the sum of edge weight of the minimum spanning tree (MST) of all the unvisited cities and the end city of the current sub-path.

$$0 \le h_2(s) = \text{MST}(\mathbb{S}_u \cup \{c_e\}) \le f^*(s) \tag{3.47}$$

Obviously, $h_2(s)$ is better than $h_1(s)$. It satisfies $h_1(s) \le h_2(s)$, so that fewer states are expanded with $h_2(s)$ than $h_1(s)$, and thus A* search can find the global optimal faster.

3.2.3.3 Monte-Carlo Tree Search

The DFS, depth-limited search, and greedy search can quickly provide a feasible solution but do not guarantee to find the global optimal solution. Some algorithms, e.g., A* search, take a long time to provide the global optimal solution, especially for the complex problem with many states, e.g., TSP with many cities. Monte-Carlo tree search is a compromising method, which tries to provide a relatively good solution within the limited time and is capable of providing a better solution as the time increases.

Monte-Carlo tree search was first proposed to solve the combinatorial game problem, such as chess game and Go game. In these problems, each state can be viewed as a chessboard arrangement. The idea of Monte-Carlo tree search combines the generality of random simulation and the accuracy of tree search. It aims to find the most promising state by random sampling in the search tree and construct the search tree with the sample results. Each sample is a feasible path from the initial state to the goal state, and the evaluation of the found goal state is pushed forward to each state along the path. The expected cost of the state is calculated based on the results of these samplings. This procedure is somewhat like a Markov process, and the expected cost of each state is more accurate with more samplings.

Figure 3.12 shows the framework of Monte-Carlo tree search [2], which consists of four main components as below. It repeats the four steps until the time runs out.

(1) Selection. The algorithm starts from the initial state and searches along the most promising state until it reaches a state with unvisited actions, which is a leaf node in the incomplete search tree. The algorithm search toward the most promising direction is also the key idea of Markov process.
(2) Expansion. The algorithm expands the selected leaf state to one of the child states c with one of the possible actions.

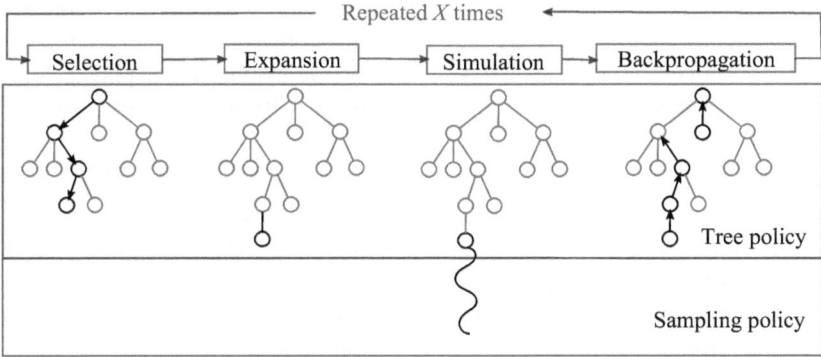

Fig. 3.12 Framework of Monte-Carlo tree search

(3) Simulation. The algorithm samples several times from the newly created state c, that is, to generate some paths from c to different goal states. The results of these samplings are used to evaluate the state c.

(4) Backpropagation. The evaluation of c is used to update the evaluation of each state along the path from c to the initial state.

One of the difficulties in selection is how to select the most promising state and maintain a good balance between exploration (the state with good performance) and exploitation (the state with few simulations). A good idea is to use the upper confidence bounds applied to trees

$$v_i + C\sqrt{\frac{\ln n_i'}{n_i}} \tag{3.48}$$

where v_i is the expected value of the current state i (normally the number of wins over the number of simulations for state i); n_i is the number of simulations for state i ; n_i' is the total number of simulations of its parent state; and C is a user-defined parameter, which is usually set as $\sqrt{2}$ in practice.

For this formulation, each node that represents each state should record two important information, the number of simulations, and the expected value of the state.

The four measurements of Monte-Carlo tree search is discussed as follows:

(1) Completeness. Monte-Carlo tree can find at least one feasible solution.

(2) Optimality. With the randomness, Monte-Carlo tree does not guarantee finding the optimal solution.

(3) Space complexity. It does not need to search the whole search tree to find a solution; its space complexity is related to the time it runs.

(4) Time complexity. The user can stop the algorithm at any time, and it will provide better solutions as time goes on.

Nowadays, the research of Monte-Carlo tree search is still very popular with its potentials to solve the complex optimization problems. In 2017, AlphaGo with Monte-Carlo tree search inside itself won the world's number one player, Kejie, in Go game. Monte-Carlo tree search is quite different from the classical tree search with advantages and disadvantages.

The advantages of Monte-Carlo tree search are significant, and it is quite useful:

(1) With simulation, it does not need any heuristic information to guide the search or an explicit evaluation function to evaluate each state. The idea of Monte-Carlo tree search is general and can be easily extended to solve complex optimization problems without any expert knowledge.
(2) The search tree grows asymmetrically as the search tends to search for the more promising subtree, which helps solve the complex problem with many states and many actions.
(3) The framework of Monte-Carlo tree search is clear and simple, so it can be easily implemented.
(4) The Monte-Carlo tree search can be terminated at any time with the best solution found.

Monte-Carlo tree search has some disadvantages, but they are very critical to its performance:

(1) Monte-Carlo tree search may miss some critical state due to the randomness in the simulation, which may lead to failure in game problems.
(2) It may need many samplings to get a relatively accurate evaluation of a state in a large-scale problem, which is time-consuming.

3.3 Single-Solution-Based Random Search

This section introduces several single-solution-based heuristic algorithms with randomness mechanism. These methods include hill climbing, simulated annealing, iterated local search, variable neighborhood search, and guided local search.

3.3.1 Hill Climbing

Hill climbing is a simple mathematical random search algorithm. It is an improvement on DFS introduced in Sect. 3.2.2.2 and employs the feedback information to create an offspring. To elaborate, it performs a loop in which the currently known best solution is adopted to generate an offspring solution. The offspring solution will replace the parent solution only if the objective value of the offspring solution is better than that of the parent. Then, it repeats the procedures all over again.

Algorithm 3.7 shows the pseudo-code of the hill climbing algorithm. At each iteration, it will adjust a single element for x and determine whether the change improves the objective value. It is worth noting that this differs from gradient descent methods introduced in Sect. 3.1, which adjust all the elements in x at each iteration according to the gradient of the hill. In hill climbing, any change that improves $f(x)$ is accepted, and the process continues until no change can be made to improve x. Then we can say that a locally optimal solution is found. Figure 3.13 gives an illustration of the hill climbing algorithm. The search starts from an initial solution x_0, and after three iterations, there is no improvement of the objective value on solution x_3, solution x_3 is then returned as a local optimum.

Algorithm 3.7: Hill climbing

Input: Initial solution x
Output: Final solution x_{best}
$x_{best} \leftarrow x$;
while *termination criterion is not satisfied* **do**
    Using x_{best} to create a new solution x, evaluate its objective value $f(x)$;
    if $f(x)$ *is better than* $f(x_{best})$ **then**
        $x_{best} \leftarrow x$;

return x_{best} and $f(x_{best})$.

From the above description, it is evident that the key issue of the hill climbing algorithm is premature convergence, which is easily stuck in some local optima. The quality of the hill climbing algorithm depends heavily on the location of the initial solution as gradient descent methods. If an initial solution is close to the global optimum, the hill climbing algorithm is more likely to find the global optimum. By contrast, if an initial solution is far away from the global optimum, a local optima solution is likely to be found.

Fig. 3.13 Diagram of hill climbing

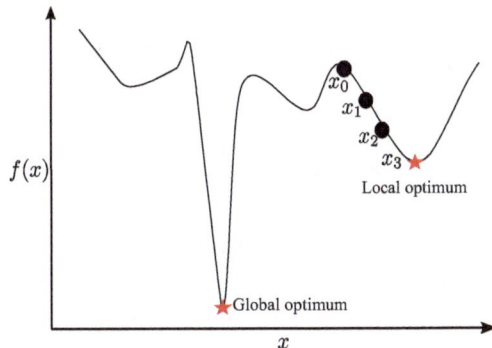

3.3.2   Simulated Annealing

Simulated annealing (SA) [5] is an optimization algorithm for addressing especially hard discrete optimization problems, e.g., the TSP. The core idea of SA is to mimic the heating and controlled cooling process of metals with the aim of increasing the ductility and reducing the hardness. To simulate the process, SA generates a new potential solution by altering the current state according to a predefined criterion at each virtual annealing temperature. The acceptance of the new state is based on the satisfaction of the Metropolis criterion, and this procedure is repeated until the termination criterion is satisfied.

Algorithm 3.8 shows the pseudo-code of the SA algorithm. At each iteration, a new solution is randomly generated within the neighbor of the current solution. The distance of the new solution from the current solution (i.e., the extent of the exploration) is typically based on a probability distribution with a scale that is proportional to the temperature. SA always accepts new solutions that are better than the current solution but also with a certain probability of selecting inferior solutions that have poorer objective values. By accepting these solutions, SA may avoid being trapped in local optima in the early iterations and thus is able to explore globally for the global optimum. In this way, SA has a much better global search capability than hill climbing and gradient-based methods. Therefore, SA is regarded as a global search method.

Algorithm 3.8: Simulated annealing

Input: Initial solution x
Output: Final solution x_{best}
$x_{cur} \leftarrow x$;
$x_{best} \leftarrow x$;
$t=0$;
while *termination criterion is not fulfilled* **do**
    Compute the energy difference $\Delta E = f(x) - f(x_{cur})$;
    if $\Delta E \leq 0$ **then**
        $x_{cur} \leftarrow x$;
        if $f(x_{cur}) < f(x_{best})$ **then**
            $x_{best} \leftarrow x_{cur}$;

    else
        $T \leftarrow$ getTemperature(t); // used in the calculation of $p(\Delta E)$
        if *rand* $< p(\Delta E)$ **then**
            $x_{cur} \leftarrow x$;

    Using x_{cur} to create a new solution x, evaluate its objective value $f(x)$;
    $t = t + 1$.
return x_{best} and $f(x_{best})$.

The laws of thermodynamics state that for a substance in a state of equilibrium at temperature T, the probability of accepting an inferior new solution is given by

$$p(\Delta E) = \exp\left(\frac{-\Delta E}{kT}\right) \tag{3.49}$$

where k is the Boltzmann constant.

Metropolis proposed a simulation model to determine the energy of a system in its ground (or frozen) state, that is, when the temperature is reduced to its limiting value. In the simulation, the material is considered as a system of atoms. For each possible configuration of atoms, the energy of the system can be calculated. At each iteration, one atom in the current configuration is subjected to a small displacement, and the increase in energy, ΔE, is calculated. If $\Delta E \leq 0$, a new configuration is accepted automatically. Otherwise, it is accepted with probability $p(\Delta E)$. This is facilitated by generating a random number, $rand$, in the interval [0,1] and accepting the new configuration if $rand < p(\Delta E)$. SA simulates the annealing process by repeatedly reducing the value of T until the system freezes at its ground state.

3.3.3 Iterated Local Search

Iterated local search [6] is a metaheuristic that tries to improve a solution by permutation and local search in each iteration. The iterative process consists of a perturbation of the current solution, obtaining an intermediate solution that is used as a new starting solution for the local search method. This simple idea has been successfully used to tackle various combinatorial optimization problems.

The iterated local search works as Algorithm 3.9. Let x be the starting solution. At each loop, a diversification process is first adopted by a *Perturbation* operator. Perturbation mutates the current best solution x and generates an intermediate solution x'. Then the local search is applied to x' and produces a new solution x''. If x'' wins x in *Apply Acceptance Criterion* operator taking the history search into account, then x will be replaced by x'' and becomes the current best solution; otherwise, the search starts again at the previous best solution x. The perturbation aims at escaping from local optima and exploring other promising regions of the search space.

Algorithm 3.9: Iterated local search

Input: Initial solution x
Output: Final solution x
while *termination criterion is not fulfilled* **do**
  $x' \leftarrow$ Perturbation(x);
  $x'' \leftarrow$ LocalSearch(x');
  $x \leftarrow$ Apply Acceptance Criterion(x, x'', history);
return x and $f(x)$.

Fig. 3.14 Diagram of the
iterated local search

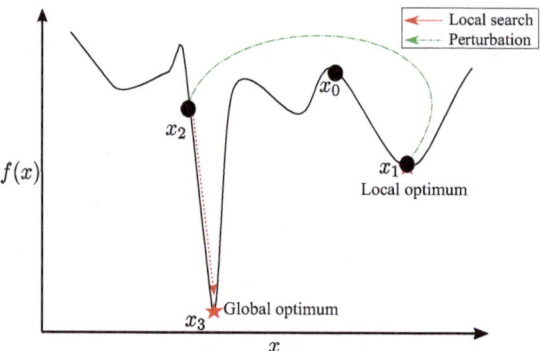

The overall iterated local search procedure is pictorially illustrated in Fig. 3.14. To be specific, after performing a local search on the initial solution x_0, a new solution x_1 is produced. Then, we apply a perturbation leading to a solution x_2. After applying the local search, a new local minimum x_3, which may be better than x_1.

3.3.4 Variable Neighborhood Search

Variable neighborhood search (VNS) [7] systematically changes the neighborhood both within a local search phase to find a local optimum and in a perturbation phase to get out of the local optima. Unlike previous metaheuristic methods, VNS explores increasingly distant neighborhoods of the current solution as the search goes on and moves to a new location if and only if an improvement was made.

The complete process of the VNS algorithm can be summarized by Algorithm 3.10.

Algorithm 3.10: Variable neighborhood search

Input: A set of neighborhood structures N_k for $k = 1, \cdots, k_{max}$ for shaking
Output: The best solution and its objective value
while *termination criterion is not fulfilled* **do**
    $k=1$;
    while $k < k_{max}$ **do**
        Pick a random solution x' from the neighborhood N_k of x;
        Apply some local search method with x' with the result of x'' ;
        if $f(x'')$ *is better than* $f(x)$ **then**
            $x \leftarrow x''$, and continue the search with N_{k+1};
        else
            $k = k + 1$.

return x and $f(x)$.

Before the iteration, a set of neighborhood structures, N_k for $k = 1, \cdots, k_{\max}$, needs to be set and will be adopted in the search. The main step of VNS mainly includes two operations: a shaking operation and a local search operation. The shaking operation is used to generate a random solution x' from the neighborhood of x ($x' \in N_k(x)$). The local search operation is used to adopt some local search approach with x' as the initial solution; the corresponding obtained local optimum is denoted by x''. If the obtained solution x'' is better than x, then x is replaced by x'', and continue the search with N_{k+1}; otherwise, set $k = k + 1$.

References

1. Barzilai, J., Borwein, J.M.: Two-point step size gradient methods. IMA J. Numer. Anal. **8**(1), 141–148 (1988)
2. Browne, C.B., Powley, E., Whitehouse, D., Lucas, S.M., Cowling, P.I., Rohlfshagen, P., Tavener, S., Perez, D., Samothrakis, S., Colton, S.: A survey of Monte Carlo tree search methods. IEEE Trans. Comput. Intell. AI Games **4**(1), 1–43 (2012)
3. Fletcher, R., Reeves, C.M.: Function minimization by conjugate gradients. Comput. J. **7**(2), 149–154 (1964)
4. Hestenes, M.R., Stiefel, E.: Methods of conjugate gradients for solving linear systems. J. Res. Natl. Bureau Stand. **29**(1), 409–436 (1952)
5. Kirkpatrick, S., Gelatt, C.D., Vecchi, M.P.: Optimization by simulated annealing. Science **220**(4598), 671–680 (1983)
6. Lourenço, H.R., Martin, O.C., Stützle, T.: Iterated local search. In: F.W. Glover, G.A. Kochenberger (eds.) Handbook of Metaheuristics, pp. 320–353. Springer International Publishing, Norwell (2003)
7. Mladenović, N., Hansen, P.: Variable neighborhood search. Comput. Oper. Res. **24**(11), 1097–1100 (1997)

Part II
Evolutionary Computation Algorithms

This part introduces the basic components of evolutionary computation algorithms including solution representation, selection, and reproduction operators. Several typical evolutionary computation algorithms are introduced, i.e., genetic algorithms, evolutionary programming, genetic programming, evolution strategy, estimation of distribution algorithms, ant colony optimization, particle swarm optimization, and differential evolution.

Chapter 4
Basics of Evolutionary Computation Algorithms

Abstract This chapter mainly introduces some basic concepts and components about evolutionary algorithms (EAs). A brief origination and history of evolutionary computation (EC) is reviewed, including some biological processes that have inspired researchers. Classic evolutionary procedures, including representation, selection, and reproduction, are described in detail.

4.1 Introduction

EAs are population-based iterative metaheuristic search methods based on biology-inspired mechanisms such as mutation, crossover, and natural selection, to reproduce new candidates and generate new populations [2]. Compared to the conventional optimization methods introduced in Chap. 3, EAs have some distinct advantages: they could work on problems called "black box" with little a priori knowledge, and even do not need an optimization model; just a performance measure could make it work. In addition, EAs do not work with a single solution but instead work in a way of parallel search with a population of solutions. This enables EAs to have much stronger global search capability than conventional optimization methods.

4.1.1 Biological Evolution

In 1859, Darwin published his book *On the Origin of Species* [6], in which he identified the principles of natural selection and survival of the fittest as the driving forces of the biological evolution. His theory can be extracted into the following observations and deductions [21].

The population size of a species would remain a dynamic balance in nature if there is no sudden interference such as food limits or disasters. Individuals in each species could reproduce an offspring, but the chances associated with them are different, depending on their fertility and robustness. Especially in sexually

C. Li et al., *Intelligent Optimization*,
https://doi.org/10.1007/978-981-97-3286-9_4

reproducing species, only two different individuals can combine and generate an offspring. During reproduction, an offspring will inherit genes or traits from their parents, especially good traits that will help them survive in the environment. After reproduction, all individuals need to compete for limited foods or adapt to environments, and only the physically fit offspring will survive and start a new reproduction process, i.e., those offspring that cannot adapt to changes or fail to compete will be eliminated eventually.

From the perspective of molecular genetics, the evolution of organisms is related not only to the external environment but also to their own genetic material. Genetic material determines the traits of organisms. In biology, the smallest genetic unit of an organism is a gene. Gene fragments make up the DNA and RNA, which are macromolecules that perform genetic functions. The characteristics and traits shown by organisms are the products of gene expression, so biologists call them genotype and phenotype. Normally, the genotype plays a vital role in the phenotype. In biological structure, DNA can exist as an alone chain like the RNA or can be combined with proteins to form chromosomes, which are widely present in eukaryotic cells.[1] Chromosomes can exist in pairs to determine the traits of organisms. Genes at the same position on a pair of chromosomes are called alleles.

In the process of biological evolution, reproduction can effectively expand the size of species and enable species to survive and genes to be preserved. In the actual reproduction process, the genetic material will undergo separation, replication, recombination, mutation, and other processes, and only one of the alleles participates in recombination. These processes help enhance the diversity of the genetic material and thus make them adapt to complex and changing environments.

In nature, reproduction of various organisms can be divided into sexual reproduction[2] and asexual reproduction.[3] For asexually reproduced organisms, the genetic material of the offspring will copy from only one parent, and genetic mutation may occur during the replication process. But for sexually reproduced organisms, genetic materials may come from multiple parents (usually two parents, as seen in nature). They can be recombined and mutated, leading to an obvious difference between offspring and their parents.

Figures 4.1 and 4.2 show the principle of gene mutation and chromosomal recombination, respectively. A gene mutation permanently changes the sequence that makes up a gene and takes effect on characteristics of a single individual. In the gene mutation process, normally, most bases of the sequence are preserved, while only a very small fragments are altered. Contrary to mutation, chromosome recombination normally involves two parents and results in the heritage of half of genetic material from each parent.

Biological evolution is a recurring process: individuals grow, reproduce offspring, and eventually die after a limited lifespan. Their genetic material is inherited

---

[1] https://en.wikipedia.org/wiki/Eukaryote

[2] https://en.wikipedia.org/wiki/Sexual_reproduction

[3] https://en.wikipedia.org/wiki/Asexual_reproduction

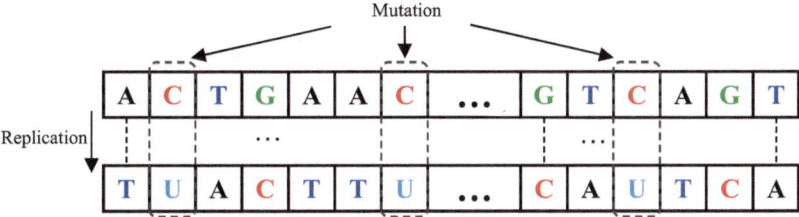

Fig. 4.1 Mutation in base complementary pair

Fig. 4.2 The process of the chromosome recombination

by their offspring and continues to propagate. Due to mutations or exchanges in genetic material during reproduction, offspring are not identical to their parents but exhibit some similar traits. This diversity is the driving force behind species' adaptation to environments and the acquisition of new traits. Mutations, in particular, introduce new genes to individuals and increase the diversity of the gene pool.

4.1.2 Origin of Evolutionary Algorithms

EAs simulate several vital processes based on biological evolution, aiming to find optima of optimization problems. The search space $\mathbb{X}$ of a problem resembles a DNA repository $\mathbb{G}$, where genetic information can be found. Each concrete element $g \in \mathbb{G}$ represents a natural genotype. Genotypes undergo a genotype-phenotype mapping, resulting in corresponding phenotypes, which are candidate solutions $x \in \mathbb{X}$ within the search space. The fitness of each phenotype is evaluated through a phenotype-fitness mapping. Generally, objective functions serve as fitness functions, and a higher fitness value increases the likelihood of a solution surviving in the current environment.

According to the similarity of the biological evolution and EAs, Table 4.1 shows some basic notions and concepts of EA that are transferred from their biological counterparts, and Fig. 4.3 shows a simple genotype of a solution.

Table 4.1 Basic evolutionary notions in biology and computer science

Notion	Biology	Computer science
Gene	A group of sequence bases has a specific function	Computational object (e.g., a bit, character, number, etc.)
Allele	Genes at the same location on chromosome	Value of a computational object
Genotype	Genetic constitution of a living organism	Encoding of a candidate solution
Phenotype	Physical appearance determined by the genotype	Implementation/application of a candidate solution
Fitness	Ability of a living organism to adapt to environments	Quality of a candidate solution
Chromosome	Combination DNA with histone proteins	Structure of computational objects
	Usually multiple chromosomes per individual	Usually only one chromosome per individual
Individual	Living organism	Candidate solution
Population	A set of living organisms	A set of candidate solutions
Reproduction	Creating offspring of one or multiple parents or organisms	Creating new candidate solutions with one or multiple parents
Generation	Count population reproduction times	Count iteration times

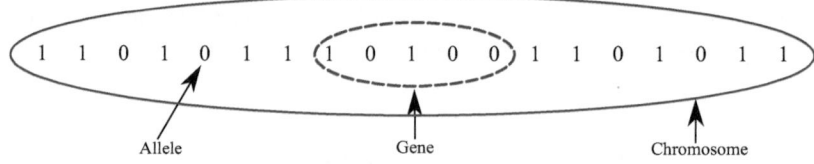

Fig. 4.3 A genotype of a candidate solution

4.1.3 Basic Evolutionary Processes

A basic structure of an evolutionary algorithm is illustrated in Fig. 4.4. An EA starts with an initial population $\mathbb{P}$ with a predefined number of individuals/solutions that are randomly generated in the search space. The objective value of each solution in $\mathbb{P}$ is calculated by the objective function, and then each solution is assigned a fitness value by a fitness function in turn. In general, objective values can also be used directly as fitness values. Afterward, a selection process filters out the solution candidates with good fitness and allows them to enter the mating pool for reproduction with a higher probability. The selected individuals recombine and cross over and even mutate with a set probability to generate new offspring. Finally, parents and offspring are put together and compete for limited resources to survive and constitute a new population. EAs repeat the above process for a new cycle of

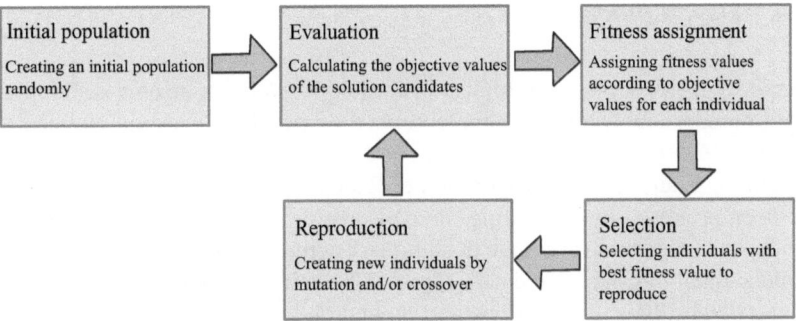

Fig. 4.4 The basic process of an evolutionary algorithms

evolution until the termination criterion is met. The solutions found $\mathbb{S}$ are returned where the number is determined by the search task.

The pseudo-code can refer to Algorithm 4.1, where t denotes the number of generations.

Algorithm 4.1: Typical procedures of a classical EA

Input: $\mathbb{P}_0$: an initial population
Output: $\mathbb{S}$: an archive for found solution(s)
$t \leftarrow 0; \mathbb{P} \leftarrow \mathbb{P}_0$;
while *not terminate* **do**

    Calculate the objective of each individual;
    Assign fitness for each individual by a fitness function;
    mating pool $\leftarrow$ *select*($\mathbb{P}$);
    offspring $\leftarrow$ *reproduce*($\mathbb{P}_t$);
    Update $\mathbb{P}_t$ and archive $\mathbb{S}$;
    $t \leftarrow t + 1$;

return $\mathbb{S}$;

There are a number of components or operators that must be designed in order to implement a particular EA. The most important components are listed below:

(1) Setting a data structure to encoding an individual or solution
(2) Design of a suitable evaluation function and/or a fitness function
(3) Selection of the parent selection mechanism and the survivor (or replacement) selection mechanism
(4) Designing variation operators, such as recombination or mutation or both

Among these operators, variation and selection operators are the most important impetus for EAs. Details of these components will be described in the following sections in this chapter.

4.1.4 Developments

Inspired by the process of biological evolution, researchers created different types of evolutionary algorithms to solve optimization problems. Several classical types of EA have emerged, such as genetic algorithm, evolutionary strategy, evolutionary programming, and genetic programming. Recently, swarm intelligence algorithms have been proposed by simulating the collective behaviors of social creatures. All these types of algorithms become evolutionary computational communities. We will introduce them in Chap. 5 in detail. Figure 4.5 depicts the development histories of the major algorithms of evolutionary computation.

Genetic algorithm (GA) is the most widely known type of EA due to its effectiveness and universality. No matter what type of optimization problem, continuous or discrete or model-based or non-model-based problems, it can work. It was at the beginning of the 1970s that John Holland first proposed the genetic algorithm in his book *Adaptation in Natural and Artificial Systems* [17]. De Jong's thesis [8] helped define what has come to be considered as the classical genetic algorithm. GAs were regarded as a function optimization method since the book of Goldberg *Genetic Algorithms in Search, Optimization and Machine Learning* [15].

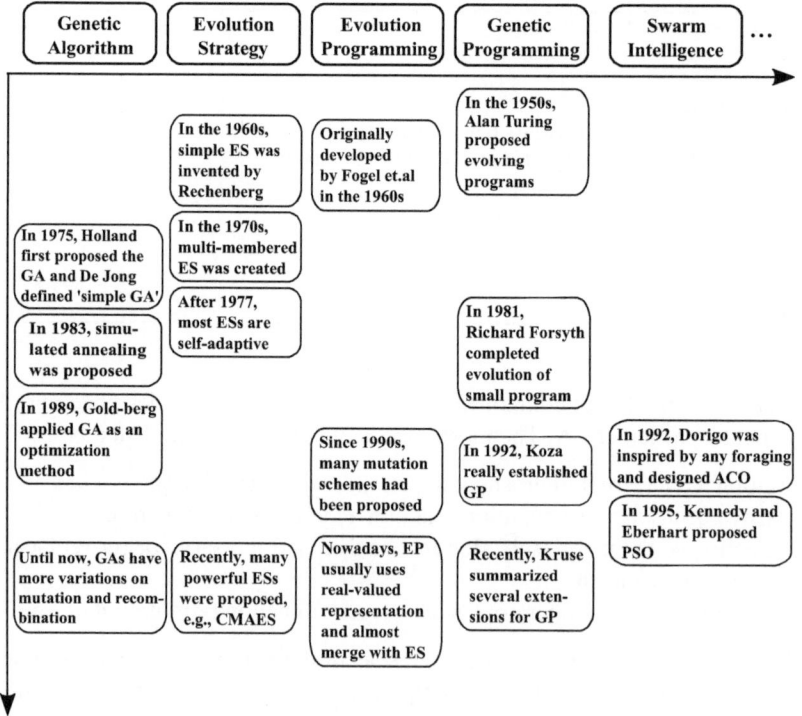

Fig. 4.5 Development histories of evolutionary computational community

Evolution strategy (ES) was invented by Rechenberg and Schwefel in the early 1960s. The first ES just had two members and is denoted $(1 + 1)$ ES or $(1, 1)$ ES, which means a parent generates an offspring and selects a better one from all individuals or just replaces the parent with children. One of the most important breakthroughs in ES was a simple mechanism proposed by Rechenberg to adjust the step size online [22]. Then in the 1970s, multimember evolution strategies were introduced, and it ruled μ individuals generated λ offspring in each generation that could be denoted as $(\mu + \lambda)$ or (μ, λ) ES. The population will have a more complex search behavior because different individuals could have different mutation step sizes, leading to the study of a self-adaptation strategy. Currently, most ESs are self-adaptive [24]. Recently, one of the state-of-the-art ES CMA-ES [16] was developed for optimizing complex real-valued problems.

Evolutionary programming (EP) was developed by Fogel et al. in the 1960s to simulate evolution [12]. In the beginning, classical EP systems used finite state machines as individuals, and nowadays, it usually uses real-valued representations and also merges with ES to optimize continuous problems. There are some differences between EP and ES. The principal difference lies in the biological concept: each individual in EP denotes a special species; therefore, there will not be any recombination operators between different individuals. Another distinction lies in the selection mechanism. In EP, each parent just generates one offspring $(\lambda = \mu)$ and then competes for the survivor together. But in ES, best parents μ are selected from the union of $\mu + \lambda$ offspring. Since the 1990s, EP variants had become more frequent to optimize real-valued optimization problems and even positioned themselves as standard EP [3], and a number of mutation schemes have been proposed. Since the proposal of meta-EP [13], controlling the step size automatically has become the normal form. A variety of schemes have been proposed, and it mainly includes mutation strategies such as Yao's fast evolutionary programming algorithm, where two different distributions were used to generate random mutations [28].

As for genetic programming (GP), the first person who proposed to evolve programs may be Alan Turing in 1950 [27]. And then there were few vital developments until 1981, when Richard Forsyth demonstrated the successful evolution of small programs, represented as trees, to perform classification [14]. In 1988, John Koza applied for a patent about his invention of a GA for program evolution [18]. And it is the series of books by Koza starting from 1992 [19] together with videos [20] that really established GP.

Apart from the above traditional EAs, there has also emerged a new type of search algorithm named swarm intelligence in the recent 20 years by the inspiration of collective behaviors of social animals; such algorithms include particle swarm optimization (PSO), ant colony optimization (ACO), artificial bee colony (ABC), and so on.

4.1.5 Related Resources

Due to the growing research interest, communities of evolutionary computation are still growing rapidly, and many worldwide journals and conferences are devoted to this domain. Here, we list some of such journals: Evolutionary Computation, IEEE Transaction on Evolutionary Computation, Swarm and Evolutionary Computation, Applied Soft Computing, Soft Computing, Journal of Global Optimization, Swarm Intelligence, Memetic Computing, Genetic Programming and Evolvable Machines.

Many other journals also accept research in this domain, such as IEEE Transaction on Cybernetics, Information Sciences, IEEE Computational Intelligence Magazine, Artificial Intelligence, Neurocomputing, Applied Intelligence, IEEE Transactions on Systems Man Cybernetics-Systems, IEEE Transactions on Emerging Topics in Computing, and ACM Computing Surveys.

Popular international conferences within this domain are Parallel Problem Solving from Nature, Genetic and Evolutionary Computation Conference, IEEE Congress on Evolutionary Computation, IEEE Symposium Series on Computational Intelligence, International Conference on Evolutionary Multi-Criterion Optimization, International Conference on Simulated Evolution and Learning, etc.

4.2 Solution Representation

In order to solve an optimization problem, the first step is to know how to denote a candidate solution. For example, in a routing optimization problem, if one needs to know the shortest distance between two cities, the solution can be represented as a sequence of cities that denotes a route. Similarly, when using EAs to solve an optimization problem, the first important step is to construct a solution. What data structure can be used to represent the genotype of a solution normally depends on the problem, but sometimes algorithms may also play an important role. The process of representation will assign genotypes to the corresponding phenotypes. Here, we introduce four common representations.

4.2.1 Binary Representation

The binary representation is the most similar to the gene sequence, and it is also one of the earliest representations. The genotype is made up of a fixed size of bit-string, and each bit is 0 or 1. It is a very popular and general choice for various problems with GAs.

In general, discrete optimization problems usually use binary representation to denote a solution. For example, in the shortest path problem introduced in Sect. 3.2.1, a solution may look like "1001010...." In this solution, "0" and "1"

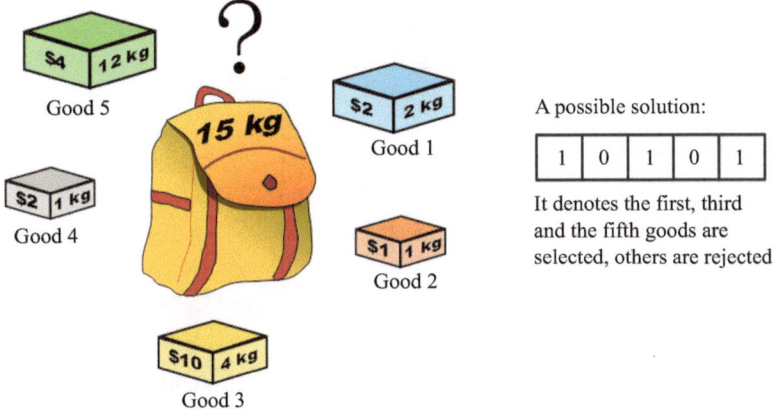

Fig. 4.6 Solution representation of a knapsack problem

denote whether a specific city is included; "1" means the city is included, and "0" means it is not. Another case is the knapsack problem, which is described in Fig. 4.6; a bag with limited volume should take as many goods as possible and each good has a volume. A possible solution could be "0100101....." Similarly, "0" means that the corresponding good is rejected, and "1" means the good is selected. We can observe that the genotype space is a set of stable-size bit-strings. If the options are countable, we can enumerate all possible solutions, where the changes in solutions are obviously not continuous.

In fact, binary representation can be used in various other situations. We all know that binary numbers can represent numbers in different bases through conversion. For example, the decimal number "10" can be represented as the binary number "1010" or "00010000" if each digit is composed of four binary digits. Although binary representation can be applied in different scenarios, there may be issues related to the algorithm used or the redundancy and validity of the representation.

4.2.2 *Integer Representation*

Integral programming problems require all variables be an integer. In these cases, integer representation is useful. In fact, a binary representation is also a special integer representation. Most problems have integer variables, and also these integers have ordinal attributes, for example, 1 is more like 2 than 4. There are also other situations where some rules can also be denoted as integers; for example, in Fig. 4.7, we want to find a path on a square grid that is the shortest from the beginning point to the end point. And we can use an integer set {1, 2, 3, 4} to represent {Up, Down, Left, Right}. Here, these numbers have cardinal attributes; they are just symbols.

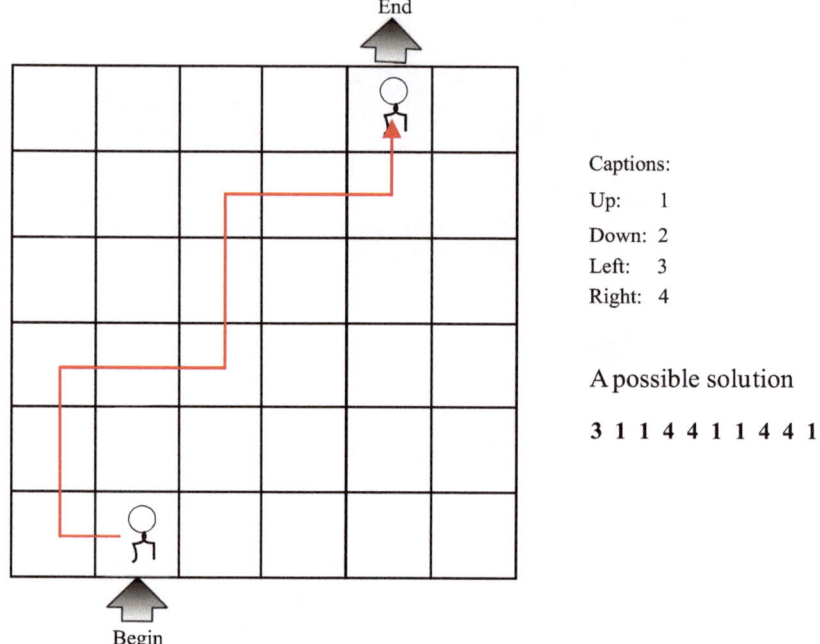

Fig. 4.7 Integer representation for a path

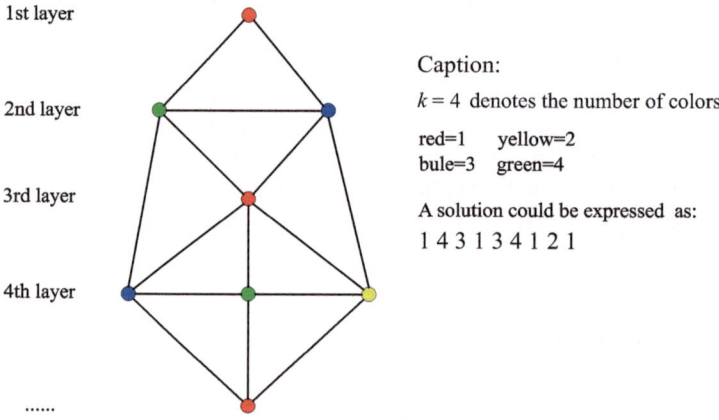

Fig. 4.8 Graph k-coloring problem

Another example is the graph k-coloring problem in which a number of points should be painted by k colors, but neighbor points (two points have a common edge) cannot have the same color as shown in Fig. 4.8. We can denote each color as an integer number; then it is an integer optimization problem, and these numbers do not have any natural ordering; they are just different symbols.

Many problems take the form of deciding the order in which a sequence of events occurs, such as the TSP introduced in Sect. 3.2.1.2. In such a permutation problem, we can also use the integer representation where the order of two neighboring integers denotes an action being taken as shown in Sect. 3.2.1.2.

4.2.3 Real-Valued Representation

Real-valued or floating-point representation is very common in continuous optimization problems because many physical quantities are real values such as length, mass, volume, voltage, speed, and so on. All these cases occur when we want to represent a variable that comes from a continuous search space, e.g., the numerical optimization problems introduced in Sect. 3.1. For example, we want to get the minimal value of equation $f(x_1, x_2) = (x_1 - 2x_1 + 2)^2 + (x_2 - 1.5)^2$, and the search space will be the whole $x_1 \perp x_2$ plane. Therefore, a pair of values (x_1, x_2) can be denoted as a candidate solution. If more variables need to be optimized, a solution will be a vector $x = (x_1, \ldots, x_D)$ with $x_i \in \mathbb{R}$.

4.2.4 Tree Representation

Trees are one of the most classical data structures in computer science and also form the basis of the GP. In general, many expressions can be denoted as (parse) trees when expressions are divided into different syntaxes, such as arithmetic formulas, logical formulas, programming codes, and so on. An expression can be divided into the terminal set and the function set. They will be located at leaves and internal nodes, respectively, in a tree. Here are three types of expressions and their tree structures:

(1) Arithmetic formula

$$5 \cdot x + ((y - 2) - \frac{z}{x + y}) \tag{4.1}$$

(2) Logical formula

$$(x \wedge true) \rightarrow ((x \vee y) \vee (z \leftrightarrow (x \wedge y))) \tag{4.2}$$

(3) C/C++ programming language

```
i = 0;
while (i <100)    i = i + 2;
```

Fig. 4.9 Tree structures for
Eq. (4.1)

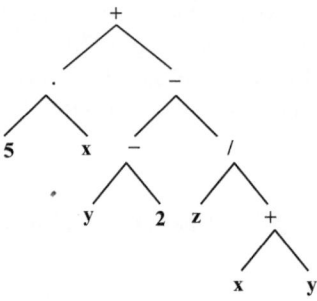

Figures 4.9, 4.10, and 4.11 show the tree structures for the above three types of expressions, respectively.

Fig. 4.10 Tree structures for
Eq. (4.2)

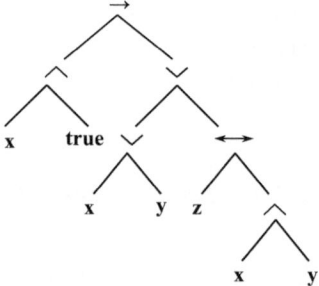

4.2.5   The Effect of Representation

The above solution representations are basic ways to represent a solution in different scenarios. Because genetic operators operate directly on the representation, the type of representation will have a significant influence on the search behavior of an EA.

Therefore, when selecting the representation for a particular problem, several issues need to be taken into account. The most important is the association of genotype and phenotype. Rothlauf listed several important properties of representations [23], and here we introduce the effect on the locality of the representation.

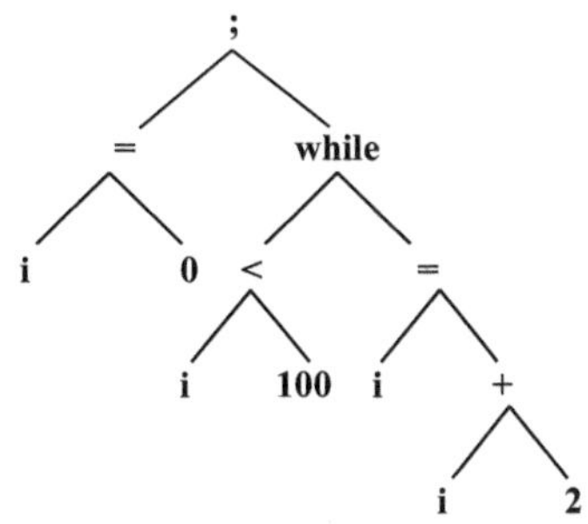

Fig. 4.11 Parse tree for the while structure of C/C++ language

Table 4.2 Different encoding schemes for binary representation

Decimal	'8421' code	Gray code
2	0010	0011
6	0110	0101
7	0111	0100
8	1000	1100
13	1101	1011

Representation is the genotype-phenotype mapping and could change the structure of the neighborhood. For example, when we use a binary string to represent an integer, there are different encoding ways to choose, such as "8421" code or gray code, and it is obvious that different bits have different significance in "8421" code. If we treat them fairly, a local mutation may result in a large jump in the phenotype space. The Hamming distance between two consecutive integers often is not equal to 1 when we use "8421" codes. For example, given an integer number 7, it has the same probabilities of changing to 8 or 6. However, when we use binary representation by "8421" codes, from "0111" to "1000" requires bit flips four times, but to "0110" just once (see Table 4.2). It is obvious that the probability of changing from 7 to 8 is much lower than that to 6, resulting in biased search. A better representation is to use the Gray code, where just one bit is different between two consecutive integral numbers. However, the Gray code brings another shortcoming that two very different integers may have a small Hamming distance, for example, "0011" (2) and "1011" (13) in Gray code. Therefore, some special measures must be taken on each bit; for example, different bits have different mutation probabilities.

Additionally, when we choose the genotype-phenotype mapping for a specific problem, we have to make sure that the encoding allows all possible genes to denote a valid solution and all possible candidate solutions can be represented. For a real-world optimization problem, a mixed representation is normally used, for example, in the mixed-integer programming problem introduced in Sect. 1.2.2.

4.3   Selection

Using EAs to solve optimization problems, we expect to get as good as possible solutions. Although there are local optima, algorithms need to jump out of these regions and explore better areas. The fundamental driven force for EAs to complete these goals are variation and selection. This section will introduce some commonly used selection rules. There are two places in the evolutionary cycle where the selection process can occur based on the fitness value: selecting individuals to take part in mating and selecting individuals to survive into the next generation.

4.3.1   Parents Selection

In EAs, the population will undergo an updating process in every generation, especially when recombination operators are adopted, in which at least two individuals are selected as parents to generate an offspring. This subsection will introduce several strategies about how to select these parents for reproduction.

Uniform Selection
Uniform means that each individual has the same chance to be selected without considering their survival ability (fitness values) during the process of evolution. This might suggest when there is no selection pressure (the difference of the selection probability of the best individual against the average selection probability of all individuals). Parent individuals will be put into a mating pool with the same probability, i.e., for a population $\mathbb{P}$, if we want to select μ parents into the mating pool, then the probability for each individual to be selected is $\mu/|\mathbb{P}|$. Because parents are chosen randomly in this way, good genes of individuals that possibly have good fitness values may not be inherited, which makes the evolutionary search process slowly.

Fitness Proportional Selection
An EA may be trapped in stagnation if there lacks select pressure, where the population does not evolve any more even if it does not converge. Adding select pressure is an effective way to address this issue, and select pressure is normally necessary for EAs to evolve with a desired speed. As described in the basic procedures in Algorithm 4.1, every individual will be assigned a fitness value in each generation. Each individual i is selected by a probability p_i, where the value of p_i is proportional to its fitness value f_i, $p_i = f_i / \sum_{j=1}^{|\mathbb{P}|} f_j$, where the sum of the selection probabilities of all individuals is equal to 1.

The easiest fitness proportional selection method is the roulette wheel selection. Figure 4.12 shows a classical single arm roulette. The first step is normalizing the fitness value of every individual (divide the sum of all fitness values). Then the probability distribution can be regarded as the area of the sector of a wheel. The

Fig. 4.12 Roulette wheel
selection

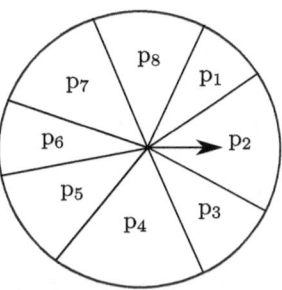

process of selecting is to spin the wheel; when it stops in a sector, the corresponding
individual is selected as a parent.

It is apparent that the larger the fitness value, the greater the chances that an
individual is chosen, especially the best one. This strategy is inherently greedy and
is also referred to as the elite selection strategy. In this way, the best individual
will have more opportunities to mate and produce offspring than others. When the
population size is not large, individuals with very low fitness values will not have
any chances to generate an offspring, so they will quickly become extinct in the
following few generations. This process mirrors the dynamics of a biosphere where
resources are limited and only the strongest individuals can obtain enough food to
survive and reproduce offspring.

When algorithms adopt the elite selection strategy, offspring are generated by
almost similar parents, so the offspring have a high probability of being similar.
After a number of generations, the whole population will move to a small region
and will not disperse, making the population converged. Although one of important
features of an algorithm is making the population converge, there are also some
shortcomings:

(1) Preponderant individuals take over the entire population quickly, resulting in
    that the genes in a population tend to be identical, losing their population
    diversity.
(2) If there is a multimodal optimization problem or there are many local optima,
    the whole population may converge to just one optimal. Other promising
    subspaces where there may exist better solutions cannot be explored. This
    phenomenon is also called premature convergence.
(3) When fitness values do not have apparent differences, there is almost no
    selection pressure. Then, the selection is almost uniformly random, and the
    improvement will become very slow. Another situation that has the same issue
    is a population with a very large number of individuals. In such a situation, good
    individuals are hard to select, as their fitness takes a very small portion to the
    total fitness due to the large population size.
(4) The probability belonging to each individual will vary greatly with the change in
    the order of magnitude of the fitness value. It will have a very different effect on
    the selection behavior when faced with problems having quite different ranges

Table 4.3 Selection probability of different orders of magnitude of objective values

Individual	Objective	Probability	Objective	Probability	Objective	Probability
A	1	0.1	11	0.275	101	0.326
B	3	0.3	13	0.325	103	0.332
C	6	0.6	16	0.400	106	0.342
Sum	10	1.0	40	1.0	310	1.0

of objective values or in dynamic environments. Table 4.3 illustrates an example where selection probabilities change with different orders of magnitude of objective functions.

Rank-Based Selection

Rank-based selection is a method inspired by the shortcomings of fitness proportionate selection [4]. The absolute fitness value has a significant effect on the selection probability. Therefore, the rank-based selection does not directly use fitness values. It sorts individuals by their fitness values and then allocates selection probabilities according to their ranks, rather than their actual values. We can assign individuals with different rank numbers. For example, the best one could have rank $|\mathbb{P}| - 1$, or the worst one has rank 0. We can even adopt some fuzzy rank methods and then map rank numbers to selection probabilities. Similarly, the sum of selection probability must be one.

Common mapping methods include linear mapping and exponential mapping. When mapping rank numbers to the selection probability linearly, a parameter named selective factor $\rho \in (1, 2]$ is set. The selection probability for an individual of rank i can be denoted as

$$p_i = \frac{(2 - \rho)}{|\mathbb{P}|} + \frac{2i(\rho - 1)}{|\mathbb{P}|(|\mathbb{P}| - 1)} \tag{4.3}$$

The probability consists of two components. The first component $\frac{(2-\rho)}{|\mathbb{P}|}$ is greater than 0, which means that every individual has a chance to be selected as a parent; it is a 'baseline' probability. Another component increases with the rank i linearly. Only one parameter ρ controls the assignment of the selection probability; we can adjust the weight of the base probability. Figure 4.13 shows the relations between ρ and the selection probability; a bigger ρ means more bias toward better individuals.

Sometimes, algorithms need more selection pressure to distinguish different individuals; linearly mapping methods might be not effective enough. Thus, using exponential mapping methods can scale the differences between individuals, such as

$$p_i = \frac{1 - e^{-i}}{|\mathbb{P}| - \sum_{i=0}^{|\mathbb{P}|-1} e^{-i}} \tag{4.4}$$

where the relationship between rank and probability is shown in Fig. 4.14.

Fig. 4.13 Linear probability

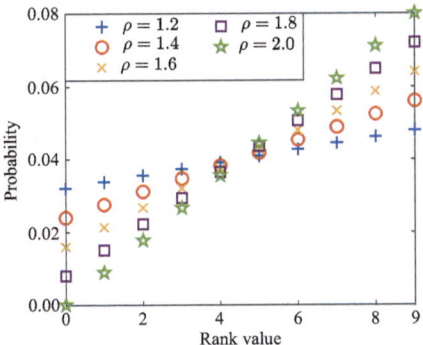

Fig. 4.14 Exponential probability

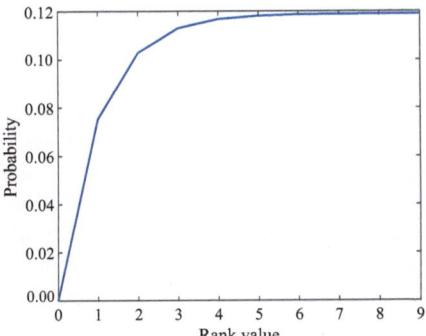

Tournament Selection

Although rank-based selection methods could increase the select pressure, they are not adjustable, that is, the population will obey a stable selection criterion during the whole evolution process. At different stages, the population may need different levels of select pressures, and the tournament selection comes into play.

Like choosing the best athlete in a competition, the tournament selection chooses the best one from a number (tournament size) of competitors, which are typically randomly selected, in each competition until the mating pool is full. Tournament selection has a property that does not need any global knowledge or information of the population nor quantifiable indicators. Instead, it relies only on relative relationships that could compare and rank any two individuals. It is very easy to implement this strategy.

The probability of an individual being selected as a parent depends on the following factors summarized in [1, 26]:

(1) Its rank makes a great influence. The better the fitness, the higher the probability.

(2) The tournament size k controls selection pressure. The more individuals take part in the tournament, the greater the chance of getting elite individuals. On the contrary, if low-fitness-value individuals need to be selected, a smaller tournament size can be used.

(3) Whether individuals are chosen repeatedly, for example, if one individual is chosen in a tournament, the remaining $k - 1$ worse members of the population will never be selected, and then those deselected individuals will have no chance for the reproduction process. But even if the worst individual has a chance to be re-selected again in the tournament, then even a very poor individual still has a chance to reproduce, which could benefit for the diversity of the population.

Tournament selection is also the most widely used selection operator in some EAs especially GAs. In addition, it is easy to control the selection pressure by adjusting the value of k. When k is equal to the population size, only the best individual will be selected, but if k is equal to one, then any individual can be chosen with the same probability as the uniform selection.

Truncation Selection

All the above selection operators are probabilistic, but the truncation selection is a deterministic selection method and is one of the strongest (most elitist) forms of selection. Individuals are sorted according to their fitness, and only the best k individuals are selected into the mating pool in each generation. Within the selected set of k individuals, all are equally able to reproduce an offspring. It is a greedy selection method and could result in a fast convergence speed.

4.3.2   Survivor Selection

In classical EAs, the size of a population will remain stable in every generation. After the reproduction process, there are more individuals than the population size. Thus, algorithms should use some strategies to decrease the size. Generally, parent selection strategies introduced above still work well here, and several special survivor selection strategies can be used.

Age-Based Replacement

This scheme is very similar to natural lives, every individual has a chance to generate offspring, and they can last for a stable generation. For example, a parameter τ is set to denote the maximum age of each individual. It means that an individual can live τ generations at most. When the number of iterations increases, a value α_i will be attached to the attribute of the individual p_i, which means that it has lived for α_i generations.

It is obvious that the reproduction strategy will affect the process of age-based replacement. If the number of offspring λ is small, after τ generations, there may be many individuals beyond τ, but just a few individuals are eliminated, leading to

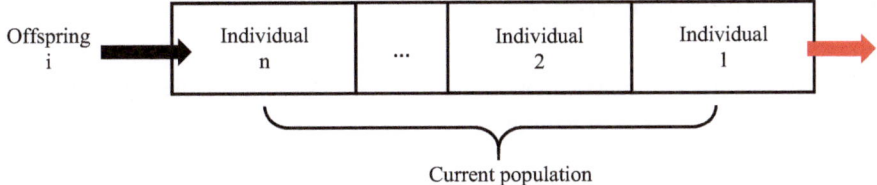

Fig. 4.15 Age-based replacement

the age rule being useless. Therefore, setting reasonable values for τ and λ is very important.

A feasible method to address this issue is to use a first-in-first-out (FIFO) queue to store individuals according to the time they are generated as described in Fig. 4.15.

Fitness-Based Replacement
There are several strategies to choose survivor individuals according to the fitness. Here we just list several commonly used ones.

(1) Replacement of the worst.
   In this strategy, for example, μ parents in the current population generate λ offspring and then merge them. The worst λ members of the population will be eliminated in order to maintain a stable size of the population. It is a greedy selection strategy and could rapidly improve the average fitness. But it faces the risk of premature convergence issue, especially when the size of population is small. It would be very useful when the population size is large.

(2) Elitism.
   This scheme is designed for preserving the best member of the population, because it has the best genes. It is usually combined with other strategies such as age-based or stochastic fitness-based schemes. No matter what parents or children are, the best individual will always survive in the population. To some extent, it is similar to the replacement of the worst strategy.

(3) Round-robin tournament.
   This mechanism is like competition in ball games, especially in football or basketball clubs. A team needs to compete with all other teams once, and then record their scores. The team with the highest scores will be the winner. Similarly, an EA uses μ parents to generate λ offspring. Initially, all the parents and offspring are merged into a single population. Then, each individual in the merged population is compared with m randomly selected members from the remaining individuals. The comparison records how many individuals are dominated by the chosen individual. After this process is completed for all members, the algorithm checks the number of individuals that each member dominates and selects the best μ members to survive to the next generation.

4.3.3 Selection Pressure

All the methods mentioned above pertain to selection, with the goal of tuning the selection pressure to a suitable level. As the saying goes, "No pressure, no diamonds." In EAs, selection is a crucial mechanism that propels the search forward and facilitates population convergence. A strong selection pressure implies significant differences in priorities among members, whereas a weak selection pressure indicates that all members have similar importance. Various measures have been proposed to quantify the selection pressure. One of the most common measures is the takeover time, which refers to the number of generations it takes for the best individual to dominate the entire population. The takeover time can be utilized as an indicator of the speed of convergence.

4.4 Reproduction

In EAs, the population undergoes updates in each generation, with new members reproduced from old ones. Reproduction is another crucial process in EAs. Currently, the most popular reproduction methods continue to draw inspiration from nature (e.g., mutation and recombination), physical processes, psychology, sociology, and more. In this section, common mutation and recombination operators are introduced.

4.4.1 Mutation

Mutation exists wildly for living beings in the natural world, as all cells or viruses may make a mistake when duplicating their genetic materials. In EAs, individuals have a mutation phenomenon when the genotype has a difference from the ancestor.

Mutation for Binary Representation
For binary encoding, the most common mutation operator is bit flipping. For example, from 1 to 0 or 0 to 1 with a probability p_m. Suppose a solution has D bits, and each bit has a probability p_m to mutate; then the expected number of mutation bits in a sequence is $D \cdot p_m$. In Fig. 4.16, the third, fourth, and eighth bits have a bit flip.

The bitwise mutation rate p_m plays an important role in EAs since it directly controls the step size (i.e., the number of bit flips). A larger value of p_m means a

Fig. 4.16 Bitwise mutation for binary encodings

larger step size to explore a wider range of the search space. Several studies have investigated how to set a suitable p_m. Here we will not discuss them in detail (please see Sect. 6.1.1), but we will only discuss the effects of the p_m. When it is at the beginning of the search, the population should explore as much as possible space; therefore, a big mutation probability is helpful. However, when the population is nearly convergent, local exploitation is necessary. Therefore, a small probability of mutation is preferred to avoid deteriorating the population.

Mutation for Integer Representations

In fact, integer encoding is similar to binary encoding. Therefore, mutation operators used in binary representation can also be used in integer representation. There are four major forms.

(1) Random resetting. Suppose that a solution is made up of D integers and the search domain is $\mathbb{X}$. Then each integer can randomly change to any other integer that belong to $\mathbb{X}$ with probability p_m. This is suitable for use when gene encoding has cardinal attributes where all values are equally to be chosen since they do not have any orders. An example shown in Fig. 4.17 is random resetting, where the second, fourth, and seventh genes are selected to mutate with a probability p_m.

(2) Creep mutation. Another situation when the encoding has an apparent order, the mutation should take different methods because it will affect the search scope. It works by adding a small value that is sampled from a distribution random to each gene with probability p. Therefore, it is a local mutation operator, and individuals will exploit local regions. Mutation probability p is an important parameter to set; it will affect the search step size. Some tests have been carried out in [7]; "big creep" and a "small creep" operators are used to compare the performance of the strategies.

(3) Swap mutation. When swap mutation is applied to binary or integer chromosomes, two genes are randomly selected to exchange their values as shown in Fig. 4.18. It is suitable for permutation-based encoding, such as the TSP.

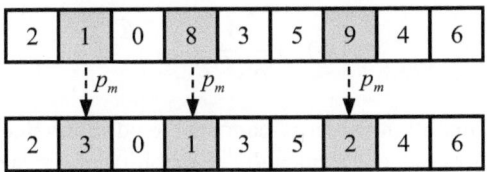

Fig. 4.17 Random resetting for integer representations

Fig. 4.18 Swap mutation for integer representations

Fig. 4.19 Inversion mutation for integer representations

(4) Inversion mutation. In contrast to swap mutation, inversion mutation selects a random gene sequence and reverses the order of genes in the sequence as shown in Fig. 4.19. Like the swap mutation, inversion mutation is also suitable for chromosomes with a sequenced list, because the new chromosome still carries the same genes as the original one.

Mutation for Real-Valued Representation

For a continuous optimization problem, any gene in a solution could change randomly. Given the lower boundary $\mathbf{l} = \{l_1, l_2, \cdots, l_D\}$ and the upper boundary $\mathbf{u} = \{u_1, u_2, \cdots, u_D\}$ of each dimension, a general mutation operator may be $(x_1, \cdots, x_D) \rightarrow (x_1', \cdots, x_D')$, where $x_i, x_i' \in [l_i, u_i]$. The uniform mutation and nonuniform mutation are commonly used mutation operators.

(1) Uniform Mutation. This is the easiest case where a uniform distribution in $[l_i, u_i]$ is applied in each dimension, and then x_i could change to any values within its boundary with equal probability. In general, the uniform mutation is good for introducing new genes and helps the population explore undiscovered regions. However, it cannot control the mutation range/step.

(2) Nonuniform Mutation. Similarly, the nonuniform mutation will adopt other types of distribution functions rather than uniform distribution functions, and at the same time some parameters will be introduced with the distribution in order to control the mutation step size. For example, for each dimension, a value sampled from Gaussian distribution with mean zero and a user-set standard deviation can be directly added to the gene value, and then the new gene value can be made within $[l_i, u_i]$ if it is out of the search domain. Through the mean value and standard deviation, the step size can be controlled in a deterministic range. Given a Gaussian distribution defined below

$$p(x_i') = \frac{1}{\sigma\sqrt{2\pi}} \cdot e^{-\frac{(x_i'-x_i)^2}{2\sigma^2}} \tag{4.5}$$

where x_i' is the mutation value, x_i denotes the current value, and σ denotes the standard deviation, which decides the probability of the size of mutation step. When σ is small, most mutation values change slightly around the baseline, the population will exploit local space. But when σ becomes larger, individuals will have greater probabilities of searching further regions, and then the population could explore in a larger area.

Another nonuniform distribution, for example, the Cauchy distribution, has different features from the Gaussian distribution. It has a sightly higher probability when x_i' is far away from the mean value with the same standard

Fig. 4.20 The circled node in the tree is selected for mutation and replaced by a randomly generated tree

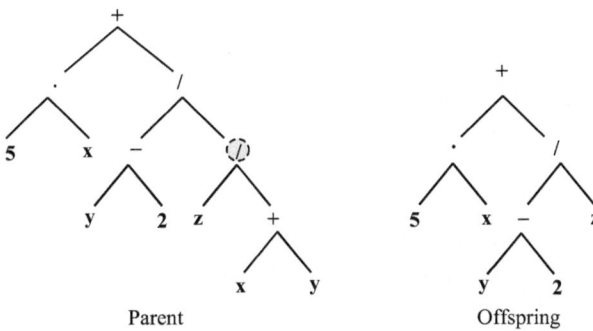

Parent   Offspring

deviation [28], which makes it suitable for exploring the search space or jumping out of local optima.

Note that the size of mutation step plays a key role in the performance. A large mutation step means exploring the areas that are far away from the current point, while a small mutation step means exploiting around the current point. How to adjust the size of mutation step is a challenging issue, and it is addressed in different ways for different algorithms. We will discuss this later in Chaps. 5 and 6.

Mutation for Tree Representation

A tree structure is made up of nodes and leaves. When a mutation occurs, a subtree will be selected randomly and replaced by a randomly generated tree subject to conditions on maximum depth and width. This newly created subtree is usually generated the same way as in the initial population and so is subject to conditions on maximum depth and width. Figure 4.20 shows a mutation in a tree, where a subtree with the circle root is simply replaced by a node z.

There are two parameters to be controlled in the tree-based mutation:

(1) The probability of choosing mutation
(2) The probability of choosing a subtree to be replaced

Tree presentation is usually used in genetic programming, and Koza's classic book [19] advised users to set the mutation rate at 0. This means that GP works without any mutation operators. More recently, Banzhaf et al. recommended that a mutation probability less than 5% is fine [5]. The current GP adopts a low mutation probability (see more in Sect. 5.3).

4.4.2 Recombination

From the knowledge of genetics, chromosome recombination occurs between different individuals. Similarly, in EAs, more than one parent will take part in the process of recombination. Recombination aims to improve parent solutions by exchanging information on the parent solutions involved.

Recombination for Binary Representation

There are three standard forms of recombination for binary representations. Although some research work on a number of parents [10], here we just introduce the way to use two parents to generate two offspring.

(1) One-point crossover. One-point crossover is the first and easiest recombination operator proposed in [17] and verified in [8]. Figure 4.21 shows a process of one-point crossover. It works as described below. Firstly, select two parents from the population using some selection strategies introduced in Sect. 4.3.1. Secondly, randomly choose a point to divide the gene sequence into two parts. Finally, the separated segments of the two parents are exchanged to generate two different offspring.

(2) k-point crossover. One-point crossover just selects one crossover point to recombine. Similarly, k-point crossover will select k crossover points to split the two gene sequences correspondingly and then exchange the segments with each other in turn. Figure 4.22 shows an instance with $k = 2$ crossover points that are randomly chosen.

(3) Uniform crossover. The feature of the previous two operators is that the recombined gene sections are all continuous, which makes the offspring very similar to their parents. Uniform crossover [25] is different from the k-point crossover. It operates on each gene location rather than a segment; the "uniform" means a gene has an equal chance from any parents. For example, in Fig. 4.23 with two

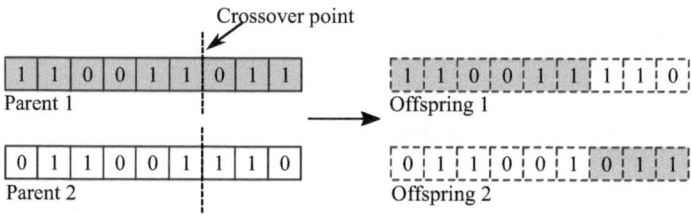

Fig. 4.21 One-point crossover

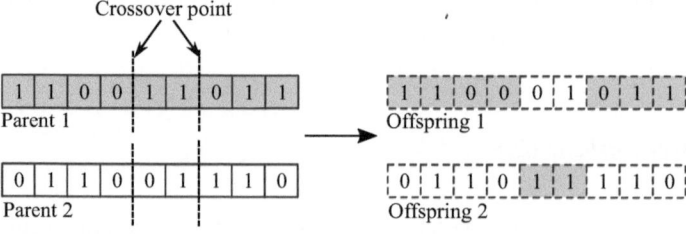

Fig. 4.22 Two-point crossover

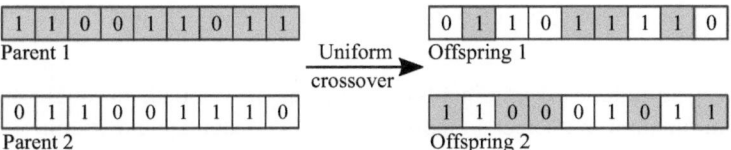

Fig. 4.23 Uniform crossover with two parents

parents, if we set a control parameter $p = 0.5$, then for each gene location i, generate a uniformly distributed random number r in $[0, 1]$. If $r < 0.5$, the gene at location i from parent 1 is given to offspring 1, and the corresponding gene from parent 2 is given to offspring 2. Repeat the process until two offspring are complete.

Recombination for Integer Representation
Recombination for integer representation is similar to that for the binary representation. In fact, we can find that these crossover methods do not introduce new genes or values during the process. All genes in offspring are directly inherited from their parents with different combination. However, this mechanism will cause the population to converge quickly and drastically lose diversity.

Recombination for Real-Valued Representation
We can use one of the following three ways to design recombination operators for solutions with the real-valued representation:

(1) We can use the same operators, which are designed for the bit strings or integer strings, for real-valued representation, i.e., all genes are directly copied from the parents by the way introduced above. This is called discrete recombination.
(2) In the continuous search space, any genes with the search range are valid. Therefore, a gene value of the offspring can be a random value between the gene values of their parents. For example, given two parents x and y, an offspring z can be obtained by

$$z_i = \alpha x_i + (1 - \alpha) y_i \qquad (4.6)$$

where α, a random number within $[0, 1]$. In this way, offspring will be located within a hyper-rectangle where their parents spanned. It is called intermediate or arithmetic recombination.
(3) We can also generate new genes around their parents as in the arithmetic recombination but do not limit the range of α. As a result, new genes can be distributed in the entire search space. This way is known as blend recombination.

Commonly used recombination operators for real-valued representation are as follows:

(1) Single arithmetic recombination. As the name implied, in single arithmetic recombination, we select a random point k and calculate the arithmetic average of the two parents at this point. The other genes are directly derived from parents as follows:

$$\text{Offspring 1: } (x_1, \ldots, x_{k-1}, \alpha \cdot x_k + (1 - \alpha) \cdot y_k, x_{k+1}, \ldots, x_D)$$

$$\text{Offspring 2: } (y_1, \ldots, y_{k-1}, \alpha \cdot y_k + (1 - \alpha) \cdot x_k, y_{k+1}, \ldots, y_D)$$

Figure 4.24 shows the process, where the fifth gene ($k = 5$) is selected to recombine and the recombination factor is $\alpha = 0.4$.

(2) Simple arithmetic recombination. Simple arithmetic recombination is based on the single arithmetic recombination where after a recombination point k is chosen, the genes after point k will all be recombined.

$$\text{Offspring 1: } (x_1, \ldots, x_k, \alpha \cdot x_{k+1} + (1 - \alpha \cdot y_{k+1}), \ldots, \alpha \cdot x_n + (1 - \alpha) \cdot y_n)$$

$$\text{Offspring 2: } (y_1, \ldots, y_k, \alpha \cdot y_{k+1} + (1 - \alpha \cdot x_{k+1}), \ldots, \alpha \cdot y_n + (1 - \alpha) \cdot x_n)$$

Figure 4.25 shows the process where the recombination point is at the seventh gene location and recombination factor is $\alpha = 0.5$.

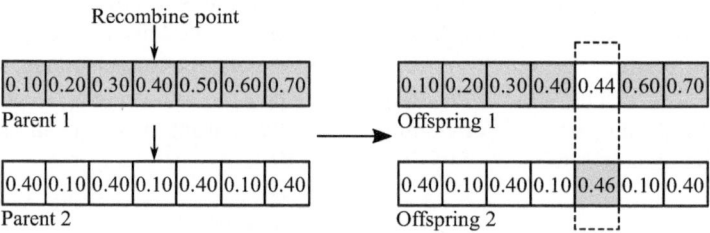

Fig. 4.24 Single arithmetic recombination with $k = 5$, $\alpha = 0.4$

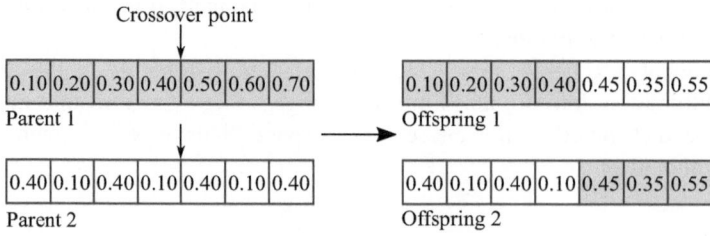

Fig. 4.25 Simple arithmetic recombination with $k = 6$, $\alpha = 0.5$

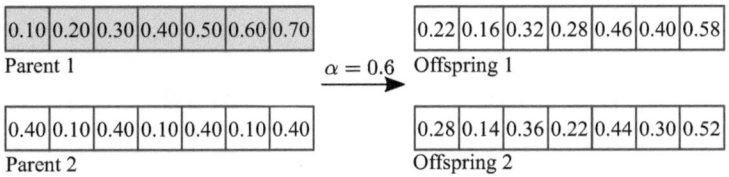

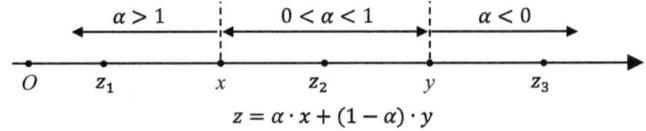

Fig. 4.26 Whole arithmetic recombination with $\alpha = 0.6$

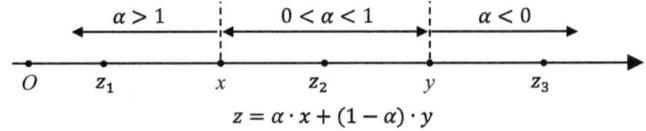

Fig. 4.27 Possible locations of z with α in different ranges in one dimension

(3) Whole arithmetic recombination. The offspring are obtained by recombination of all genes of parents in whole arithmetic recombination as follows:

$$\text{Offspring 1} = \alpha \cdot x + (1 - \alpha) \cdot y$$

$$\text{Offspring 2} = \alpha \cdot y + (1 - \alpha) \cdot x$$

Figure 4.26 shows two children that are recombined from two parents with recombination factor $\alpha = 0.6$.

(4) Blend crossover. Blend crossover was introduced in [11]. The biggest difference between it and the arithmetic methods mentioned above is that it allows offspring to distribute everywhere in the search space rather than in a hyper-rectangle. Similarly, a parameter α is set to control the location of offspring, but the value of α does not have any limitation in Eq. (4.6). Figure 4.27 shows the possible location of z on a one-dimensional line as α changes.

By controlling the value of α, we can control the search scope of offspring. But only a stable α will limit the search direction during the whole process. Here, we introduce a random number $u \in (0, 1)$ when calculating a gene value of the offspring in each dimension.

$$\gamma = k \cdot (2u - 1), k = 1, 2, \cdots$$

$$z_i = \gamma \cdot x_i + (1 - \gamma) \cdot y_i$$

Figure 4.28 shows that different γ will generate offspring in different regions in a two-dimensional space, where γ_1 and γ_2 denote the values in the two dimensions, respectively.

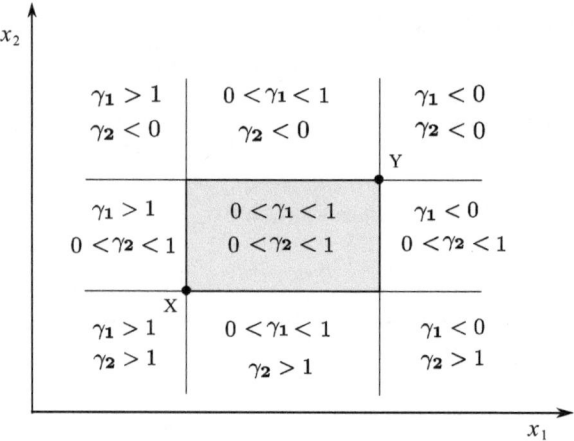

Fig. 4.28 Controlling the search space with blend crossover

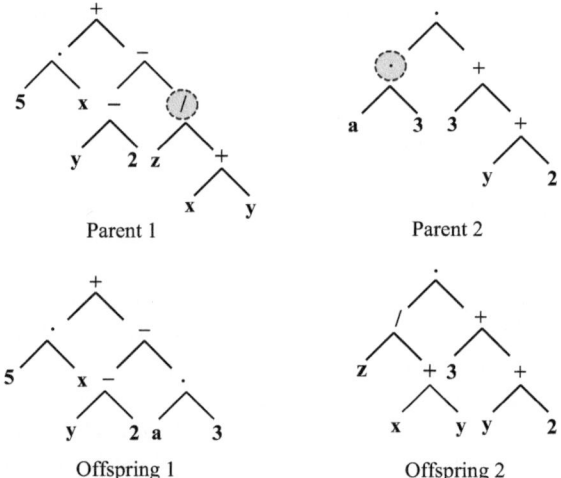

Fig. 4.29 Tree-based crossover

More recent methods such as a simulated binary crossover [9] were built on blend crossover. It does not generate offspring uniformly; it uses a distribution that is likely to generate small changes, and the parameters of the distribution are controlled by the distance between the parents.

Recombination Operators for Tree Representation

Tree-based recombination creates offspring by swapping subtrees from the selected parent trees. It works by exchanging the subtrees starting at two randomly selected nodes in the given parents as shown in Fig. 4.29. Randomly selecting a crossover point with the tree-based crossover operator may cause an increase in the depth of the tree without any control. As a result, gene blocks with good semantics may be destroyed. Therefore, controlling the depth of the crossover points should be taken into account for tree-based crossover operators.

References

1. Bäck, T.: Generalized convergence models for tournament-and (μ, lambda)-selection. In: Proceedings of the Sixth International Conference on Genetic Algorithms, pp. 2–8. Morgan Kaufmann, San Francisco (1995)
2. Back, T.: Evolutionary Algorithms in Theory and Practice: Evolution Strategies, Evolutionary Programming, Genetic Algorithms. Oxford University Press, New York (1996)
3. Back, T., Schwefel, H.: An overview of evolutionary algorithms for parameter optimization. Electron. Commer. **1**(1), 1–23 (1993)
4. Baker, J.E.: Reducing bias and inefficiency in the selection algorithm. In: Proceedings of the Second International Conference on Genetic Algorithms, pp. 14–21. Lawrence Erlbaum, New Jersey (1987)
5. Banzhaf, W., Nordin, P., Keller, R.E., Francone, F.D.: Genetic Programming: An Introduction. Morgan Kaufmann San Francisco, San Francisco (1998)
6. Darwin, C.: On the Origin of Species, 1859. Routledge, Abingdon (2003)
7. Davis, L.: Handbook of Genetic Algorithms. Van Nostrand Reinhold Company, New York (1991)
8. De Jong, K.A.: An analysis of the behavior of a class of genetic adaptive systems. Ph.D. thesis, Ann Arbor (1975)
9. Deb, K., Beyer, H.: Self-adaptive genetic algorithms with simulated binary crossover. Electron. Commer. **9**(2), 197–221 (2001)
10. Eiben, Á.E., van Kemenade, C.H., Kok, J.N.: Orgy in the computer: multi-parent reproduction in genetic algorithms. In: Proceedings of the Third European Conference on Artificial Life, pp. 934–945. Springer, Berlin (1995)
11. Eshelman, L.J., Schaffer, J.D.: Real-coded genetic algorithms and interval-schemata. In: Whitley, D. (ed.) Foundations of Genetic Algorithms, pp. 187–202. Morgan Kaufmann, San Mateo (1993)
12. Fogarty, T.C., Vavak, F., Cheng, P.: Use of the genetic algorithm for load balancing of sugar beet presses. In: Proceedings of the 6th International Conference on Genetic Algorithms, pp. 617–624. Morgan Kaufmann, San Francisco (1995)
13. Fogel, D.B.: Evolving artificial intelligence. Trends Cogn. Sci. **5**(11), 500 (2001)
14. Forsyth, R.S.: Beagle-a Darwinian approach to pattern recognition. Kybernetes **10**(3), 159–166 (1981)
15. Goldberg, D.E.: Genetic Algorithms in Search, Optimization and Machine Learning. Addison-Wesley Longman Publishing Company, Boston (1989)
16. Hansen, N., Ostermeier, A.: Completely derandomized self-adaptation in evolution strategies. Evol. Comput. **9**(2), 159–195 (2001)
17. Holland, J.H.: Adaptation in Natural and Artificial Systems: An Introductory Analysis with Applications to Biology, Control, and Artificial Intelligence. MIT Press, Cambridge (1992)
18. Koza, J.R.: Non-linear genetic algorithms for solving problems (1990)
19. Koza, J.R.: Genetic Programming: On the Programming of Computers by Means of Natural Selection. MIT Press, Cambridge (1992)
20. Koza, J.R.: Genetic programming II: automatic discovery of reusable subprograms. Cambridge **13**(8), 32 (1994)
21. Mayr, E.: The Growth of Biological Thought: Diversity, Evolution, and Inheritance. Harvard University Press, Cambridge (1982)

22. Rechenberg, I.: Evolution strategies-Optimierung technisher Systeme nach Prinzipien der biologischen Evolution. Frommann–Holzboog, Stuttgart (1973)
23. Rothlauf, F.: Representations for Genetic and Evolutionary Algorithms. Springer, Berlin (2006)
24. Schwefel, H.P.: Numerische Optimierung von Computer-Modellen mittels der Evolutionsstrategie. Birkhäuser, Basel (1977)
25. Syswerda, G.: Uniform crossover in genetic algorithms. In: Proceedings of the 3rd International Conference on Genetic Algorithms, pp. 2–9. Morgan Kaufmann, San Mateo (1989)
26. Thiele, L., Blickle, T.: A comparison of selection schemes used in genetic algorithms. Technical report, Swiss Federal Institute of Technology Zurich, Computer Engineering and Communications Networks Lab, Federal (1995)
27. Turing, A.M.: Computing Machinery and Intelligence. Springer, Dordrecht (2009)
28. Yao, X., Liu, Y., Lin, G.: Evolutionary programming made faster. IEEE Trans. Evol. Comput. **3**(2), 82–102 (1999)

Chapter 5
Popular Evolutionary Computation Algorithms

Abstract This chapter introduces several popular EAs, including the four classical paradigms of GA, GP, EP, and ES and also including the emerging mainstream of PSO, DE, EDA, and ACO. GA, GP, PSO, and DE generate offsprings mainly by means of recombining individuals, while EP, ES, EDA, and ACO use specific probability distribution models to generate offsprings. ES, EP, DE, and PSO are mainly used to solve continuous optimization problems, while EDA and ACO are proposed to solve combinatorial optimization problems. However, GA can solve both types of optimization problems by using different solution representations.

5.1 Genetic Algorithm

According to Darwin's theory of evolution and Mendel's theory of heredity, biologists have known the role of genes in natural evolution in the 1950s to the 1960s. Many researchers hope to simulate this process on the emerging computer to attempt to quantitatively study the relationship between genes and evolution.

Based on what is mentioned above, John Holland, the pioneer of complex theory and nonlinear science, realized the similar relationship between biological genetic and natural evolution phenomena and artificial adaptive systems. Furthermore, he proposed to learn from the mechanism of biological genetics for designing artificial adaptive system, i.e., he developed a method of adaptive search with a population of individuals and found that the operational strategies of crossover and mutation are adaptive in the system. In 1975, Holland first systematically discussed the genetic algorithm in the adaptation of natural and artificial systems [42].

GA is the earliest and mostly well-known evolutionary algorithm proposed by imitating the iteration process of biological evolution, including environmental selection, chromosome crossover, and gene mutation. Its essence is an intelligent optimization method of parallel, efficient, and global search. It can automatically obtain the knowledge about the search space during the evolutionary process and self-adaptive control the search direction to obtain the optimal solution.

Since the 1990s, as an efficient optimization technology, GA has been widely used in the fields of machine learning, pattern recognition, neural networks, and control systems.

5.1.1 Basic Principle and Framework

As one of the typical evolutionary algorithms, GAs use parallel search technology, i.e., the population represents a group of solutions and evolves into a state of containing an approximate optimal solution through a series of genetic operations, which are selection, crossover, and mutation, to generate a new generation of population.

The workflow of a standard GA is given in Algorithm 5.1. Initially, randomly generate N individuals as the initial population $\mathbb{P}_0$. Then it goes into a loop to perform the following procedures: calculate the fitness of newly generated individuals, and select some individuals with high fitness. These individuals undergo crossover to generate some new individuals; thereafter, a mutation operation may be taken. The selection methods are used to select some excellent individuals from $\mathbb{P}_t$ as parents to perform reproduction operations. The reproduction of new individual mainly includes crossover and mutation introduced in Sect. 4.4.2. Finally, check whether the termination condition is satisfied. If the stopping condition is not fulfilled, then continue the above evolutionary processes; otherwise, terminate the algorithm.

Algorithm 5.1: Genetic algorithm

Input: Maximum Generation $G_{\max}$, population Size N
Output: Best solution x_{best}
Initialization: Randomly generate initial population $\mathbb{P}_0$, generation counter $t = 0$;
while $t < G_{\max}$ **do**
     Calculate the fitness of individuals;
     Perform selection;
     Perform crossover on good candidates;
     Perform mutation on good candidates;
     Update the best solution x_{best};
     $t = t + 1$;
Return x_{best};

The GA has the following characteristics:

(1) GA takes the coding of decision variables as the operation object, so that it can use the concepts of chromosome and gene in biology for reference in the optimization calculation, and standard genetic algorithm (SGA) uses binary coding.

(2) GA drives the evolution process by fitness, in which the fitness function is generally positively correlated with the objective function, that is, the better the objective value of a solution, the higher the fitness.
(3) GA has parallel search characteristics, that is, the search process of the most optimal solution starts from a population containing multiple solutions.
(4) The selection, crossover, and mutation operations in GA improve its search flexibility.

Note that the occurrence of crossover and mutation depends on a certain probability, i.e., the crossover probability and mutation probability. Generally, the probability of crossover is high (above 0.6), while the probability of mutation is very low (close to zero). In fact, the suitable values for these two parameters are problem-dependent and may even change during the evolutionary progress. Therefore, self-adaptation techniques come into play; for example, the mutation probability is encoded into the chromosome and allowed to evolve (see Sect. 6.1.1.2).

5.1.2 Applications of Genetic Algorithms

As one of the most popular evolutionary algorithms, GAs have been widely applied in different fields since they were proposed. The application fields of GAs include engineering design, agriculture, medicine, etc.

GAs have emerged as a promising tool with far-reaching implications across various medical specialties, including disease screening, diagnosis, healthcare management, and medical image segmentation. To illustrate, GAs play a vital role in breast cancer screening using mammography and are instrumental in the establishment of healthcare management models aimed at enhancing patient service and satisfaction [35].

Furthermore, GAs find extensive applications in the realm of economic management, particularly in areas such as logistics scheduling [79], flow-shop scheduling [59], and resource-constrained project scheduling [40]. Modern agricultural development planning and rural infrastructure transformation also benefit significantly from GAs, with applications ranging from irrigation planning [64] to the spatial optimization of intensive utilization of biological resources in farmland [54].

In the field of engineering, GAs have been instrumental in tasks such as optimizing the design of spindle structures for computer numerical control machine tools, fine-tuning motor parameters, and optimizing geological drilling trajectories [27]. Noteworthy achievements in engineering, like the steel structure of Beijing Bird's Nest stadium, the design of the new-generation high-speed train head in Japan [77], and the development of the ST5 antenna for NASA [43], have all been accomplished through the utilization of GAs.

5.2   Evolutionary Programming

Evolutionary programming (EP) was firstly proposed to simulate the evolution process and generate artificial intelligence and then has been applied in optimization domains.

5.2.1   The Emergence of Evolutionary Programming

There is a consistent pursuing for our human to obtain the artificial intelligence. Even about 60 years ago, some researchers attempted to build the model of natural evolution to realize automatic programming, sequence prediction, and so on. Friedberg, the pioneer of automatic programming, aimed to design the algorithm to find the program with certain inputs and outputs [32]. It proved too early for him to accomplish the arduous task, due to the poor performance of computers at that time. Furthermore, he initialized the program with binary codes; then set the counter to record the success number of expected outputs during a number of executions; finally, random instructions were interchanged with the original program to obtain the new better one. Unfortunately, the mechanism proved worse even than a pure random search, resulting mainly from the absence of effective selection pressure and extreme disassociation (small changes in program syntax usually cause large changes in the input-output behavior of the program).

In the 1960s, having the knowledge of Friedberg's disappointing results, Bremermann focused on the work of relatively simple optimization problems [9], especially for linear programming and convex programming. But it is too limited for real optimization applications. In 1964, a kind of evolutionary algorithm, called evolutionary programming (EP) [31], was formally proposed by L.J. Fogel. In his scheme, the individual, on behalf of a transition table of finite-state-machine (FSM), mutates to reproduce new FSMs; meanwhile, whether the individual was able to mutate in the next generation depended on the performance in the evaluation testing. Compared with Friedberg's and Bremermann's algorithms, Fogel applied evolutionary programming algorithm to more sophisticated optimization problems, with appropriate selective pressure provided. Nevertheless, the refined work did not receive the remarkable attention in the field, until genetic algorithm and evolution strategies were fully accepted in the 1970s. The limitation in Friedberg and Bremermann's experiments caused the ignorance of Fogel's works in almost 30 years. The initial contribution of L.J. Fogel was extended for applications in continuous parameter optimization by D.B. Fogel (L.J. Fogel's son). Details of elaborated EP algorithms can be checked in [29, 30].

5.2.2 The Classical Evolutionary Programming

Compared with other evolutionary algorithms, the initial EP distinguished itself by its fixed chromosome structure and its changing numerical parameters, which were allowed to evolve along with decision variables. The specific mechanism was brought by many other EAs and obtained a formal term, "self-adaptation." Initial EP algorithms can be divided into the following categories according to their main characteristics:

(1) Standard EP, which has no self-adaptation specialties
(2) Continuous standard EP, different from generation-based algorithms, in which the individual is evaluated and added into the population.
(3) Meta-EP, in which the variance of mutation step size is cooperated
(4) Continuous meta-EP, in which the individual is evaluated and added into the population due to the variance of the mutation operator
(5) Rmeta-EP, which cooperates covariance and standard deviations for adaptation

In the following section, the key elements in the process of EP algorithms will be represented in the sequence of representation of solutions, mutation operators, and selection mechanisms. It should be remained that though EP and ES are similar in many perspectives, their respective development is independent with each other.

5.2.2.1 Representation

The EP used in continuous parameter optimization adopts the optimization domain $x \in [l, u]$. What is different from many other algorithms is that the domain is only considered in the population initialization, and mutation operators may disobey them. Essentially, the search space of the algorithm is unconstrained, which is the space of D dimensions. Similarly, the variances v in meta-EP is initialized in a range $[0, b]$, but they can change in the domain of real numbers $\mathbb{R}$. The number of variances in meta-EP is equal to D. In short, individuals of standard EP can be described as $\mathbf{p} = (x, v)$, where $x \in \mathbb{R}^D$, $v = 0$. Slightly different from standard EP, individuals in meta-EP represent $\mathbf{p} = (x, v)$, where $x \in \mathbb{R}^D$, $v \in \mathbb{R}_+^D$.

5.2.2.2 Mutation

In standard EP, Gaussian mutation operator is applied, and it can be described in the form of $\mathbf{m} = (\{\beta_1, \ldots, \beta_D\}, \{\gamma_1, \ldots, \gamma_D\})$, where proportionality constant vector β and offset γ are parameters that must be tuned for a particular task. However, often the constants in β and γ are set to one and zero, respectively. The operator works with a standard deviation that is determined by the square root of a linear transformation of the fitness value $f(x)$ for each element in the variable vector.

Mutating with Gaussian mutation operator, the element x_i in variable vector x updated by

$$x_i' \leftarrow x_i + \sqrt{\beta_i \cdot f(x) + \gamma_i} \cdot \mathcal{N}_i(0, 1) \tag{5.1}$$

where $\beta_i = 1$, $\gamma_i = 0$ are commonly used; therefore, the mutation operator can be simplified

$$x_i' \leftarrow x_i + \sqrt{f(x)} \cdot \mathcal{N}_i(0, 1) \tag{5.2}$$

The mutation operator used in standard EP obviously incurs several difficulties:

(1) When the fitness value is large, the step size in the search will be large, which results in almost like random search.
(2) It needs a lot of effort to tune the extra 2D parameters.
(3) If the fitness value of global optimum is not zero, approaching the global optimum is not possible.

Compared with standard EP, meta-EP has a main specialty of adapting variances to overcome tuning difficulties. Different mutation operators combined with the variances vector v are applied

$$x_i' \leftarrow x_i + \sqrt{v_i} \cdot \mathcal{N}_i(0, 1) \tag{5.3}$$

$$v_i \leftarrow v_i + \sqrt{\zeta \cdot v_i} \cdot \mathcal{N}_i(0, 1) \tag{5.4}$$

where the parameter ζ is introduced to assure v to be positive all the time.

It should be noted that EP relies only on mutation operators but not on any recombination operators. It is still an argument with reservation whether mutation alone is a more effective way to evolution than that including recombination.

5.2.2.3 Selection

Carrying out the mutation policy, all individuals produce their offspring, resulting in a larger population with the size of 2N. To reduce the size from 2N back to N, selection mechanism is requisite. For example, the tournament selection is a popular method to weed out N individuals based on fitness value in the pairwise comparison. Details about selection can also be found in Sect. 4.3.1.

5.2.3 Framework and Parameter Settings

Using the above components, the framework of a meta-EP algorithm is shown in Algorithm 5.2 below.

Table 5.1 Parameter settings for standard EP and meta-EP

Paramter	EP	meta-EP	Default
Domain range l_i, u_i	×	×	$l_i = -50, u_i = 50$
Upper bound c of σ_i		×	$c = 25$
Proportionality constants β_i	×		$\beta_i = 1$
Offset constants γ_i	×		$\gamma_i = 0$
Meta-parameter ζ for adaptation		×	$\zeta = 6$
Tournament size q	×	×	$q = 10$
Population size N	×	×	$N = 200$

Algorithm 5.2: meta-EP algorithm

Input: Maximum Generation $G_{\max}$
Initialization: $t = 0$, $\mathbb{P}(t) = \{\mathbf{p}_i(t)\}$,$\mathbf{p}_i = (\mathbf{x}_i, \mathbf{v}_i, f(\mathbf{x}_i)))$, $i = 1, \cdots, N$;
Evaluate all individuals;
while $t < G_{\max}$ **do**
$\quad$ **Mutate**: $\mathbf{x}_i(t + 1) = \mathbf{x}_i(t) + \sqrt{\mathbf{v}_i(t)} \cdot \mathcal{N}(0, 1)$;
$\quad$ **Evaluate**: $f(\mathbf{x}(t + 1))$;
$\quad$ **Select**: $\mathbb{P}(t + 1) = \mathbb{P}(t) \bigcup New\ Generated$;
$\quad$ $t := t + 1$;
end

The algorithm starts with a set of solutions with randomly generated positions and mutation step sizes. Then the population undergoes mutation, evaluation, and selection procedures until the stop criterion is satisfied. The default parameter settings of standard EP and meta-EP can be seen in Table 5.1.[1]

5.2.4   Recent Advances in Evolutionary Programming

Ever since EP was applied to continuous optimization problems, adaptive and hybrid mechanisms have become the norm for the EP framework. The main reasons for the trend are:

(1) Step size is closely related to the find of the global optimum. All traditional EPs, including the above ones, rely on Gaussian mutation to produce offsprings. With a small step size, individuals may lose the capability to jump out of local optima. On the contrary, the mutation with a large step size will make it easy to miss the global optimum.
(2) With more applications to the optimization problems with different characteristics, it is impossible to handle all situations with a single mutation strategy. So it

---

[1] The default values were suggested by Fogel for low-dimensional problems; they may change for high-dimensional problems.

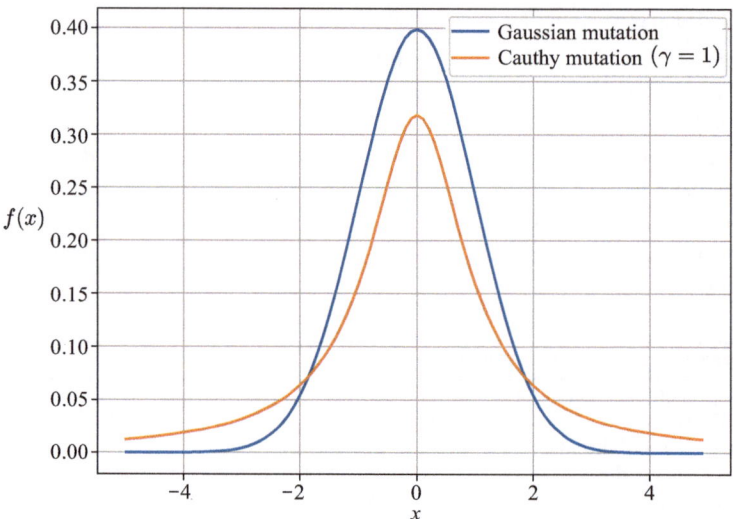

Fig. 5.1 Probability distribution of mutation step size for Gaussian and Cauchy mutation

is a natural way to construct multiple mutation strategies to deal with complex situations.

The classic evolutionary programming (CEP) with Gaussian mutation tends to converge quite slowly; on account of which, a faster evolutionary programming (FEP) was proposed [84, 85]. Different from the CEP, FEP uses Cauchy mutation as the primary search operator, which has larger jumps than Gaussian mutation. Both analytical and experimental studies show that FEP is skilled in multimodal problems and CEP is good at unimodal problem with only a few local optima.

Figure 5.1 shows the probability distribution of the step size in the Gaussian mutation and the Cauchy mutation, respectively. It is obvious that Cauchy mutation has more chances to get a jump with bigger step size than Gaussian mutation.

Based on the fact that FEP and CEP have benefits in different situations, the hybrid method can be utilized to incorporate the strength of both. Generally speaking, it would be ideal to use Cauchy mutation when the individual is far away from the optimum and to use Gaussian mutation when near to the optimum. However, global optimum is not actually known. So some algorithms use the self-adaptation technique to learn to "change" by itself. For example, in the meta-EP, the standard deviation (step size) will be modified in every iteration and the decision variable of the individual update too. In 2004, a generalized EP with Lévy distribution [52] was put forward, which can adjust the variation in the mutation with different parameters. Because Cauchy distribution is just a special case of Lévy distribution, the self-adaptive algorithm can derive abundant mutation density with different parameters. The most popular probability distributions are listed in Table 5.2.

Table 5.2 Popular probability distributions used in EP

Type	Parameter	Probability density function
Gaussian mutation	μ, σ	$\frac{1}{\sqrt{2\pi\sigma^2}} e^{-\frac{(x-\mu)^2}{2\sigma^2}}$
Cauchy mutation	x_0, γ	$\frac{1}{\pi\gamma}\left[\frac{\gamma^2}{(x-x_0)^2+\gamma^2}\right]$
Lévy mutation	μ, c	$\sqrt{\frac{c}{2\pi}} \frac{e^{-\frac{c}{2(x-\mu)}}}{(x-\mu)^{3/2}}$

Apart from the adaptation method to obtain multiple search strategies, some researchers also try to "switch" the policy in the search process for an individual. An improved FEP (IFEP) [52] generates two offsprings by the Cauchy mutation and the Gaussian mutation; then the better one will be selected. Furthermore, the mutation strategy with probability is controlled [41]. Each individual selects mutation operator m from a mutation operator set according to a selection probability p. The probability of Gaussian mutation (m_1) and Cauchy mutation (m_2) can be calculated as follows. Let $\boldsymbol{x}_i(k)$ and $\boldsymbol{x}'_i(k)$ be the parent and its offspring in k-th dimension, respectively. The output $o(\boldsymbol{x}_i, m_j)$ of an individual $\boldsymbol{x}_i$ is defined by how far an individual $\boldsymbol{x}_i$ moves during a successful mutation by m_j.

$$o(\boldsymbol{x}_i, m_j) = \max_{1 \leq k \leq D} o(\boldsymbol{x}'_i(k)) \tag{5.5}$$

$$o(\boldsymbol{x}'_i(k)) = \begin{cases} |\boldsymbol{x}_i(k) - \boldsymbol{x}'_i(k)| & f(\boldsymbol{x}'_i) < f(\boldsymbol{x}_i) \\ 0 & \text{otherwise} \end{cases} \tag{5.6}$$

The output of the mutation strategy $o(m_j)$ can be defined as

$$o(m_j) = \max_{1 \leq i \leq N} o(\boldsymbol{x}_i, m_j) \tag{5.7}$$

Then the payoff $\pi(m_j)$ and the selection probability p_j of the mutation operator $\pi(m_j)$ can be described as follows:

$$\pi(m_1) = \frac{o(m_1)}{o(m_2)}, \pi(m_2) = \frac{o(m_2)}{o(m_1)} \tag{5.8}$$

$$p_1 = \frac{\pi(m_1)}{\pi(m_1) + \pi(m_2)}, p_2 = \frac{\pi(m_2)}{\pi(m_1) + \pi(m_2)} \tag{5.9}$$

In short, not only different mutation operators but also the different parameter settings in the same operator can make big differences in the search process. Self-adaptation of parameters and the selection control of the operators are the main methods to improve algorithms' performance. In Chap. 6, methods of parameter control and policy adaptation will be covered in detail.

5.3   Genetic Programming

Genetic programming (GP) describes a general term for all EAs that use tree data structures to represent solutions or the set of all EAs that breed programs, algorithms, and similar constructs. We will focus more on the latter one and discuss tree-shaped genomes.

5.3.1   Introduction

Figure 5.2 shows a conventional input-processing-output model, which takes input data to compute and return output. In GP, we usually know some inputs or situations, and the corresponding outputs are known or can be simulated or produced. The goal is to find the model to produce desired outputs according to the specific inputs or situations. From this point of view, it has something in common with machine learning. Both of them want to build a model, rather than hold a model to find the best input—like what genetic algorithm does.

The history of GP can be traced back to 1950 with the Turing test [76]. The pioneer Friedberg [32] used a learning algorithm to improve a program for a theoretical computer called Herman. Around the same time, Samuel created the world's first self-learning program to the game of checkers.

In the mid-1980s, the next generation of scientists tried to find a way to evolve programs, and a genetic algorithm was used to evolve programs in some programming languages such as PL [15], Proglog [20], and LISP [33]. The mutation operation on tree genome was used [5]. Then Koza pushed this area forward for a big step by using tree structures (see Sect. 4.2.4) to employ genomes, which is called standard GP. It brought the possibility of evolving more complex programs.

```
list<individual> CreatePop(int s) {
    list<individual> pop;
    for (int i = 0; i < s; i++){
        pop.push_back(Create());
    }
    return pop;
}
```

Program for Algorithm 5.3 in C++

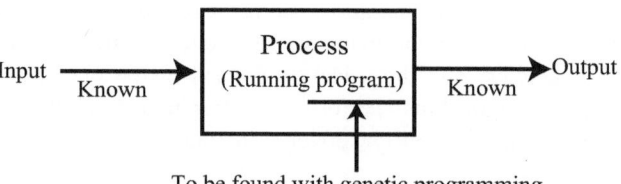

Fig. 5.2 Genetic programming in the context of the input-process-output model

---

Algorithm 5.3: POP = CreatePop(s)

---

input : *s* the size of the population to be created
output: *pop* the new random population
pop() ← ();
i ← 0;
while *i* < *s* **do**
    pop ← AppendList (pop, Create ());
    i ← i +1;
return pop

Fig. 5.3 The AST representation of algorithms/programs

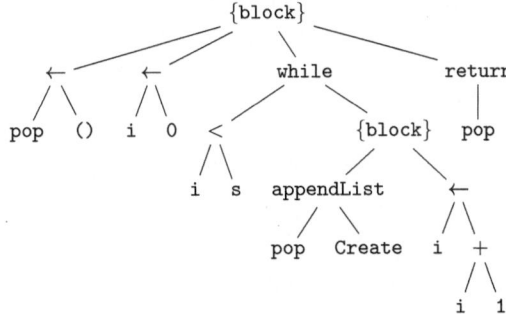

Depending on tasks, a tree can represent a rule set [56], a decision tree [48], or the blueprint of an electrical circuit [45]. Furthermore, the structure of trees is very close to the structure of algorithms and programs, not only because of the hierarchy structure of a program but also because compilers use tree representation internally. When reading a program, they first split it into tokens, parse these tokens, and finally create an abstract syntax tree (AST) [44], where the internal nodes represent operators and the leaf nodes contain operands. In theory, we can represent any program into an AST, e.g., Algorithm 5.3 can be implemented in C++ language and represented using an AST shown in Fig. 5.3.

Tree-based GP has some special properties compared with artificial neural networks (ANN). It is direct and intuitive since the tree structure is initially used to represent mathematical expressions. It often offers white-box solutions that are human interpretable, while ANN generates black-box outputs, which are highly complicated to fully understand. Furthermore, the structure of GP can be evolved internally during the optimization process, while neural architecture search (NAS) techniques, which are optimization methods, are needed to optimize the architecture of ANN.

5.3.2   *Genotype-Phenotype Mapping*

Genotype-phenotype mapping (GPM) methods are used in many different GP methods. Here, we take the integer string approach and the gene expression programming as two examples.

5.3.2.1   Integer String Approach

Cramer's GP approach [15] is one of the earliest GP methods, which is based on genotype-phenotype mapping, to evolve programs in a variation of the programming language PL. Two simple examples of PL language from his work are shown in Listing 5.1.

Listing 5.1 Two examples for the PL dialect used by Cramer for GP

```
(: ZERO v0)
(: LOOP v1 (: INC v0))
(: ZERO v5)
(: LOOP v3 (: LOOP v4 (: INC v5)))
```

where (: ZERO var) means to set variable var to 0; (: INC var) means to increase var by 1; (: LOOP var STAT) means to perform the statement STAT var times. So the first two lines will set the value of v0 to v1's value. The two bottom lines will multiply v3 to v4 and store the result in v5.

This algorithm uses integer strings as genomes and two methods to convert them into program trees. The first one is JB mapping, which uses a tuple of a fixed-length integer string to represent arbitrary instruction. In the example in Listing 5.2, there are some triplets where the first integer identifies the operation and the two following integers represent parameters (could be statements and variables).

Listing 5.2 JB Mapping example

```
(0 4  2) → (: BLOCK AS4 AS2 )
(1 6  0) → (: LOOP v6 AS0 )
(2 1  9) → (: SET v1 v9)
(3 17 8) → (: ZERO v17 ) ;;the 8 is ignored
(4 0  5) → (: INC v0) ;;the 5 is ignored
```

The first statement defines the main statement, which is the enter point of the program (the BLOCK here means the parameters of it will be performed sequentially). The statements after the main statement are auxiliary statements. In Listing 5.2, the numbers appearing in vn and ASn represent variables and auxiliary statements, respectively. Based on this translation rule, a program can be encoded as (0 0 1 3 5 8 1 3 2 1 4 3 4 5 9 9 2) and the translation is shown in Listing 5.3.

Listing 5.3 Another JB Mapping example

```
(0  0  1)  ;;main statement      → (: BLOCK AS0 AS1 )
(3  5  8)  ;;auxiliary statement 0 → (: ZERO v5)
(1  3  2)  ;;auxiliary statement 1 → (: LOOP v3 AS2 )
(1  4  3)  ;;auxiliary statement 2 → (: LOOP v4 AS3 )
(4  5  9)  ;;auxiliary statement 3 → (: INC v5)
```

To reduce the influence of strong positional epistasis—the strong relation of the meaning of an instruction to its position—Cramer introduced TB mapping (a modified version of the JB mapping), which uses a treelike structure and decodes recursively. Thus, the program tree in Listing 5.3 will be represented in string (0 (3 5) (1 3 (1 4 (4 5))))) in TB mapping.

5.3.2.2   Gene Expression Programming

Like CGP, gene expression programming (GEP) proposed by Ferreira [28] also translates fixed-length chromosomes into tree phenotypes. However, it introduces a new way to handle the remaining unsatisfied function arguments.

A special gene structure is used to achieve this. A gene in GEP can be divided into two parts, that is, the head part and the tail part. The head part of the term contains codons representing arbitrary expressions, and the tail part contains only instance parameters. For each problem, the length of the head h is fixed, and the length of the tail is defined by Eq. (5.10), where n is the number of arguments of the function with the most arguments.

$$t = h(n - 1) + 1 \qquad\qquad (5.10)$$

Since we have h expressions, and the largest number of arguments required by arbitrary one of them is n, then the upper bound of the number of arguments is $h * n$. In addition, there have been $h - 1$ arguments that have been satisfied by $h - 1$ expressions after the first expression since any expression can be an argument for the expression instantiated before. Therefore, the number of remaining unsatisfied parameters is $h * n - (h - 1) = h * n - h + 1 = h(n - 1) + 1$.

An example illustrated in Fig. 5.4 from [28] shows the genotype-phenotype structure and the translation between them. A phenotype is built by interpreting the gene as a level-order traversal of the nodes of the expression tree.

5.3.3   Other Genome Structures

There are other genome structures used in GP besides the tree structure. Some of them will be discussed in this part.

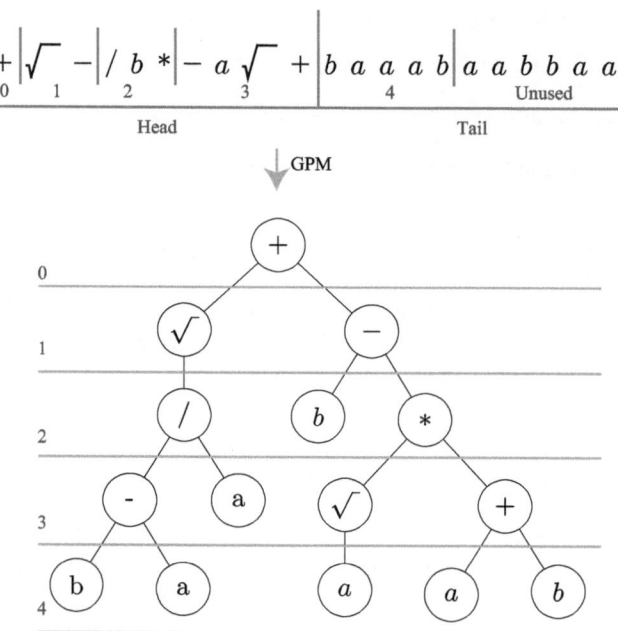

Fig. 5.4 A GPM example of GEP

5.3.3.1   Linear GP

One of the main goals of GP is to generate programs automatically. The tree structure is not the only way to achieve that. In fact, programs are processed as sequences of instructions in a computer. GP methods based on the instruction sequence are called linear GP (LGP).

Since LGP does not use a tree as an intermediate representation, it needs less time to evaluate each individual. However, conventional operators cannot be applied directly to it. This unavailability can be more serious with an LGP method that allows arbitrary jumps to call instructions. Take Fig. 5.5a, for instance. This is a part of a memory address in a program. If we use relative address shift to represent a jump (e.g., a −4 jump from an address that contains 83 to an address that contains 9A), then an insertion of a new command without any further corrections (e.g., insert 23 after 9A) may lead to an invalid jump and then destroy the whole program after that. Tree-based methods are less vulnerable with this operation (see the insertion of $n \leftarrow i$ in Fig. 5.5b).

There are two ways to address the influence of this issue. One way is to use some intelligent operators, which can analyze the control flow. The other way is to use the operations that can only have the minimal effect.

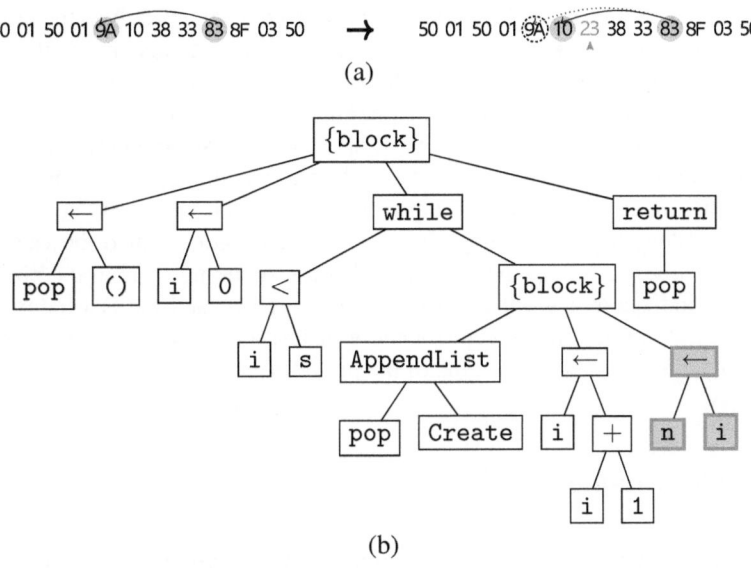

(a)

(b)

Fig. 5.5 Insertion operation in both tree-based GP and LGP. (**a**) Insertion into an instruction string. (**b**) Insertion in a tree representation

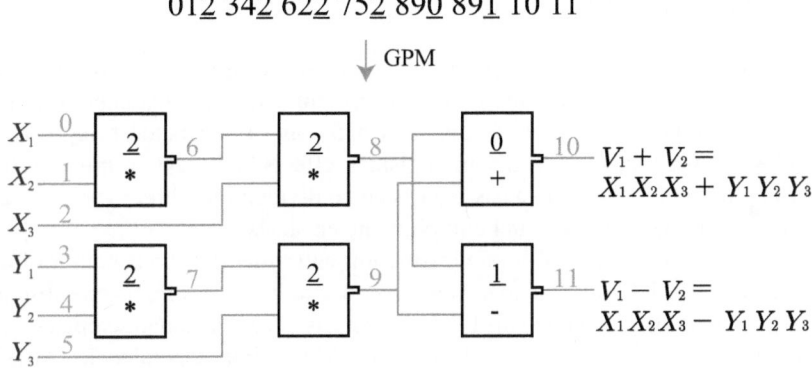

Fig. 5.6 An example of GMP of CGP

5.3.3.2   Graph-Based GP

If you think about the flowchart of a program, you will see that in addition to tree representation and linear representation, a program can also be represented as a graph. Figure 5.6 shows an example of cartesian GP (CGP)—one of the best known graph-based GP approaches.

This program has $o = 2$ outputs, which are the difference and the sum of volume of two boxes $V_1 = X_1X_2X_3$ and $V_2 = Y_1Y_2Y_3$. The $i = 6$ inputs are side lengths $X_1 \cdots X_3$ and $Y_1 \cdots Y_3$, which are numbered from 0 to 5. And the predefined function set are $\{+, -, *, /, \vee, \wedge, \oplus, \neg\}$, where eight elements are numbered from 0 to 7, respectively. The grid size, which is $n = 3$ cells wide and $m = 2$ cells deep, is fixed. All the cells have the same input size and output size (e.g., $i' = 2$ and $o' = 1$ in this example). Outputs of all cells are also numbered in ascending order beginning with i. Since there are 6 cells and we already have 6 inputs for the program, the index of the 6 cells' outputs is $6, 7, \cdots, 11$.

The fixed-length integer strings genotype of this program is also shown at the top of Fig. 5.6. $m * n$ genes in the front encode corresponding cells in the phenotype. The first i' indices indicate the inputs the cell receives, and the last index denotes the function it carries out. Except for these genes, the last o genes indicate which outputs of cells are the final outputs of the program.

5.3.4   Open Issues

Several open issues should be taken into account with GP. Here we would like to discuss the following aspects.

5.3.4.1   Epistasis in GP

We have introduced some GP approaches that can be used to evolve "real algorithms" in the previous sections. However, in applications, researchers often find that the fitness landscape is very rugged, which can be attributed to epistasis. In the following section, we will discuss epistatic effects and ways to mitigate them. According to their causes, epistasis in GP can be divided into three types: semantic epistasis, positional epistasis, and embryogenic epistasis.

In an algorithm, the result of one instruction will influence the behavior of those executed afterward. If we change $a = b + c$ into $a = b - c$, its effects on subsequent instructions will also change. This type of epistasis is called semantic epistasis.

Another epistasis occurs in the form of *positional* interdependencies. It is one of the causes that many of the phenotypic and genotypic representations in GP are rather fragile in terms of insertion and crossover points. It can be simply understood as the vulnerability of a program when statements are swapped from their positions.

The third epistasis is *embryogenic* epistasis. It occurs in grammar-guided GP approaches, which translate string genome to trees. Ryan et al. [68] think it can very well be compared to artificial embryology: the translation of the DNA into proteins. This process depends very much on the proteins already produced: if a certain piece of DNA has created a protein X and is transcribed again, a protein Y may be produced for the presence of X.

5.3.4.2 Correctness

For the application of GP to evolve programs and algorithms, methods to evaluate both the functional and non-functional requirements are required to guide the direction of the evolutionary process.

According to the Entscheidungsproblem[2] and halting problem,[3] we cannot determine whether arbitrary evolved programs will provide correct results or whether they will finish running by using some kind of algorithm. In addition, exhaustive testing is unacceptable in most cases. So, how to measure the correctness of an evolved program is an open issue.

5.3.4.3 All-or-Nothing

The evolution of algorithms is often a special instance of the needle-in-a-haystack problem (see Sect. 2.2.2). For example, an algorithm computing the greatest common divisor of two numbers is either correct or wrong. If you choose only one objective function that results right or wrong, you will get the fitness landscape with a few steep spikes of equal height distributed over a large plane of infeasible solution candidates with equally bad fitness.

5.3.4.4 Non-functional Features of Algorithms

In addition to evaluating the functionality of an algorithm, some non-functional features should also be considered. In GP, a problem is code *bloat*, which is the uncontrolled growth in the size of the individuals during the course of the evolution. It is often used in conjunction with code *introns*. It can cause the evolved programs to become unnecessarily big, make operators like mutation and recombination useless, slow down both the evaluation and breeding processes, and increase memory consumption.

Tree depth limitation and an additional objective to evaluate the bloat degree [6, 18, 25] could be useful to deal with it:

(1) *Runtime and memory consumption.* Compared with code size, the existence of loops and recursion in programs invalidates a more direct relation to runtime consumption. The number of variables and memory cells needed by programs in order to perform the task should be as small as possible.
(2) *Errors.* In the application of symbolic regression, some non-functional errors such as division by zero could occur. Then the division operator can be rendered

---

[2] https://en.wikipedia.org/wiki/Entscheidungsproblem
[3] https://en.wikipedia.org/wiki/Halting_problem

to a nop or yields a default value such as 1. However, you can also set error numbers as an objective and then minimize it.

(3) *Transmission count.* In distributed algorithms, message transmission is time-consuming and costly, since the network is very slow compared with the computing speed. So, the number of transmission should be as low as possible.

5.4 Particle Swarm Optimization

Particle swarm optimization (PSO) is a kind of population-based algorithm inspired by simulating the behaviors of natural bird flocks and fish schools.

5.4.1 The Rise of Particle Swarm Optimization

Though evolutionary algorithms are the mainstream of evolutionary computation, there are another genres of algorithms for addressing complex real-world problems that cannot be solved by traditional methodologies, called swarm intelligence (SI). SI focuses on the collective behavior of the natural or artificial agent systems. To simulate the social and collective behavior of bird flocks, Reynolds built a simple model called "BOIDS" [67] in 1987. In the model, all agents in the swarm obey just three ordinary rules, separation, alignment, and cohesion, and then dramatic collective behavior will take into shape soon. As shown in Fig. 5.7, a particle will tend to separate with others by the repulsive force when other particles get too close to itself (move into the circle), so overcrowded crash can be avoided. To make the swarm stable, the particle will have the tendency to move according to the swarm average velocity. To avoid the swarm to be scattered, all particles will get centripetal force to the current swarm center.

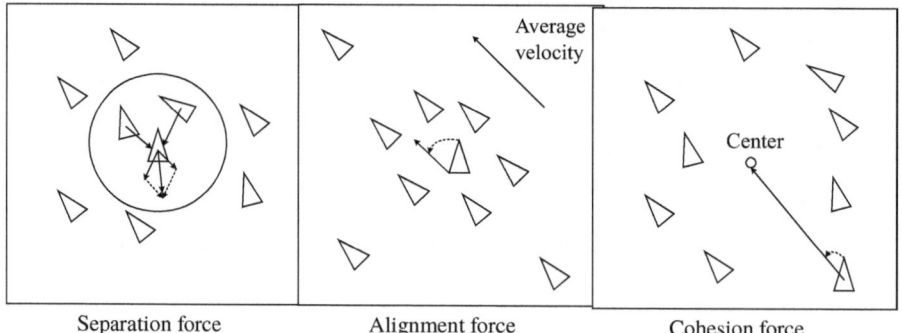

| Separation force | Alignment force | Cohesion force |

Fig. 5.7 The free-body diagram of the particle

A simulation-inspired algorithm PSO was proposed to solve nonlinear function optimization problems by Kennedy and Eberhart [24] several years later in 1995. It is not like the GA and other traditional EAs. Every particle in the swarm/population has a velocity vector along with the potential solution. The production of a new solution relies on the "flying" according to the velocity vector in hyperspace. Meanwhile, the velocity will update in line with the previous best solution and the global best solution. That is, all agents will record their own historical position with the best fitness value as well as the global best position. In that, most of the SI algorithms are inspired by social creatures such as ants, birds, etc. There is no obvious evolutionary process in the algorithm. However, from the perspective of algorithm analysis, they can be classified into the general EC framework as shown in Fig. 4.5 in Sect. 4.1.4.

5.4.2 Original Particle Swarm Optimization

As mentioned earlier, the original PSO was inspired by the social behavior of animals, especially the ability of animals to locate the position such as the food source or predators. Many individuals work as a group to improve the probability of survival. The locating behavior can be associated with the search process. In the original PSO, particles move in the search space with the attraction toward the best solution found by themselves and the best solution found in their neighborhood. The particles in the neighborhood can learn from each other, like the biological models where birds and fishes are just able to communicate in the vicinity.

In the PSO framework, a particle i is made up of three vectors: a position vector in D-dimensional space $x_i = [x_{i,1}, x_{i,2}, \cdots, x_{i,D}]$, a velocity vector $v_i = [v_{i,1}, v_{i,2}, \cdots, v_{i,D}]$, and a personal best position vector found so far $pbest_i = [pbest_{i,1}, pbest_{i,2}, \cdots, pbest_{i,D}]$. The position vectors and velocity vectors are set to uniformly generate random numbers for the swarm initialization; then particles move in the hyperspace according to the velocity vector, and the velocity vector updates toward the personal best position and the best solution found by all neighbors.

In every iteration, the velocity and position of particle i will be updated by

$$v_{i,d} = v_{i,d} + \eta_1(pbest_{i,d} - x_{i,d}) \cdot \mathcal{U}_1(0, 1) + \eta_2(lbest_{i,d} - x_{i,d}) \cdot \mathcal{U}_2(0, 1) \tag{5.11}$$

$$x_{i,d} = x_{i,d} + v_{i,d} \tag{5.12}$$

where $\mathcal{U}_1(0, 1)$ and $\mathcal{U}_2(0, 1)$ are constant values with 2.0; η_1 and η_2 are independent random numbers generated within $[0, 1]$ for every dimension update; and $lbest_i$ is the best position found by the neighbors of particle i.

The velocity update contains three components, which are the momentum component, the cognitive component, and the social component. The momentum will keep the particle in its current direction, the cognitive component will force the particle to return to its best position found so far, and the social component will attract the particle to the best position found so far from its neighborhood, which is called population topology defined by users. The PSO algorithm can be described in Algorithm 5.4.

Algorithm 5.4: Original PSO algorithm

Input: Maximum Generation $G_{\max}$, Population Size N
Initialization: $t = 0$, $\mathbb{P}(t) = \{\mathbf{p}_1(t), \mathbf{p}_2(t), \ldots, \mathbf{p}_N(t)\}$,
      where $\mathbf{p}_i(t) = (\mathbf{x}_i(t), \mathbf{\textit{pbest}}_i(t), \mathbf{v}_i(t)\}$, $i \in \{1, 2, \cdots, N\}$;
while $t < G_{\max}$ **do**
   foreach *particle* $\mathbf{p}_i(t)$ *in swarm* **do**
     Update $\mathbf{x}_i(t + 1)$ according to position update equation ;
     Update $\mathbf{v}_i(t + 1)$ according to velocity update equation ;
     Calculate particle fitness $f(\mathbf{x}_i(t + 1))$;
     Update $\mathbf{p}_i(t + 1)$ and $\mathbf{\textit{gbest}}(t + 1)$;
   end
   $t=t+1$;
end

Many variants of PSO have been proposed, and some standard PSO series, which are widely accepted, are SPSO 2006, SPSO 2007, and SPSO 2011. The source code of the three successive versions of the standard PSO can be obtained on the Particle Swarm Central.[4]

5.4.3 Standard Particle Swarm Optimization

Once the original PSO was proposed, many researchers from all over the world committed to modify the PSO from different perspectives. A standard PSO (SPSO) algorithm [8] was proposed to unify the previous work and made it easier to compare the different PSO algorithms. The key components of SPSO will be present in a comparative way, which can allow the differences between them easy to find.

5.4.3.1 Initialization

All the SPSO algorithms initialize their particles randomly in the search spaces for both the location and velocity. For the population size, SPSO2006 and SPSO 2007 set the swarm size according to the number of dimensions $10 + \sqrt{D}$, while SPSO 2011 suggests the value to be 40 by experiences.

---

[4] www.particleswarm.info

5.4.3.2 Population Topology

The swarm topology defines the particles to communicate for a given particle. Therefore, it determines the choice of the **pbest** of a particle, which is an adaptive random neighborhood topology in the three SPSO versions. At the initialization or for each unsuccessful iteration (i.e., no fitness improvement of the best solution found of the swarm), each particle will be assigned K randomly chosen particles as its neighbors (including itself) for sharing their previous best position.

5.4.3.3 Confinement

When a particle i moves away from the problem boundary, SPSO 2006 and SPSO 2007 regard the boundary as a "wall," and the position and velocity are modified according to Eq. (5.13) and Eq. (5.14), respectively. The position will be set to the boundary values, and the velocity will be set to zero accordingly.

$$x_{i,d}(t+1) = \begin{cases} l_d & x_{i,d}(t) < l_d \\ u_d & x_{i,d}(t) > u_d \end{cases} \tag{5.13}$$

$$v_{i,d}(t+1) = \begin{cases} 0 & x_{i,d}(t) < l_d \text{ or } x_{i,d}(t) > u_d \\ \text{update as normal} & \text{otherwise} \end{cases} \tag{5.14}$$

In SPSO 2011, the particles serve as a bouncy "ball" in a confined room, so they bounce back from the "punky wall." The velocity updates according to Eq. (5.15)

$$v_{i,d}(t+1) = -0.5 v_{i,d}(t+1) \tag{5.15}$$

5.4.3.4 Velocity Updating

In terms of the velocity updating, SPSO2006 and SPSO2007 follow the same rule proposed in [70], where an inertia weight is introduced in the momentum component to control the velocity for swarm convergence as

$$v_{i,d} = \omega v_{i,d} + c_1 \eta_1 (pbest_{i,d} - x_{i,d}) + c_2 \eta_2 (lbest_{i,d} - x_{i,d}) \tag{5.16}$$

To have a deeper sight into the search mechanism, Fig. 5.8a illustrates the search behavior in a 2-D problem search space, where Δv_i of particle i can cover the shadow range from the theoretical analysis aspect. The velocity is updated dimension by dimension, which results in the bias of parallel search along one of the coordinate axes [72]. This behavior is too naive for most optimization problems, so SPSO2011 replaces the update velocity equation dimension by dimension.

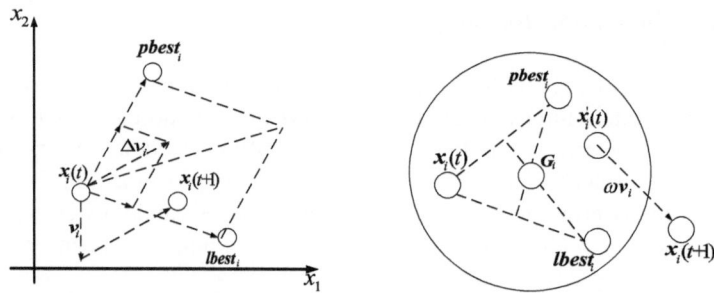

(a) Velocity updated in SPSO2006 and SPSO2007 (b) Velocity updated in SPSO2011

Fig. 5.8 Velocity update of SPSO2006 and SPSO2007 (**a**) and SPSO2011 (**b**)

Unlike the canonical velocity update, where Δv_i tends to the point on the line of *pbest* and *lbest*, SPSO2011 constructs a virtual geometrical point G_i. The constructed point can be defined by Eq. (5.17), which is similar to the "barycenter" of the current position of particle i, *pbest*, and *lbest* as shown in the right graph in Fig. 5.8. Then random position x_i' is generated around the center G_i of radius $\|G_i - x_i\|$. So the corresponding velocity and position updating equations become Eqs. (5.18) and (5.19).

$$G_i = \frac{x_i + (x_i + c(pbest_i - x_i)) + (x_i + c(lbest_i - x_i))}{3}$$

$$= x_i + c\frac{pbest_i + lbest_i - 2x_i}{3} \tag{5.17}$$

$$v_i(t+1) = \omega \cdot v_i(t) + x_i'(t) - x_i(t) \tag{5.18}$$

$$x_i(t+1) = \omega \cdot v_i(t) + x_i'(t) \tag{5.19}$$

5.4.4 Recent Advances in Particle Swarm Optimization

PSO has prominent performance on continuous function optimization problem; as a result, many variants have been proposed in the literature in different fields. From the perspective of improvement, variants can be classified mainly into four aspects: communication topology, PSO with diversity maintenance, adaptive PSO, and hybrid PSO.

5.4.4.1 Communication Topology

As mentioned above, one of the most obvious differences between PSO and other EAs is that particles in PSO record the best positions found by themselves. For

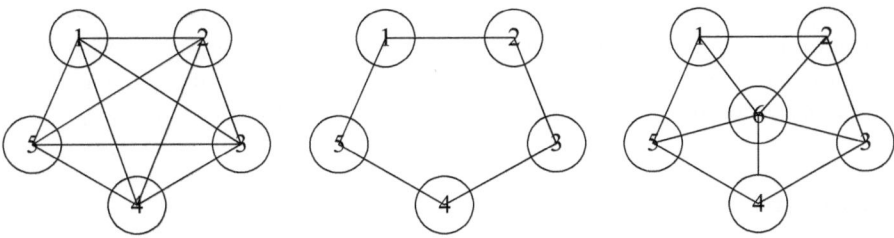

Fig. 5.9 Topology sketch map

the guarantee of convergence of the whole swarm, particles also learn from the best positions found by their neighbors. In addition, using historical best positions will help particles revise searching behaviors. The topology of the swarm can be divided into two categories: static topologies and dynamic topologies. After the original proposed PSO, circles, stars, and other communication topology methods were introduced, showing different search effectiveness with various problems [47]. Some topology sketch maps are shown in Fig. 5.9, in which the number means the index of the particle and the lines between them means the two can know the previous best solution of each other.

Furthermore, some dynamic communication structures were also built to solve various optimization problems. For example, a dynamic swarm topology [75] changes from the local best model to the global best model, which smoothens the search process from exploration to exploitation (the balance between exploration and exploitation can be referred to Chap. 7).

5.4.4.2 PSO with Diversity Maintenance

Diversity maintenance is a vital factor for the balance between exploration and exploitation. Premature convergence is always an undesirable situation that happens in EAs. To avoid this, negative entropy [80] was applied to the original PSO, enabling the algorithm keep a relative balance on exploration and exploitation. Similarly, a niching mechanism was also cooperated in the initial PSO model [78]. Furthermore, multi-swarm techniques recently proposed can be added into the algorithm.

5.4.4.3 Adaptive PSO

Besides communication topologies, there are many non-negligible aspects that determine the performance of the algorithm directly. Settings such as parameters, velocity updating equations, and so on always vary with the search process and different optimization problems. So, the adaptive mechanism for PSO becomes the common choice for many researchers. Adaptive PSO [66] added time-varying acceleration coefficients, where η_1 and η_2 are set to a large and a small value, respectively, at the beginning and are gradually reversed during the search. In recent years, some PSO algorithms applied different velocity and position updating equations, according to the distribution of population and particles' fitness [53].

5.4.4.4 Hybrid PSO

On account of the fact that different algorithms yield various applicability to the optimization problems, hybrid EAs become more and more popular. A selection scheme was added to the PSO, and the first hybrid PSO algorithm was introduced [2]. Then more techniques derived from other EAs were incorporated into PSO algorithms, such as the hybrid PSO applied with genetic programming [62], PSO model integrated with recombination operators [12], and so on.

In general, since the proposal of the original PSO, a great number of methods have been published on the design of algorithms for general purposes. In addition, many variants have been utilized in many kinds of applications, e.g., electromagnetics, economic dispatching, power flow optimization, etc. The PSO model has become one of the most popular optimization tools in the world.

5.5 Differential Evolution

Differential evolution (DE) is one of the most classical EC methods, which was proposed by Storn and Price [73] in 1995. Since then, DE and its variants have emerged as one of the most versatile family of EC methods. Meanwhile, DE variants have been successfully applied to solve real-world problems, especially in the domain of numerical optimization problems.

5.5.1 Introduction of Differential Evolution

The canonical DE algorithm is simple but effective. A standard DE algorithm contains four basic steps, which are initialization, mutation, crossover (recombination), and selection. The initialization is a conventional population initialization as in other EAs.

DE is a parallel and stochastic search method. In the algorithm, a population is composed of N D-dimensional solution vectors x_i; $i = 1, 2, \cdots, N$. For each generation t, the population size remains the same during the evolutionary process.

5.5.1.1   Mutation

The original mutation operation is used to generate donor solution vectors by adding a weighted difference between two solution vectors to a third vector. For each vector $x_i(t), i = 1, 2, \cdots, N$ at iteration t, a mutant vector is generated according to Eq. (5.20) as follows:

$$v_i(t+1) = x_{r_1}(t) + F(x_{r_2}(t) - x_{r_3}(t)) \tag{5.20}$$

where r_1, r_2, r_3 are random indexes chosen within $1, 2, \cdots, N$ and $F \in [0, 2]$ is a scale factor that controls the amplification of the differential variation $(x_{r_2}(t) - x_{r_3}(t))$. Note that the randomly chosen integers r_1, r_2, and r_3 are all different from the selected index i, so N must be greater than or equal to 4 to allow for this condition.

The mutation operation of Eq. (5.20) proposed in the original DE algorithm is called DE/rand/1, which is almost the most credible. Figure 5.10 illustrates how DE/rand/1 generates a target vector in a two-dimensional search space. From the figure, we can see that the target vector generated by this method has a large uncertainty since the donor vector x_{r_1} and differential vectors x_{r_2} and x_{r_3} all are selected randomly from the current population.

Another commonly used mutation operation is the DE/best/1:

$$\text{DE/best/1} : v_i(t+1) = x_{best}(t) + F(x_{r_2}(t) - x_{r_3}(t)) \tag{5.21}$$

Comparing to DE/rand/1, we can see that the only difference is the base vector to be mutated. Here, the base vector is $x_{best}(t)$, which represents the solution vector with the best objective at time t. From the example in Eq. (5.21), it is not hard to find that the newly generated target vector has a large possibility of locating near

Fig. 5.10 An example of the process for generating a target vector by DE/rand/1 in a two-dimensional search space

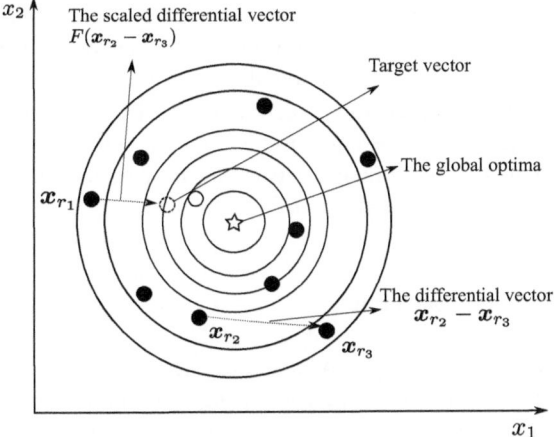

the best solution vector in the current population, especially when the scale factor F is small.

Besides the above two mutation operators, there are some other mutation operators that are also referred frequently:

DE/rand/2:

$$v_i(t+1) = x_{r_1}(t) + F(x_{r_2}(t) - x_{r_3}(t)) + F(x_{r_4}(t) - x_{r_5}(t)) \qquad (5.22)$$

DE/best/2 :

$$v_i(t+1) = x_{best}(t) + F(x_{r_1}(t) - x_{r_2}(t)) + F(x_{r_3}(t) - x_{r_4}(t)) \qquad (5.23)$$

DE/target-to-best/1 :

$$v_i(t+1) = x_i(t) + F(x_{best}(t) - x_i(t)) + F(x_{r_1}(t) - x_{r_2}(t)) \qquad (5.24)$$

5.5.1.2 Crossover

In order to increase the diversity of the perturbed parameter vectors, there is usually a crossover operator after the mutation operation. The DE family of algorithms commonly uses a binomial (or uniform) crossover method. Binomial crossover is performed on each of the D variables whenever a randomly generated number between 0 and 1 is less than or equal to a predefined crossover rate CR. The scheme can be expressed as:

$$u_{ij}(t+1) = \begin{cases} v_{ij}(t+1) & \text{if } (\mathcal{U}_{ij}(0,1)) \le CR) \text{ or } j = K \\ x_{ij}(t) & \text{otherwise} \end{cases} \qquad (5.25)$$

where K is a randomly chosen natural number in $\{1, 2, \cdots, D\}$ with the aim to ensure that $u_i(t+1)$ gets at least one component from $v_i(t+1)$; $\mathcal{U}_{ij}(0,1)$ is instantiated for every component of each vector per iteration.

Figure 5.11 shows how the crossover operator affects the newly generated vector in a two-dimensional search space. If the crossover probability CR is 1, the new vector will all be equal to the target vector generated by mutation, and if CR is less than 1, some dimensional values may be inherited from the parent individual, e.g., the vector u_i' or vector u_i'' in Fig. 5.11.

5.5.1.3 Selection

When a vector is produced at generation t, we should decide whether the new trail vector can be saved for the next iteration. The trial vector $u_i(t+1)$ is simply compared to the target vector $x_i(t)$ as in Eq. (5.26). If vector $u_i(t+1)$ has better

Fig. 5.11 An example of binomial crossover a two-dimensional search space

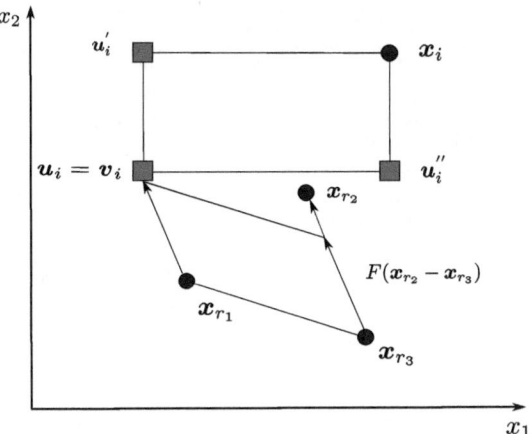

fitness value than $x_i(t)$, $x_i(t)$ will be replaced by $u_i(t+1)$; otherwise, $x_i(t)$ is retained.

$$x_i(t+1) = \begin{cases} u_i(t+1) & \text{if } f(u_i(t+1)) \le f(x_i(t)) \\ x_i(t) & \text{otherwise} \end{cases} \qquad (5.26)$$

5.5.2   Framework and Parameter Settings

Algorithm 5.5 shows the pseudocode of the differential evolution with DE/rand/1 mutation operator. Several key factors, including the mutation operator, the mutation scale factor F, and the crossover probability CR, can affect the performance of DE. Varying mutation operators, the values of CR or F will obtain different search behaviors of the algorithm.

DE/best/1 utilizes the current best solution to guide the mutation according to Eq. (5.21) and can thus converge very fast on simple unimodal problems. However, such a greedy strategy may suffer from premature convergence when dealing with complex multimodal problems. DE/rand/1 generates a mutated solution guided by a random solution in a current population, which can maintain diversity of population. It does not easily fall into local optima, so DE/rand/1 performs better for complex multimodal problems but converges not as fast as DE/best/1 for simple problems. New solutions will be near around the base solution when F is small or CR is big, so the population will converge quickly.

How to choose the mutation operator and values of F and CR is a problem to balance the exploration and exploitation, which is a great challenge for optimization. In this book, Chap. 7 will discuss parameter settings in EAs. Generally, for DE, DE/rand/1 is more robust than DE/best/1 on the whole, when solving global optimization problems according to experience.

Algorithm 5.5: Differential evolution

Initialization: Set the generation counter $t = 0$, the maximum generation G, the population size NP, and the dimensions of problem D, the mutation factor F and the crossover probability CR, and initializing a parent population;

while $t \leq G$ **do**

    for $i = 1 \cdots N$ **do**

        Select three indexes r_1, r_2, and r_3 which are different from i;

        Select an integer $K \in [1, D]$;

        for $j = 1 \cdots D$ **do**

            if $randn() \leq CR$ *or* $K == D$ **then**

                $u_{ij}(t+1) = x_{r_1}j(t) + F(x_{r_2}j(t) - x_{r_3}j(t))$;

            else

                $u_{ij}(t+1) = x_{ij}(t)$;

        if $f(u_i(t+1)) \leq f(x_i(t))$ **then**

            $x_i(t+1) = u_i(t)$;

        else

            $x_i(t+1) = x_i(t)$;

    $t = t + 1$;

5.5.3 Some Advances in Differential Evolution

Due to the excellent performance, DE has been widely studied since it was first proposed. Many improved algorithms have been proposed, which can be divided into two major categories. Some algorithms focus on parameter adaptation, and the others deal with adaptation of mutation operators. Below, we present brief ideas about the two types algorithms.

5.5.3.1 Parameter Adaptation

To solve complex optimization problems, researchers build adaptive DE algorithms by controlling the parameters of DE. One typical method called jDE was proposed by Brest et al. [10] based on the classic DE/rand/1 operator. Each individual in the algorithm has different choices for the parameters F_i and CR_i. In the beginning, all individuals use the same initial values of $F_i = 0.5$ and $CR_i = 0.9$; then new values for F_i and CR_i are generated according to uniform distributions with probabilities of $\tau_1 = \tau_2 = 0.1$ at each generation, respectively, as shown below:

$$F_i(t+1) = \begin{cases} 0.1 + 0.9 \cdot \mathcal{U}_1(0,1) & \text{if } \mathcal{U}_2(0,1) < \tau_1 \\ F_i(t) & \text{otherwise} \end{cases} \tag{5.27}$$

$$CR_i(t+1) = \begin{cases} \mathcal{U}_3(0,1) & \text{if } \mathcal{U}_4(0,1) < \tau_2 \\ CR_i(t) & \text{otherwise} \end{cases} \tag{5.28}$$

where $\tau_1 = \tau_2 = 0.1$ represent the probabilities of adjusting the control parameters. Recently used parameter values will replace the old ones if the offspring can survive in the next iteration. Then parameters can be controlled since the selection mechanism will get rid of individuals with unsuitable parameter settings.

5.5.3.2 Adaptation of Mutation Operators

The selection of optimal mutation operator is another factor that will have an important impact on the performance of DE on different problems. To address this issue, some adaptive operator selection mechanisms have been proposed.

One of the most representative method is self-adaptive DE (SaDE) [63]. In SaDE, four mutation operators, DE/rand/1, DE/rand-to-best/2, DE/rand/2, and DE/current-to-rand/1, are implemented. The selection probability of each mutation strategy p_k ($k = 1, 2, .., 4$) is the same in the first LP generations. For each individual at each iteration, a mutation operator is selected from the four operator by the stochastic universal selection method. The success information of each used operators is recorded (an operator is successful when the tried vector generated by the operator is better than the base vector) during the evolutionary progress. After the initial LP generations, the probabilities of choosing different operators will be updated at each subsequent generation based on the success information in the last LP generations. The probability of choosing the k-th strategy is updated by

$$p_k(t) = \frac{S_k(t)}{\sum_{k=1}^{4} S_k(t)} \tag{5.29}$$

where

$$S_k(t) = \frac{\sum_{g=t-LP}^{t-1} n_{suc_k}(g)}{\sum_{g=t-LP}^{t-1} n_{sel_k}(g)} \tag{5.30}$$

is the success rate of operator k in the last LP generations and $n_{suc_k}(g)$ and $n_{sel_k}(g)$ are the number of successful iterations and the total number of being selected in past LP generations, respectively, for operator k.

5.5.3.3 New Mutation Operators

Apart from optimizing the selection of different mutation operators, other researchers also made several attempts to design new mutation schemes (involving difference vectors in various forms) that can induce better performance in complex

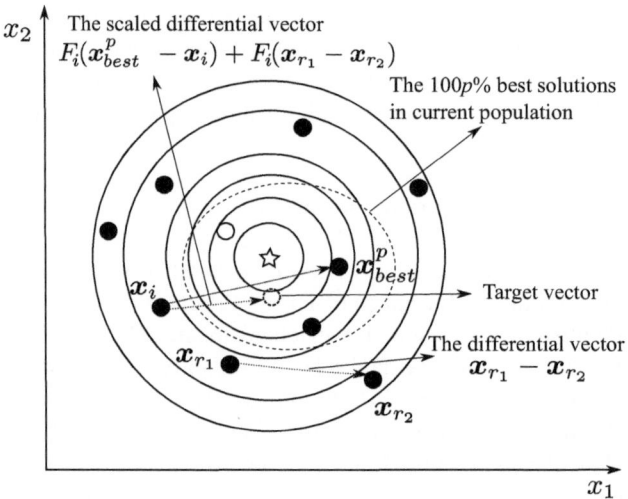

Fig. 5.12 An example of the process for generating target vectors by DE/current-to-pbest in a two-dimensional search space

optimization problems. For example, DE/current-to-pbest [86] is a competitive new mutation strategy, which is shown as follows:

$$v_i(t+1) = x_i(t) + F_i(x_{best}^p(t) - x_i(t)) + F_i(x_{r_1}(t) - x_{r_2}(t)) \qquad (5.31)$$

where $x_{best}^p(t)$ is randomly chosen as one of the top $100p\%$ individuals in the current population with $p \in (0, 1]$, which are shown as points within the dotted circle in Fig. 5.12. Any of the top $100p\%$ solutions can be randomly chosen to play the role of the best solution.

To achieve a faster and better convergence speed than classical DE, a 2-Opt algorithm is proposed by [13], which uses a better vector rather than a random vector as base vector for mutating. In [26], a proximity is defined as modification of mutation scheme for DE, where neighbors of a parent vector, but not random ones, will be used to generate the donor vector.

Besides the methods presented above, there are also some advanced algorithms on other aspects. The population size is adjusted depending on the population diversity in [81]. To address the issues of premature convergence and population stagnation, the population diversity is automatically enhanced at the dimensional level whenever the population convergence or stagnation is detected during the evolution [82]. For more advanced DE algorithms, please refer to [16].

5.6 Evolution Strategy

As introduced in Chap. 4, evolution strategy (ES) is one of the four major traditional evolutionary computation paradigms. The first traceable version of ES is bionics-inspired schemes, which were proposed by Bienert, Rechenberg, and Schwefel at the Technical University of Berlin, Germany, for evolving optimal shapes of minimal drag bodies in a wind tunnel.

At the beginning of the proposal of ES, discrete binomial distributed mutations was used. Later on, it used a normal distribution to mutate real-valued decision variables to address the premature convergence issue of discrete mutations, which becomes the core feature of ES and distinguishes it from other EAs.[5]

5.6.1 Basic Evolution Strategy Paradigm

A basic ES paradigm is often denoted by

$$(\mu/\rho \overset{+}{,} \lambda)\text{-ES}$$

where μ denotes the size of population; $\rho \leq \mu$ means the number of parent individuals that participant in each offspring reproduction; λ is the number of offsprings reproduced in each iteration; symbol "$\overset{+}{,}$" refers to the type of selection used, i.e., "+" and "," -selection, respectively (see the survivor selection below).

5.6.1.1 Parent Selection

In some early ES versions, mutation was the only way to reproduce offspring. In this way $\rho = 1$, and the parent individual is randomly selected from the population. Many later versions will include recombination in the reproduction process. In this way, $\rho > 1$ parents are randomly selected from the population.

5.6.1.2 Mutation

The earliest ES is an (1+1)-ES version, and in this case, mutation is the only feasible reproduction method. As mentioned before, the normal distribution is adopted in the mutation process

$$x' = x + \sigma \cdot \mathcal{N}_D(\mathbf{0}, \mathbf{I}) \tag{5.32}$$

---

[5] Although some other EAs such as the continuous EDA [7] also use a normal distribution to evolve real-valued decision variables.

where x' and x denote the decision variables of the offspring and parent, respectively; σ is the mutation step size.

5.6.1.3 Mutation Step Size Control

The famous 1/5 success rule is proposed to adjust the mutation step size. This rule states that if less than one-fifth of mutations fails ($f(x') < f(x)$), then the mutation encourages exploration too much and σ needs to be decreased; otherwise, the mutation is too exploitative and σ needs to be increased.

To adaptively control the mutation step size, σ is added into the representation of individual and co-evolves with the decision variables x

$$p_i = (x^{p_i}, \sigma^{p_i}) \tag{5.33}$$

The mutation step size is modified by equation

$$\sigma' = \sigma \cdot \exp(\tau \cdot \mathcal{N}(0, 1)) \tag{5.34}$$

where σ' and σ denote the mutation step sizes of the offspring and the parent, respectively, and τ is the learning parameter that decides the rate of self-adaptation.

5.6.1.4 Recombination

After multi-membered ES ($\mu > 1$) was introduced, the recombination was also introduced into the reproduction of ES individuals. But unlike the GA where the recombination (crossover) plays the leading role, the recombination in ES is a pre-operation of the mutation, i.e., the parent individual is replaced with the recombination of ρ randomly selected parents.

There are two ways to recombine parents, namely, discrete recombination and intermediate recombination. In the previous one, each allele is randomly chosen from either parents. In the latter one, each allele is the average allele of the parents. Discrete recombination is more appropriate for recombining x, while intermediate recombination is more appropriate for recombining σ

$$\langle x \rangle_j = x_j^{p\lceil \mathcal{U}(0,\rho)\rceil}, \, j \in \{1, \cdots, D\} \tag{5.35}$$

$$\langle \sigma \rangle = \frac{\sum_{i=1}^{\rho} \sigma^{p_i}}{\rho} \tag{5.36}$$

where $\langle x \rangle$ and $\langle \sigma \rangle$ denote x and σ of the recombination parent, respectively, and $\{p_i, \cdots, p_\rho\}$ denotes the ρ randomly selected parent individuals.

5.6.1.5 Survivor Selection

The differences between (μ, λ) and $(\mu + \lambda)$ are their survivor selection schemes. In (μ, λ) (termed as *comma-selection*), $\mu < \lambda$ must be hold, and the next generation population is selected only from the offspring. In $(\mu + \lambda)$ (referred to as *plus-selection*), both the offspring and the parent population compete to survive to the next generation.

5.6.1.6 Canonical Self-Adaptation Evolution Strategy

Canonical self-adaptation ES is shown in Algorithm 5.6. Firstly, generate μ individuals as the initial population. Then in each iteration, generate λ offspring, and use them to update the population.

Algorithm 5.6: $(\mu/\rho \overset{+}{,} \lambda)$ self-adaptation ES

Initialize population $\mathbb{P} = \{p_1, \cdots, p_\mu\}$;
while *termination criterion is not fulfilled* **do**
    for $o_i \in \mathbb{O} = \{o_1, \cdots, o_\lambda\}$ **do**
        Randomly select ρ parents from $\mathbb{P}$;
        Recombine the ρ parents by using Eqs. (5.35) and (5.36) to generate $\langle x \rangle$ and $\langle \sigma \rangle$;
        Update σ^{o_i} by mutating $\langle \sigma \rangle$ with Eq. (5.34);
        Update x^{o_i} by mutating $\langle x \rangle$ with Eq. (5.32);
    if *use plus-selection* **then**
        $\mathbb{P} :=$ best μ individuals in $\mathbb{P} \cup \mathbb{O}$;
    else
        $\mathbb{P} :=$ best μ individuals in $\mathbb{O}$;

To generate an offspring, first use Eqs. (5.35) and (5.36) to combine the decision variables and mutation step size of ρ with randomly selected parents. Next, use Eq. (5.34) to mutate the combined mutation step size $\langle \sigma \rangle$, thus obtaining the mutation step size of the offspring σ^{o_i}. Then use Eq. (5.32) and σ' to mutate the combined decision variables $\langle x \rangle$, thus obtaining decision variables of the offspring x^{o_i}. Finally, the fitness value of this offspring will be evaluated. After λ offsprings are generated, use the *plus-selection* scheme or the *comma-selection* scheme to update the population.

5.6.2 Covariance Matrix Adaptation Evolution Strategy

In ES, parameters that parameterize the normal distribution of the decision variable's mutation are called strategy parameters. So far, only one strategy parameter, i.e., the

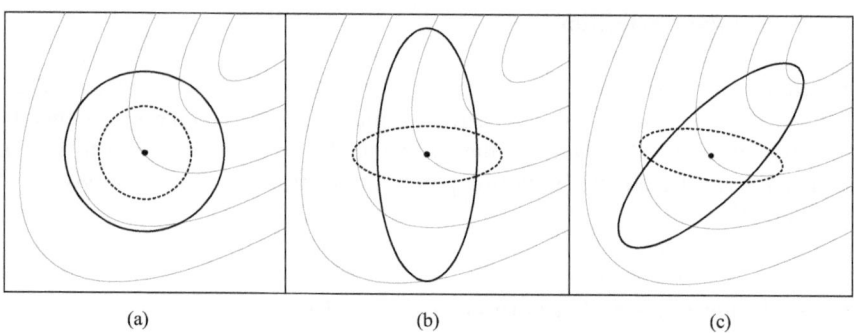

Fig. 5.13 Ellipsoids (thick full lines and thick dashed lines) depicting one-σ lines of equal probability density of six normal distributions; thin gray lines depict objective function contour lines (This figure is abstracted from [36]). (**a**) $\mathcal{N}(\mathbf{0}, \sigma^2\mathbf{I})$. (**b**) $\mathcal{N}(\mathbf{0}, \text{diag}^2(\boldsymbol{d}))$. (**c**) $\mathcal{N}(\mathbf{0}, \boldsymbol{C})$

mutation step size σ, has been introduced. As mentioned in Eq. (5.32), in this case, mutation distribution can be described as $\sigma \cdot \mathcal{N}(\mathbf{0}, \mathbf{I})$ or $\mathcal{N}(\mathbf{0}, \sigma^2\mathbf{I})$. As shown in Fig. 5.13a, this kind of mutation distribution is isotropic.

To make the mutation distribution more flexible, we can replace the scale parameter σ in $\sigma \cdot \mathcal{N}(\mathbf{0}, \mathbf{I})$ with a diagonal matrix diag$(\boldsymbol{d})$. Figure 5.13b shows the shape of the resulting normal distribution diag$(\boldsymbol{d})\mathcal{N}(\mathbf{0}, \mathbf{I})$, which also equals to $\mathcal{N}(\mathbf{0}, \text{diag}^2(\boldsymbol{d}))$. The two ellipsoids are axis parallel oriented. In this case, the D diagonal elements in diag$(\boldsymbol{d})$ denote the mutation step sizes in each decision variable.

However, the mutation distributions with anisotropy in Fig. 5.13b are still not consistent with the gradient direction. So a rotation is added to the mutation distribution in advance. The resulting normal distribution can be described as $\mathcal{N}(\mathbf{0}, \boldsymbol{C})$, where $\boldsymbol{C}$ denotes the covariance matrix. The eigendecomposition of $\boldsymbol{C}$ obeys

$$\boldsymbol{C} = \boldsymbol{B} \cdot \text{diag}^2(\boldsymbol{d}) \cdot \boldsymbol{B}^{\mathrm{T}} \tag{5.37}$$

where $\boldsymbol{B}$ is an orthogonal matrix whose columns form an orthonormal basis of eigenvectors; diag$(\boldsymbol{d})$ is a diagonal matrix with square roots of eigenvalues of $\boldsymbol{C}$ as diagonal elements.

$\boldsymbol{B}$ defines the orientation of the ellipsoid, i.e., the new principal axes of the ellipsoid correspond to the columns of $\boldsymbol{B}$, and diag$(\boldsymbol{d})$ can be interpreted as step size matrix, i.e., its diagonal entries are the standard variations along each principal axes of the ellipsoid.

5.6.2.1 Reproduction with Covariance Matrix

Since the eigendecomposition of a covariance matrix C costs a lot, it is impractical to attach a covariance matrix as a component of the strategy parameter to each individual. So, the reproduction with covariance matrix turns out to be

$$\forall\, o_i \in \mathbb{O} = \{o_1, \cdots, o_\lambda\}: \quad x^{o_i} = m + \sigma \cdot \mathcal{N}(0, C) \tag{5.38}$$

$$m = \frac{1}{\mu}\sum_{i=1}^{\mu} x^{p_i} \tag{5.39}$$

$$C = \frac{1}{\mu}\sum_{i=1}^{\mu}(x^{p_i} - m)(x^{p_i} - m)^{\mathrm{T}} \tag{5.40}$$

where m denotes the mean decision variables of the parents and C is the covariance matrix of the parents. In this case, all offsprings share the same mutation distribution $\mathcal{N}(m, \sigma^2 C)$, and the strategy parameters C, m, and σ are updated only once per generation.

5.6.2.2 Covariance Matrix Adaptation

The mutation in the form of a normal distribution with covariance matrix is first used in the estimation of multivariate normal algorithm (EMNA) [51]. In the mid-1990s, Gawelczyk, Hansen, and Ostermeier proposed the covariance matrix adaptation ES (CMA-ES) [37], which improves the multivariate normal distribution's sampling efficiency. The main feature which distinguishes CMA-ES from EMNA is that CMA-ES uses the information from previous generations to update strategy parameters[6]

$$\forall\, i \in \{1, \cdots, \lambda\}: \ b_i(t) = \begin{cases} 1 & \text{if } o_i(t) \in \mathbb{P}(t+1) \\ 0 & \text{if } o_i(t) \ni \mathbb{P}(t+1) \end{cases} \tag{5.41}$$

$$w(t) = \frac{1}{\mu} \cdot \sum_{i=1}^{\lambda} b_i(t)(x^{o_i}(t) - m(t)) \tag{5.42}$$

$$m(t+1) = m(t) + w(t) \tag{5.43}$$

$$s(t+1) = (1-\tau) \cdot s(t) + \sqrt{\tau(2-\tau)\mu} \cdot \frac{w(t)}{\sigma(t)} \tag{5.44}$$

---

[6] For the convenience of instruction, here we adopt the simplest other than the most efficient way to update w (average) and C (only the rank-one update).

$$C(t+1) = (1-\tau_c) \cdot C(t) + \tau_c \cdot s(t+1)s(t+1)^{\mathrm{T}} \tag{5.45}$$

$$s_\sigma(t+1) = (1-\tau_\sigma) \cdot s_\sigma(t) + \sqrt{\tau_\sigma(2-\tau_\sigma)\mu} \cdot C(t)^{-\frac{1}{2}} \frac{w(t)}{\sigma(t)} \tag{5.46}$$

$$\sigma(t+1) = \sigma(t) \cdot \exp\left(\frac{\tau_\sigma}{d_\sigma}\left(\frac{\|s_\sigma(t+1)\|}{\mathsf{E}\|\mathcal{N}(0,I)\|} - 1\right)\right) \tag{5.47}$$

where (t) denotes the number of generations; b_i denotes whether the i-th offspring is saved to the next generation.

Since CMA-ES uses *comma-selection*, if the i-th offspring belongs to the best μ offspring, then $b_i = 1$. Therefore, w denotes the average vector of all successful mutations in the last generation. In Eq. (5.42), the center point of the normal distribution is updated by adding w and the center point in the last generation. In Eqs. (5.44) and (5.46), the covariance matrix and the step size are also updated by utilizing w and the corresponding parameters in the last generation. s and s_σ are intermediate variants for the convenience of calculation. τ, τ_c, and τ_σ are all time horizons, deciding how much the strategy parameters inherited from the last generation. $C(t)^{-\frac{1}{2}}$ is an eigendecomposition of $C(t)$. $d_\sigma \approx 1$ is the damping parameter that scales the change magnitude of $\sigma(t)$. $\mathsf{E}\|\mathcal{N}(0,I)\|$ is an expectation of the Euclidean norm of a $\mathcal{N}(0,I)$ distributed random vector.

5.6.2.3   Complete Process of CMA-ES

The complete process of CMA-ES is shown in Algorithm 5.7. In each generation, all offsprings are reproduced by the same normal distribution. This is quite different from the canonical ES. And as a result, the moving track of the normal distribution can be viewed as the evolution path of the population. For example, Fig. 5.14 shows the process of CMA-ES searching for the minimum of the 2-D Rosenbrock function. The colored curves depict objective function contour lines, where the red line denotes small values and the purple line denotes large values. The red plus denotes the minimum point $x_{\min} = [1.0, 1.0]^{\mathrm{T}}$. The dashed black ellipses depict the normal distribution of populations of four different generations.

Algorithm 5.7: CMA-ES

Initialize population $\mathbb{P} = \{p_1, \cdots, p_\mu\}$;
Initialize strategy parameters $C, m, \sigma, s, s_\sigma, \tau, \tau_c$ and τ_σ;
while *termination criterion is not fulfilled* **do**
  Generate offsprings $\mathbb{O} = \{o_1, \cdots, o_\lambda\}$ by Eq. (5.38);
  Evaluate all individuals in $\mathbb{O}$;
  $\mathbb{P} :=$ best μ individuals in $\mathbb{O}$;
  Update the strategy parameters by Eqs. (5.41) ~(5.47);

Fig. 5.14 The process of CMA-ES is searching for the minimum of the 2-D Rosenbrock function

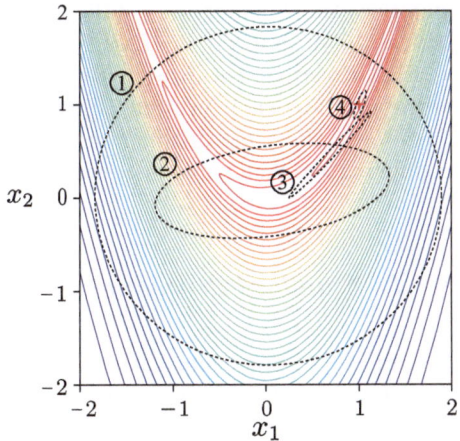

5.6.2.4 Niching CMA-ES

CMA-ES has a great performance on solving non-linear, non-separable, and ill-conditioned continuous optimization problems. However, its principle mechanism determines that the population of CMA-ES always converges to one optima, while the population of most other EAs (e.g., GA, DE, and PSO, even canonical ES, etc.) can converge to more than one peak. This capability is necessarily needed when solving multimodal optimization problems, which have more than one global optima, and the decision-maker wants as many optimal solutions as possible (for more details, see Chap. 8). Faced with this limitation, some niching methods are integrated into CMA-ES [1, 71], thus helping the population to converge to many global optima or some high-fitness local optima.

5.7 Estimation of Distribution Algorithm

The estimation of distribution algorithm (EDA) is an emerging stochastic optimization algorithm based on statistical principles. EDA and GA have obvious differences. GA uses crossover and mutation operations to generate new individuals, and EDAs use search space sampling and statistical learning to predict the best search area and then produce new individuals. Compared with GA-based micro-level evolution of genes, EDAs use a macro-level evolution method, with a stronger global search capability and faster convergence speed.

The concept of EDA was first proposed in 1996 [58]. It developed rapidly around 2000 and became the research frontier in the field of evolutionary computation. In 2005, the journal of Evolutionary Computation published the EDA. In recent years, major academic conferences in the field of evolutionary computation in the world, such as ACM SIGEVO, IEEE CEC, etc., have discussed EDA as an important topic.

EDA describes the distribution of candidate solutions in search space with a probability model. It uses statistical learning methods to establish a probability model describing the distribution of solutions from the macro perspective of the population and then randomly generate new solutions based on the model.

5.7.1   Standard Procedures

General EDA procedures are shown in Algorithm 5.8. A simple example is introduced below for users to have an intuitive understanding of the algorithm.

Algorithm 5.8: General EDA

Initialize population $\mathbb{P} = \{\mathbf{p}_1, \cdots, \mathbf{p}_\mu\}$;
while *termination criterion is not fulfilled* **do**
  Select promising solutions from $\mathbb{P}$;
  Build probabilistic model;
  Randomly sample new solutions;
  Generate new population;

Suppose one finds the maximum value of the function $f(x) = \sum_{i=1}^{D} x_i$, $x \in \{0, 1\}^D$, $D = 3$. In this example, the probability model describing the distribution of solutions is represented by a simple probability vector $\boldsymbol{\rho} = (\rho_1, \rho_2, \cdots, \rho_D)$. $\boldsymbol{\rho}$ represents the probability distribution of the population, $\rho_i \in [0, 1]$ represents the probability of taking the value of 1 for the gene at location i, and $1 - \rho_i$ means the probability of taking 0 for the gene at location i.

The first step is to initialize the population $\mathbb{P}_0$. Table 5.3 presents an initial population with eight solutions, where each solution is randomly generated in the solution space according to a uniform distribution $\boldsymbol{\rho} = (0.5, 0.5, 0.5)$. The fitness value of each individual is calculated by the function $f(x) = \sum_{i=1}^{D} x_i$.

In the second step, we select half solutions with higher fitness values to create a dominant population $\mathbb{P}'$ and then update the probability vector as shown in Table 5.4. The probability vector $\boldsymbol{\rho}$ is updated by the formula $\rho_i = P\left(x_i = 1 | \mathbb{P}'\right)$, for example, $\rho_1 = P\left(x_1 = 1 | \mathbb{P}'\right) = 0.75$, so that the probability vector is updated by $\boldsymbol{\rho} = (0.75, 0.75, 0.75)$.

The third step is to generate a new generation of population based on the probability vector $\boldsymbol{\rho}$. The probability of generating any solution $\boldsymbol{b} = (b_1, b_2, \cdots, b_D)$ is

$$P(\boldsymbol{b}) = P\left(x_1 = b_1, x_2 = b_2, \cdots, x_D = b_D\right) = \prod_{i=1}^{D} P\left(x_i = b_i\right) \tag{5.48}$$

Table 5.3 An initial
population with eight
solutions

#	x_1	x_2	x_3	f
1	0	0	1	1
2	1	1	0	2
3	0	0	0	0
4	0	1	1	2
5	0	1	0	1
6	1	0	0	1
7	1	0	1	2
8	1	1	1	3

Table 5.4 A selected
dominant population

#	x_1	x_2	x_3	f
2	1	1	0	2
4	0	1	1	2
7	1	0	1	2
8	1	1	1	3

Table 5.5 A new population
by random sampling

#	x_1	x_2	x_3	f
1	1	1	1	3
2	1	1	0	2
3	1	1	0	2
4	0	1	1	2
5	1	1	0	2
6	1	0	1	2
7	1	1	1	3
8	1	0	0	1

For example, if $b = (1, 0, 1)$, then $P(1, 0, 1) = 0.75 \times 0.25 \times 0.75 \approx 0.14$. Table 5.5 presents a new population generated by random sampling. It can be found that the new individuals of the population are significantly improved.

In this example, it can be found that with the progress of the EDA, the distribution probability of the global optimum $(1, 1, 1)$ is from the initial 0.125 ($0.5 \times 0.5 \times 0.5$) to 0.42 ($0.75 \times 0.75 \times 0.75$). By updating the probability distribution of individuals in the solution space, the probability of distribution of individuals with high fitness values becomes larger, and the probability of distribution of individuals with low fitness values becomes smaller, and repeated evolution will eventually produce the optimal solution to the problem.

Through the above simple examples, in the EDA, there is no operation such as crossover and mutation in GAs. Instead, the probability distribution and sampling operation are used to make the distribution of the population evolve toward the direction of excellent individuals. From the perspective of biological evolution, the GA simulates microscopic changes between individuals, and the EDA is the modeling and simulation of the overall distribution of biological population.

5.7.2   Discrete Versions

EDAs use probabilistic models to describe the relationship between variables, so they can solve problems that traditional GAs are difficult to solve, especially for high-dimensional problems that are nonlinear and variable coupled. According to the complexity of the optimization problem, researchers have designed many different probability models to express the relationship between variables. In this section, the EDA is divided into three types: univariate factorizations, bivariate factorizations, and multivariate factorizations. This section only discusses the algorithms of discrete space, and the related algorithms of continuous space will be introduced in the next section.

5.7.2.1   Univariate Factorizations

For the optimization problems, the simplest case is that there is no relationship between variables, i.e., decision variables of problems are separable (see Sect. 2.2.5). In this case, the distribution of the solution can generally be expressed by a simple probability vector as shown in the example above. For a D-dimensional problem of solutions with finite cardinality (usually binary strings), the independence between variables makes the probability of any solution expressed by

$$\rho\left(x_1, x_2, \cdots, x_D\right) = \prod_{i=1}^{D} \rho\left(x_i\right) \tag{5.49}$$

The earliest EDAs were proposed for univariate factorizations problems. The representative algorithms include population based incremental learning algorithm (PBIL) [3], univariate marginal distribution algorithm (UMDA) [57], compact genetic algorithm (cGA) [39], etc.

The PBIL algorithm is used to solve the optimization problem of binary coding. Although the concept of EDA has not been proposed in academia in 1994, the PBIL algorithm is recognized as the earliest model of distribution estimation algorithm. In the PBIL algorithm, the probability model representing the distribution of the solution space is a probability vector. The process of PBIL algorithm is as follows: in each generation, M individuals are randomly generated by the probability vector, then the fitness values of the M individuals are calculated, and the best N ($N \leq M$) individuals are selected to update the probability vector. The rule of probability vector adopts the Heb rule in machine learning [3], $\rho(\boldsymbol{x}(t))$ is used to represent the probability vector of the t-th generation, and $\boldsymbol{x}^1(t), \boldsymbol{x}^2(t), \cdots, \boldsymbol{x}^N(t)$ is the selected N individuals. The update process is as follows:

$$\rho(\boldsymbol{x}(t+1)) = (1 - \alpha)\rho(\boldsymbol{x}(t)) + \alpha \frac{1}{N} \sum_{k=1}^{N} \boldsymbol{x}^k(t) \tag{5.50}$$

where α is the learning rate.

The difference between the UMDA algorithm and the PBIL algorithm is the update algorithm of the probability vector. In the simple example in the previous section, the UMDA was used. The cGA differs from the PBIL and UMBA not only in the update algorithm of the probability model but also in the small size of the cGA population. In the cGA, only two individuals are randomly generated from the probability vector each time, and then compare two individuals and update the probability vector according to a certain strategy.

5.7.2.2 Bivariate Factorizations

PBIL, UMDA, and cGA do not consider the relationship between variables. The joint probability density of any solution vector in the algorithm can be obtained by multiplying the edge probability density of each independent component.

In practical problems, variables are not completely independent. In the field of EDA, the first algorithm to consider the correlation of variables is to assume pairwise interactions between variables. Such algorithms include mutual-information-maximization for input clustering (MIMIC) [17] and combining optimizers with mutual information trees (COMIT) [4].

In the MIMIC algorithm, the mutual relationship between variables is a chain relationship, and the probability model describing the solution space is written as

$$\rho^\pi (x) = \rho \left(x_{i_1} | x_{i_2} \right) \rho \left(x_{i_2} | x_{i_2} \right) \cdots \rho \left(x_{i_{n-1}} | x_{i_n} \right) \rho \left(x_{i_n} \right) \tag{5.51}$$

where $\pi = (i_1, i_2, \cdots, i_D)$ represents an arrangement of variables $(x_1, x_2, \cdots, x_D)$; $\rho_t (x_{i_j} | x_{i_{j+1}})$ represents the conditional probability of x_{i_j} given $x_{i_{j+1}}$.

When constructing a probabilistic model in the MIMIC algorithm, the optimal arrangement was expected so that ρ^π is closest to the probability distribution $\rho(x)$ of each generation of dominant groups obtained in the experiment. Due to the complexity of the model, the sampling method is also different from the univariate factorization distribution estimation algorithm. The basic idea is to sample the variables $i_n, i_{n-1}, \cdots, i_1$ in the reverse order of π to construct a complete solution.

The biggest difference between the COMIT algorithm and the MIMIC algorithm is that the probability model of the COMIT algorithm is a tree structure. The COMIT algorithm maintains an array containing a number for every pair of variables to estimate how many recently generated "good" solutions having bad gene $x_i = a$ and gene $x_j = b$ and then create the first-order probabilities and second-order probabilities $\rho(x_i)$ and $\rho(x_i, x_j)$ from the information stored in the array. Finally, a dependency tree is created to contain an optimum set of D − 1 first-order dependencies.

5.7.2.3   Multivariate Factorizations

In multivariate factorizations EDA, the relationship between variables is more complicated, and a more complex probability model is required to describe the solution space of the problem; hence, a more complex learning algorithm is also required to construct the corresponding probability model. Two representatives are the extended compact GA (ECGA) [38] and Bayesian Optimization Algorithm (BOA) [61].

The ECGA algorithm is an extension of cGA with the aim of learning the linkage between variables. In the ECGA algorithm, variables are divided into several groups, and each group of variables has nothing to do with other groups of variables. Initially, each group contains only one variable and then performs a marginal product model search by attempting to merge all pairs of groups into larger groups. If the combination leads to a decrease in the combined complexity of model and compressed population, the merging is carried out until no further pairs of groups can be merged.

In BOA, a Bayesian network is constructed from the selected dominant population as a sample set, and then a Bayesian model is sampled to generate a new generation of population. The Bayesian network is a directed acyclic graph. This graph can characterize the relationship between random variables. The most important thing in BOA is the learning algorithm and sampling algorithm. Bayesian network learning includes structure learning and parameter learning. Structure learning refers to learning the topology of the network, and parameter learning refers to learning after a given topology, the conditional distribution probability of each node in the network.

5.7.3   Continuous Versions

Similar to GAs, EDAs start with binary encoding for discrete optimization problems at the beginning and later extended to real encoding for continuous optimization problems. Due to the infinite search space, the models introduced above are no longer applicable to the continuous domain. One of the most straightforward way to apply EDAs to the continuous domain is to discretize the problem and use discrete EDAs. Another type of approach is to build probability models based on the continuous domain. In this section, we focus on the latter type of EDAs.

The UMDAc algorithm [49] is an extension of the UMDA algorithm, and the PBILc algorithm [69] is an extension of the PBIL algorithm. These two algorithms are relatively representative variable-independent distribution estimation algorithms, and both use Gaussian distribution as a probability model to describe the continuous solution space. The difference between the UMDAc algorithm and the PBILc algorithm is that different construction methods are used to update the Gaussian distribution model.

Continuous EDAs with multivariate factorizations mainly have the following work. In the estimation of multivariate normal algorithm (EMNA) [50, 51], a multivariate Gaussian model is used to represent the probability distribution of the solution, and each generation in the evolution process uses the maximum likelihood estimation method to estimate the mean vector and covariance matrix of the multivariate Gaussian distribution. The estimation of Gaussian networks algorithm (EGNA) [50, 51] is a distribution estimation algorithm based on Gaussian graph model. The directed edges in the Gaussian graph model represent the relationship between variables. Each variable is distributed by a Gaussian distribution. In the evolution process, the Gaussian graph network structure needs to be reconstructed according to the current population. Both EMNA and EGNA use a single-peak probability model. For optimization problems with complex shapes, the single-peak Gaussian distribution model cannot effectively describe the distribution of solutions in space. The iterated density EA (IDEA) [7] is a distribution estimation algorithm based on mixed Gaussian distribution and Gaussian kernel function. Although this method can overcome the shortcomings of EMNA and EGNA to a certain extent in solving the multimodal problem, the IDEA algorithm also does not fully consider the relationship between variables.

The design of the continuous domain EDA algorithm still faces great difficulties. The main reason is that each continuous variable has infinite values, which makes the search space of the optimization algorithm very large. Second, it is a difficult problem to construct a continuous space probability model with limited samples. As the dimensions increase, it becomes more difficult to construct a continuous domain probability model. With a limited sample set, constructing a probabilistic graphical model representing the coupling relationship between continuous variables still needs further study.

5.8 Ant Colony Optimization

Inspired by ants' foraging behaviors, Dorigo [21] proposed ant colony optimization (ACO) to solve the TSP (see the introduction in Sect. 3.2.1) in his doctoral dissertation in 1992. The classic TSP can be described as: Given a graph with a set of nodes and the distances between every two nodes, the objective is to find the shortest Manhattan circle, which contains all the nodes and each node is visited only once. The TSP has become a standard benchmark for testing the performance of ACO variants.

Similar to ants' foraging behaviors, ACO uses pheromone as a media for information exchange and transfers the hard combinatorial optimization problem into the problem of ants searching for the shortest path from the nest to the food source. ACO is a typical probabilistic sample-based feedback learning algorithm with the characteristics of distributed computing and robust performance.

5.8.1 Biological Inspiration

ACO takes inspiration from ants' foraging behaviors, and the behaviors were comprehensively studied in a famous experiment—double bridge experiment. In the experiment, Deneubourg et al. [19] designed two paths of different lengths from the nest of the ant colony to the food source. At the beginning, the pheromone concentration of the two paths was the same (both were zero). Ants chose the two paths with the same probability, but ants that chose the shorter path returned to the nest quicker and released more pheromone trail; slowly, the pheromone concentration on the shorter path becomes higher and higher. Ants choose shorter paths with greater probability. Finally, fewer and fewer ants choose the longer path. After a period of time, ants will only choose the shorter path. From the experiment, we can see that pheromone mechanism plays an important role to help ants find the shortest path.

5.8.2 ACO Framework

Algorithm 5.9 shows the basic framework of an ACO algorithm. Based on the connection between nodes, the algorithm constructs a pheromone matrix (modeling). When the algorithm does not satisfy the termination condition, each ant constructs a solution based on the pheromone matrix and a heuristic value related to the length between two nodes. The algorithm can apply a local search based on the generated solution to further improve the quality of the solution. The algorithm then chooses some elite solutions to update the probability matrix. The algorithm is described in detail below.

Algorithm 5.9: ACO framework

Initialization;
while *termination criterion is not met* **do**
  Construct ant solutions ;
  Apply local search (optional);
  Update pheromone;

5.8.2.1 Initialization

The algorithm chooses a value of τ_0 to initialize the pheromone matrix, which is a parameter of the algorithm. Different versions of ACO may have different initialization values.

5.8.2.2 Solution Construction

An ant starts with a node, which is randomly chosen, and constructs a complete solution by extending a partial solution $\mathbf{x}_p$ (each partial solution belongs to a solution state; see Sect. 3.2.1). Each pair of nodes i and j (can be regarded as a solution component denoted by e_{ij}) is associated with a pheromone trail $\tau(e_{ij})$ and a heuristic value $\eta(e_{ij})$ (if there exists), which are used to calculate the probability ρ of taking e_{ij} for $\mathbf{x}_p$ (i.e., the probability of moving from node i to node j for the ant).

Suppose $\mathbb{N}_{\mathbf{x}_p}$ is the feasible set of solution components that $\mathbf{x}_p$ can take. The calculation of the probability of taking e_{ij} for $\mathbf{x}_p$ is as follows:

$$\rho(e_{ij}|\mathbf{x}_p) = \frac{\left[\tau(e_{ij})\right]^\alpha \left[\eta(e_{ij})\right]^\beta}{\sum_{e_{il}\in\mathbb{N}_{\mathbf{x}_p}} \left[\tau(e_{il})\right]^\alpha \left[\eta(e_{il})\right]^\beta}, \forall e_{ij} \in \mathbb{N}_{\mathbf{x}_p} \tag{5.52}$$

where $\eta(e_{ij})$ is a heuristic function indicating the cost of moving from node i to node j (i.e., the cost of taking e_{ij} for $\mathbf{x}_p$) and α and β are factors to control the relative influence of the pheromone trail and the heuristic information, respectively.

If there is no heuristic information for the problem, we can set $\eta(e_{ij}) = 1.0$ and $\beta = 1.0$, which means that heuristic information is not considered in the probability function.

5.8.2.3 Local Search

One of the famous local search operator for TSP is the 2-opt operator. For example, in a TSP route in Fig. 5.15, node a is connected to node d and b is connected to c. The 2-opt operator disconnects a, d and b, c and attempts to connect a, b and c, d by judging whether there is improvement. The solution is gradually improved with 2-opt operator until there is no improvement. For different problems, we can apply different local search operators to improve the solution.

Fig. 5.15 2-opt search operator for the TSP

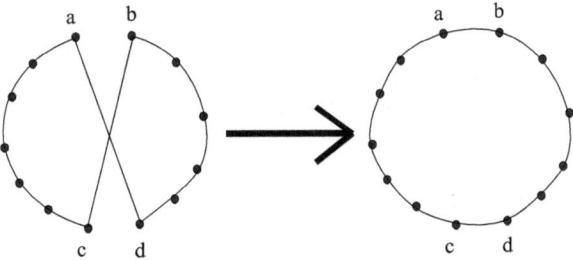

5.8.2.4   Pheromone Update

The pheromone update aims to help components of good solutions to be more attractive to ants, so that ants have a larger probability to construct good solutions. The pheromone update consists of two mechanism: pheromone deposit and pheromone trail evaporation.

In pheromone deposit, a set of good solutions will deposit pheromone along the solution components that they take. The pheromone deposit is a positive feedback loop to strengthen the choice for ants and pushes the colony to produce a result. Therefore, it aims to quickly exploit good solutions and favors the convergence of the colony.

However, in pheromone trail evaporation, the pheromone trail decreases over time so as to erase the previously deposited pheromone. The pheromone evaporation is a negative feedback loop for ants and makes the colony forget some choices that may not be good for the future or are made based on incomplete information. The pheromone evaporation mechanism enables ants to quickly adapt to changes in dynamic environments. From a practical point of view, the pheromone trail evaporation mechanism can slow down the convergence speed to prevent the algorithm from converging to the local optimum.

The pheromone update is implemented as follows:

$$\tau_{ij} = (1 - \alpha)\tau_{ij} + \sum_{\mathbf{x} \in \mathbb{S}_{upd} | e_{ij} \in \mathbf{x}} g(\mathbf{x}) \tag{5.53}$$

where $\mathbb{S}_{upd}$ is a set of good solutions that is allowed to deposit pheromone; $0 < \alpha < 1$ is a pheromone decay parameter; and $g(*)$ is a function to evaluate the solution and determine how much pheromone a solution should deposit.

Generally $g(*)$ should satisfy that $g(*) : f(\mathbf{x}) < f(\mathbf{x}') \Rightarrow g(\mathbf{x}) \geq g(\mathbf{x}')$ so that better solutions deposit more pheromone.

5.8.3   ACO Variants

ACO variants generally differ in the way of updating the pheromone. These ACO variants can be further divided into two classes according to changes in canonical ACO: set of update solutions and pheromone update equation.

5.8.3.1   Set of Update Solutions

Some ACO variants distinguish themselves from others mainly in the definition of the set of update solutions. Although ACO variants in this type may also have different pheromone update equations, the difference in the set of update solutions makes the algorithm different in performance from other ACO variants.

Ant system (AS) was firstly proposed in [21] and improved in [23]. It is the first version of ACO. Although it is quite simple and cannot compete with other ACO variants, it inspires the design of the later versions. In AS, the heuristic information $\eta(e_{ij}) = 1/l(e_{ij})$, $g(\mathbf{x}) = 1/f(\mathbf{x})$, and $\mathbb{S}_{upd} = \mathbb{S}_{iter}$, where $l(e_{ij})$ is the length of e_{ij} and S_{iter} is the set of all solutions generated by the current iterations.

Elitist ant system (EAS) [23] is a direct improved version of AS. Unlike AS, the best solutions found so far deposit pheromone on every iteration along with all other ants, i.e., $\mathbb{S}_{upd} = \mathbb{S}_{iter} \cup \{\mathbf{x}_b\}$, where $\mathbf{x}_b$ indicates the best solution found so far. Also, a strong weight is given to $\mathbf{x}_b$ for its pheromone deposit: $g(\mathbf{x}_b) = e/f(\mathbf{x}_b)$, where e is the number of elitist ants defined by users. Compared with AS, EAS has a stronger exploitation ability with the help of the best solution found so far.

Best-worst AS (BWAS) [14] considers the best global and current worst solutions to perform positive and negative pheromone update, i.e., $\mathbb{S}_{upd} = \{\mathbf{x}_b, \mathbf{x}_w\}$, where $\mathbf{x}_w$ denotes the worst solution at the current iteration and includes a restart mechanism when the algorithm is in stagnation. The pheromone update implemented in BWAS is as follows:

$$\tau_{ij} = (1 - \alpha)\tau_{ij} + \sum_{\mathbf{x} \in \mathbb{S}_{upd} | e_{ij} \in \mathbf{x}} [g(\mathbf{x})h(e_{ij})]$$

$$h(e_{ij}) = \begin{cases} 1, & e_{ij} \in \mathbf{x}_b \\ 0, & \text{else} \end{cases} \tag{5.54}$$

In BWAS, only the global best solution will deposit pheromone, and the pheromone tail on the edges that only belong to the current worst solution will decrease. BWAS favors the exploitation ability by using the information from the current worst solution, but sometimes, it may converge so fast to local optima.

5.8.3.2 Pheromone Update

Ant-Q combines reinforcement learning and ACO [34]. Gambardella found that ACO shared many similarities with Q-learning and tried to combine these two algorithms. Experimental results showed that Ant-Q outperforms AS in TSP instances. But it is not very popular because its pheromone update strategy is quite complex. Later in 1996, the ant colony system (ACS) [22] simplified the pheromone update strategy and achieved better results. ACS uses only the globally best solution to update the global pheromone trail. It also employs a local update rule when construct a solution for an ant, which is defined as follows:

$$\tau_{ij} = (1 - \alpha)\tau_{ij} + \alpha \Delta_{ij}, \pi_{ij} \in \mathbf{x} \tag{5.55}$$

where Δ_{ij} is a parameter in ACS and experimental results show that $\Delta_{ij} = \tau_0$ enables the ACS to achieve a competitive performance.

ACS also introduces a new way to choose a solution component by the so-called pseudo-random proportional rule with a user-defined probability q, that is, an ant chooses the solution component that has the largest product of the pheromone trail and the heuristic information and with probability $1 - q$ based on Eq. (5.52).

MAX-MIN AS (MMAS) [74] is one of the most effective and popular ACO variants. Unlike other ACO variants, MMAS includes pheromone trail limits and uses the globally best or iteration-best solution to update the pheromone so that it has a strong exploration ability. Combined with appropriate local search operators that focus on exploitation, MMAS can obtain world-class results on TSP instances. The pheromone trail in MMAS is constrained within $[\tau_{\min}, \tau_{\max}]$ for each τ_{ij}, where $\tau_{\min}$ and $\tau_{\max}$ are parameters, which will be updated at each iteration according to the pheromone matrix, to control the convergence speed of the algorithm.

Rank-based AS [11] derives from EAS. For most of the ACO variants, the objective value of the solutions is used to calculate the evaluation function $g(*)$. But in the late evolution, the survivals have similar objective values, and thus the guiding role of pheromone matrix is not strong as it was in the early evolution stage, which deduce the exploitation ability of the algorithm. To overcome this shortcoming in EAS, rank-based AS uses the rank of each solution in the solution set as their influence factor to increase the diversity of the algorithm. The pheromone update implemented in Rank-based AS is as follows:

$$\tau_{ij} = \alpha \tau_{ij} + \sum_{\mathbf{x} \in \mathbb{S}_{iter} \cup \{\mathbf{x}_b\} | e_{ij} \in \mathbf{x}} g(\mathbf{x})$$

$$g(\mathbf{x}) = (\sigma - r_{\mathbf{x}})/f(\mathbf{x}) \tag{5.56}$$

where σ is the number of solutions in the set of update solutions and $r_{\mathbf{x}}$ is the rank of $\mathbf{x}$ in the set. Similarly to EAS, it uses the best global solution found so far and $\sigma -1$ best solutions at the current iteration to deposit the pheromone.

5.8.4   Recent Advances

In recent years, research on the design of the ACO algorithm has become increasingly mature, and researchers have paid more attention to the extended applications and theoretical analysis of ACO algorithms. The main research progress of ACO algorithms includes the following: (1) dynamic optimization, (2) multi-objective optimization, (3) multi-model optimization, and (4) the parallel implementation. The idea of ACO algorithms is simple and easy to be extended to different types of problems. ACO appropriately combined with some local search operators and some problem-related strategies can greatly improve its performance.

5.8.4.1   Dynamic Optimization

The goal for dynamic optimization is not only to find the global optimum but also to track the changing optimum over time. When a traditional EA has converged, it hardly finds new good solutions due to the loss of diversity. Actually, ACO is an effective optimizer to solve the dynamic optimization problems since ACO variants generate the solutions through a pheromone matrix, which can be regarded as a model to store previous knowledge, and each item in the matrix has a non-zero probability; thus, it not only can reuse the knowledge of previous environments but also can quickly adapt to new environments by the pheromone updated mechanism. However, to track the changing global optimum more effectively, several dynamism handling strategies are needed to be combined with ACO, which can be divided into five classes: adaptive, diversity, prediction, memory, and multi-population strategies [55, 60] (see the detail in Chap. 11).

5.8.4.2   Multi-objective Optimization

Multi-objective optimization aims to find non-dominated solutions that are evenly distributed on the Pareto front, and thus it requires higher population diversity (see the detail in Chap. 9). With the global pheromone matrix, ACO can maintain the population diversity during the whole evolution. Combined with multi-objective optimization techniques, ACO can obtain competitive performance, such as the work for solving the multi-objective 0-1 knapsack problem and multi-objective TSP [46].

5.8.4.3   Multimodal Optimization

Similar to the multi-objective optimization, the strong global search ability of ACO variants also benefits to multimodal optimization, which aims to find multiple global optima. Some research [83] show that combined with ACO variants, the multimodal optimization algorithm well performs several state-of-the-art multimodal algorithms.

5.8.4.4   Parallel ACO Implementations

With more CPUs and GPUs integrated in the computer, parallelization of algorithms makes algorithms work more efficiently. ACO variants are quite suitable for parallelization because the behavior of each ant is independent from each other. Parallel ACO algorithms can be divided into two classes: coarse grained and fine grained. In the former ones, many ants share the same CUP, and there are a small amount of communications between CPUs, while in the second ones, only a few ants use the same CPU, and there are a great amount of communications [65].

References

1. Ahrari, A., Deb, K., Preuss, M.: Multimodal optimization by covariance matrix self-adaptation evolution strategy with repelling subpopulations. Evol. Comput. **25**(3), 439–471 (2017)
2. Angeline, P.J.: Evolutionary optimization versus particle swarm optimization: Philosophy and performance differences. In: International Conference on Evolutionary Programming, pp. 601–610. IEEE Neural Networks Council, Springer, Heidelberg (1998)
3. Baluja, S.: Population-based incremental learning. a method for integrating genetic search based function optimization and competitive learning. Tech. rep., Computer Science Department, Carnegie Mellon University (1994)
4. Baluja, S., Davies, S.: Using optimal dependency-trees for combinatorial optimization: Learning the structure of the search space. Tech. rep., Computer Science Department, Carnegie Mellon University (1997)
5. Bickel, A.S., Bickel, R.W.: Tree structured rules genetic algorithms. In: Proceedings of the Second International Conference on Genetic Algorithms, pp. 77–81. L. Erlbaum Associates, United States (1987)
6. Bleuler, S., Brack, M., Thiele, L., Zitzler, E.: Multiobjective genetic programming: Reducing bloat using SPEA2. In: Proceedings of the 2001 Congress on Evolutionary Computation, pp. 536–543. IEEE, New York (2001)
7. Bosman, P.A., Thierens, D.: Expanding from discrete to continuous estimation of distribution algorithms: The IDEA. In: S. Marc (ed.) Parallel Problem Solving from Nature - PPSN VI, pp. 767–776. Springer, Berlin (2000)
8. Bratton, D., Kennedy, J.: Defining a standard for particle swarm optimization. In: 2007 IEEE Swarm Intelligence Symposium, pp. 120–127. IEEE Computational Intelligence Society, IEEE, Honolulu (2007)
9. Bremermann, H.J.: Optimization through evolution and recombination. Self Organiz. Syst. **93**, 106–117 (1962)
10. Brest, J., Greiner, S., Boskovic, B., Mernik, M., Zumer, V.: Self-adapting control parameters in differential evolution: A comparative study on numerical benchmark problems. IEEE Trans. Evol. Comput. **10**(6), 646–657 (2006)
11. Bullnheimer, B., Hartl, R.F., Strauß, C.: A new rank based version of the ant system - a computational study. Cent. Eur. J. Oper. Res. Econ. **7**, 25–38 (1997)
12. Chen, Y.P., Peng, W.C., Jian, M.C.: Particle swarm optimization with recombination and dynamic linkage discovery. IEEE Trans. Syst. Man Cybern. **37**(6), 1460–1470 (2007)
13. Chiang, C.W., Lee, W.P., Heh, J.S.: A 2-opt based differential evolution for global optimization. Appl. Soft Comput. **10**(4), 1200–1207 (2010)
14. Cordón, O., de Viana, I.F., Herrera, F., Moreno, L.: A new ACO model integrating evolutionary computation concepts: The best-worst ant system. In: Proceedings of 2nd International Workshop on Ant Algorithms, pp. 22–29, Brussels (2000)
15. Cramer, N.L.: A representation for the adaptive generation of simple sequential programs. In: Proceedings of the First International Conference on Genetic Algorithms, pp. 183–187. L. Erlbaum Associates, United States (1985)
16. Das, S., Mullick, S.S., Suganthan, P.N.: Recent advances in differential evolution–an updated survey. Swarm Evol. Comput. **27**, 1–30 (2016)
17. De Bonet, J.S., Isbell Jr, C.L., Viola, P.A.: MIMIC: Finding optima by estimating probability densities. In: Proceedings of the 9th International Conference on Neural Information Processing Systems, pp. 424–431. MIT Press, Cambridge (1997)
18. De Jong, E.D., Watson, R.A., Pollack, J.B.: Reducing bloat and promoting diversity using multi-objective methods. In: Proceedings of the 3rd Annual Conference on Genetic and Evolutionary Computation, pp. 11–18. Morgan Kaufmann, San Francisco, CA, United States (2001)
19. Deneubourg, J.L., Aron, S., Goss, S., Pasteels, J.M.: The self-organizing exploratory pattern of the Argentine ant. J. Insect Behav. **3**(2), 159–168 (1990)

20. Dickmanns, D., Schmidhuber, J., Winklhofer, A.: Der genetische algorithmus: Eine imple-mentierung in prolog. Tech. rep., Institution of Informatics, Technical University of Munich, Munich (1987)
21. Dorigo, M.: Optimization, learning and natural algorithms. Ph.D. thesis, Politecnico di Milano, Milan (1992)
22. Dorigo, M., Gambardella, L.M.: Ant colony system: a cooperative learning approach to the traveling salesman problem. IEEE Trans. Evol. Comput. **1**(1), 53–66 (1997)
23. Dorigo, M., Maniezzo, V., Colorni, A.: Ant system: optimization by a colony of cooperating agents. IEEE Trans. Syst. Man Cybern. **26**(1), 29–41 (1996)
24. Eberhart, R., Kennedy, J.: A new optimizer using particle swarm theory. In: Proceedings of the Sixth International Symposium on Micro Machine and Human Science, pp. 39–43. IEEE, Nagoya (1995)
25. Ekárt, A., Nemeth, S.Z.: Selection based on the pareto nondomination criterion for controlling code growth in genetic programming. Genet. Program. Evolvable Mach. **2**(1), 61–73 (2001)
26. Epitropakis, M.G., Tasoulis, D.K., Pavlidis, N.G., Plagianakos, V.P., Vrahatis, M.N.: Enhanc-ing differential evolution utilizing proximity-based mutation operators. IEEE Trans. Evol. Comput. **15**(1), 99–119 (2011)
27. Fang, X.: Engineering design using genetic algorithms. Ph.D. thesis, Iowa State University, Ames (2007)
28. Ferreira, C.: Gene expression programming: a new adaptive algorithm for solving problems. Complex Syst. **13**(2), 87–129 (2001)
29. Fogel, D.B.: System Identification Through Simulated Evolution: A Machine Learning Approach to Modeling. Ginn Press, Needham Heights (1991)
30. Fogel, D.B.: Evolutionary Computation: Toward a New Philosophy of Machine Intelligence. Wiley-IEEE Press, New York (1995)
31. Fogel, L.J., Owens, A.J., Walsh, M.J.: Artificial Intelligence Through Simulated Evolution. Wiley, New York (1966)
32. Friedberg, R.M.: A learning machine: Part I. IBM J. Res. Dev. **2**(1), 2–13 (1958)
33. Fujiki, C.: Using the genetic algorithm to generate lisp source code to solve the prisoner's dilemma. In: Proceedings of the Second International Conference on Genetic Algorithms, pp. 236–240. L. Erlbaum Associates, United States (1987)
34. Gambardella, L.M., Dorigo, M.: Ant-Q: A reinforcement learning approach to the traveling salesman problem. In: Machine Learning Proceedings 1995, pp. 252–260. Morgan Kaufmann, Palo Alto (1995)
35. Ghaheri, A., Shoar, S., Naderan, M., Hoseini, S.S.: The applications of genetic algorithms in medicine. Oman Med. J. **30**(6), 406 (2015)
36. Hansen, N.: The CMA evolution strategy: A tutorial. arXiv preprint arXiv:1604.00772 (2016)
37. Hansen, N., Ostermeier, A.: Completely derandomized self-adaptation in evolution strategies. Evol. Comput. **9**(2), 159–195 (2001)
38. Harik, G.: Linkage learning via probabilistic modeling in the ECGA. Tech. rep., Illinois Genetic Algorithms Laboratory, Department of General Engineering, University of Illinois at Urbana-Champaign (1999)
39. Harik, G.R., Lobo, F.G., Goldberg, D.E.: The compact genetic algorithm. IEEE Trans. Evol. Comput. **3**(4), 287–297 (1999)
40. Hartmann, S.: A competitive genetic algorithm for resource-constrained project scheduling. Nav. Res. Logist. **45**(7), 733–750 (1998)
41. He, J., Yao, X.: A game-theoretic approach for designing mixed mutation strategies. In: International Conference on Natural Computation, pp. 279–288. Springer, Berlin (2005)
42. Holland, J.H.: Adaptation in Natural and Artificial Systems. University of Michigan Press, Ann Arbor (1975)
43. Hornby, G.S., Lohn, J.D., Linden, D.S.: Computer-automated evolution of an X-band antenna for NASA's space technology 5 mission. Evol. Comput. **19**(1), 1–23 (2011)
44. Jones, J.: Abstract syntax tree implementation idioms. In: Proceedings of the 10th Conference on Pattern Languages of Programs, pp. 26–35. ACM, Illinois (2003)

45. Kalganova, T., Miller, J.: Evolving more efficient digital circuits by allowing circuit layout evolution and multi-objective fitness. In: Proceedings of the First NASA/DoD Workshop on Evolvable Hardware, pp. 54–63. IEEE, New York (1999)

46. Ke, L., Zhang, Q., Battiti, R.: Moea/d-aco: A multiobjective evolutionary algorithm using decomposition and ant colony. IEEE Trans. Cybern. **43**(6), 1845–1859 (2013)

47. Kennedy, J.: Small worlds and mega-minds: effects of neighborhood topology on particle swarm performance. In: Proceedings of the 1999 Congress on Evolutionary Computation, pp. 1931–1938. IEEE Computational Intelligence Society, IEEE, Washington, D.C (1999)

48. Koza, J.R.: Concept formation and decision tree induction using the genetic programming paradigm. In: International Conference on Parallel Problem Solving from Nature, pp. 124–128. Springer, Berlin, Heidelberg (1990)

49. Larrañaga, P., Etxeberria, R., Lozano, J.A., Pena, J.M.: Optimization in continuous domains by learning and simulation of gaussian networks. In: Proceedings of the Genetic and Evolutionary Computation Conference, pp. 201–204. Association for Computing Machinery, San Francisco (2000)

50. Larrañaga, P., Lozano, J.A.: Estimation of Distribution Algorithms: A New Tool for Evolutionary Computation. Springer Science & Business Media, Norwell (2001)

51. Larranaga, P., Lozano, J.A., Bengoetxea, E.: Estimation of distribution algorithms based on multivariate normal and gaussian networks. Tech. rep., Department of Computer Science and Artificial Intelligence, University of the Basque Country, Boston (2001)

52. Lee, C.Y., Yao, X.: Evolutionary programming using mutations based on the lévy probability distribution. IEEE Trans. Evol. Comput. **8**(1), 1–13 (2004)

53. Li, C., Yang, S.: An adaptive learning particle swarm optimizer for function optimization. In: IEEE Congress on Evolutionary Computation, pp. 381–388. IEEE Computational Intelligence Society, IEEE, Washington, D.C (2009)

54. Ma, C., Ma, C., Ye, Q., He, R., Song, J.: An improved genetic algorithm for the large-scale rural highway network layout. Math. Probl. Eng. **2014** (2014)

55. Mavrovouniotis, M., Li, C., Yang, S.: A survey of swarm intelligence for dynamic optimization: Algorithms and applications. Swarm Evol. Comput. **33**, 1–17 (2017)

56. Mendes, R.R., de Voznika, F.B., Freitas, A.A., Nievola, J.C.: Discovering fuzzy classification rules with genetic programming and co-evolution. In: European Conference on Principles of Data Mining and Knowledge Discovery, pp. 314–325. Springer, Berlin, Heidelberg (2001)

57. Mühlenbein, H.: The equation for response to selection and its use for prediction. Evol. Comput. **5**(3), 303–346 (1997)

58. Mühlenbein, H., Paass, G.: From recombination of genes to the estimation of distributions i. binary parameters. In: V. HM., E. W., R. I., S. HP. (eds.) Parallel Problem Solving from Nature - PPSN IV, pp. 178–187. Springer, Berlin (1996)

59. Murata, T., Ishibuchi, H., Tanaka, H.: Genetic algorithms for flowshop scheduling problems. Comput. Ind. Eng. **30**(4), 1061–1071 (1996)

60. Nguyen, T.T., Yang, S., Branke, J.: Evolutionary dynamic optimization: A survey of the state of the art. Swarm Evol. Comput. **6**, 1–24 (2012)

61. Pelikan, M., Goldberg, D.E., Cantú-Paz, E., et al.: Boa: The Bayesian optimization algorithm. In: Proceedings of the Genetic and Evolutionary Computation Conference, pp. 525–532. Morgan-Kaufmann Publishers, San Fransisco (1999)

62. Poli, R., Di Chio, C., Langdon, W.B.: Exploring extended particle swarms: a genetic programming approach. In: Proceedings of the 7th Annual Conference on Genetic and Evolutionary Computation, pp. 169–176. ACM SIGEVO, ACM, New York (2005)

63. Qin, A.K., Huang, V.L., Suganthan, P.N.: Differential evolution algorithm with strategy adaptation for global numerical optimization. IEEE Trans. Evol. Comput. **13**(2), 398–417 (2008)

64. Raju, K.S., Kumar, D.N.: Irrigation planning using genetic algorithms. Water Resour. Manag. **18**(2), 163–176 (2004)

65. Randall, M., Lewis, A.: A parallel implementation of ant colony optimization. J. Parallel Distrib. Comput. **62**(9), 1421–1432 (2002)

66. Ratnaweera, A., Halgamuge, S.K., Watson, H.C.: Self-organizing hierarchical particle swarm optimizer with time-varying acceleration coefficients. IEEE Trans. Evol. Comput. **8**(3), 240–255 (2004)
67. Reynolds, C.W.: Flocks, herds, and schools: A distributed behavioral model. Comput. Graph. **21**(4), 25–34 (1987)
68. Ryan, C., O'Neill, M., Collins, J.: Grammatical evolution: Solving trigonometric identities. In: Proceedings of Mendel '98: 4th International Conference on Genetic Algorithms, Optimization Problems, Fuzzy Logic, Neural Networks and Rough Sets, pp. 111–119. Technical University of Brno, Brno, Czech Republic (1998)
69. Sebag, M., Ducoulombier, A.: Extending population-based incremental learning to continuous search spaces. In: A.E. Eiben, T. Bäck, M. Schoenauer, H.-P. Schwefel (eds.) Parallel Problem Solving from Nature - PPSN V, pp. 418–427. Springer, Berlin (1998)
70. Shi, Y., Eberhart, R.: A modified particle swarm optimizer. In: IEEE World Congress on Computational Intelligence, pp. 69–73. IEEE Computational Intelligence Society, IEEE, Anchorage (1998)
71. Shir, O.M., Emmerich, M., Bäck, T.: Adaptive niche radii and niche shapes approaches for niching with the CMA-ES. Evol. Comput. **18**(1), 97–126 (2010)
72. Spears, W.M., Green, D.T., Spears, D.F.: Biases in particle swarm optimization. In: Y. Shi (ed.) Innovations and Developments of Swarm Intelligence Applications, pp. 20–43. IGI Global, Hershey (2012)
73. Storn, R., Price, K.: Differential evolution–a simple and efficient heuristic for global optimization over continuous spaces. J. Global Optim. **11**(4), 341–359 (1997)
74. Stützle, T., Hoos, H.H.: Max–min ant system. Future Gener. Comput. Syst. **16**(8), 889–914 (2000)
75. Suganthan, P.N.: Particle swarm optimiser with neighbourhood operator. In: Proceedings of the 1999 Congress on Evolutionary Computation, pp. 1958–1962. IEEE Computational Intelligence Society, IEEE, Washington, D.C (1999)
76. Turing, A.M.: Computing machinery and intelligence. Mind **59**(236), 433–460 (1950)
77. Ueno, M., Usui, S., Tanaka, H., Watanabe, A.: Technological overview of the next generation shinkansen high-speed train series N700. In: Proceedings of the Elevator, Escalator and Amusement Rides Conference, pp. 1–4. Central Japan Railway Company, Tokyo (2008)
78. Van Den Bergh, F., et al.: An analysis of particle swarm optimizers. Ph.D. thesis, University of Pretoria South Africa, Pretoria (2001)
79. Wen, C., Eberhart, R.C.: Genetic algorithm for logistics scheduling problem. In: Proceedings of the 2002 Congress on Evolutionary Computation, pp. 512–516. IEEE Press, Honolulu (2002)
80. Xie, X.F., Zhang, W.J., Yang, Z.L.: Dissipative particle swarm optimization. In: Proceedings of the 2002 Congress on Evolutionary Computation, pp. 1456–1461. IEEE Computational Intelligence Society, IEEE Computer Society, Washington, DC (2002)
81. Yang, M., Cai, Z., Li, C., Guan, J.: An improved adaptive differential evolution algorithm with population adaptation. In: Proceedings of the 15th Annual Conference on Genetic and Evolutionary computation, pp. 145–152. Association for Computing Machinery, New York, United States (2013)
82. Yang, M., Li, C., Cai, Z., Guan, J.: Differential evolution with auto-enhanced population diversity. IEEE Trans. Cybern. **45**(2), 302–315 (2014)
83. Yang, Q., Chen, W.N., Yu, Z., Gu, T., Li, Y., Zhang, H., Zhang, J.: Adaptive multimodal continuous ant colony optimization. IEEE Trans. Evol. Comput. **21**(2), 191–205 (2016)
84. Yao, X., Liu, Y.: Fast evolutionary programming. Evol. Program. **3**, 451–460 (1996)
85. Yao, X., Liu, Y., Lin, G.: Evolutionary programming made faster. IEEE Trans. Evol. Comput. **3**(2), 82–102 (1999)
86. Zhang, J., Sanderson, A.C.: Jade: adaptive differential evolution with optional external archive. IEEE Trans. Evol. Comput. **13**(5), 945–958 (2009)

Part III
Optimization Techniques

This part first introduces techniques for the control of the values of parameters and the policy control regarding the operator selection of evolutionary computation algorithms introduced in previous chapters. The trade-off between exploitation and exploration is then discussed. Finally, techniques for multimodal optimization are provided at the end of this part.

Chapter 6
Parameter Control and Policy Control

Abstract This chapter aims to explain some basic methods for parameter control and strategy control, which have a significant impact on the performance of EAs. Single-parameter-specific methods and multi-parameter ensemble methods are introduced. The commonly used policy control frame is explained, and methods for operator selection control and hyper-heuristics control are discussed. Finally, promising trends in parameter and policy control are provided.

6.1 Parameter Control

In all kinds of algorithms introduced in previous chapters, they all incorporate some numeric parameters. Parameter settings have a great impact on the performance of algorithms. Parameter control refers to the mechanism that changes the values of the parameters (start with initial values) during the runtime of an algorithm. Furthermore, tuning the values of parameters can be traceable to the originality of EAs, in which the parameter control problem itself can also be considered as a challenging coupled optimization problem.

To bring about a satisfying performance, a series of parameters have to be adjusted precisely according to the evolving status of the algorithm and/or the features of optimization problems. Furthermore, as the search process goes on to solve a particular problem, the optimal parameter values may change accordingly. Therefore, the parameter control problem is a dynamic optimization problem (see the detail of dynamic optimization in Chap. 11).

From a "bird's-eye view," the important and challenging problem raises much concern among scholars in EC fields. There are many works on parameter control, most of which can be classified into two types based on controlled objects: parameter specific and parameter ensembles [7, 11]. Parameter-specific methods aim to control a specific parameter or component of an algorithm, while parameter ensembles control multiple parameters, which may couple with each other. The following section will introduce the representative works.

6.1.1   Unary Parameter Control

Parameter-specific methods are intended to control specific parameters in an algorithm. Generally, the parameter to be controlled can mainly be categorized as population size, operator parameters, and problem-related parameters.

6.1.1.1   Population Size Control

Since the GA was proposed, the question of how to set the size of the population of an EA has become inevitable due to its population-based search mechanism. Some works focused on the theoretical aspects but did not gain the attention deserved. Goldberg [9] tried to obtain the minimum population size with probability derivation on building sufficient blocks. However, it is difficult to analyze the performance of the parameters in such an uncertain situation.

Some researchers attempt to control the population size indirectly. The concept of "age" was proposed to control the survival time for each individual according to its fitness value, in which when the age of the individual reaches the calculated lifetime, the individual will be removed from the population [2]. Proportional allocation and linear allocation strategies are shown in Eqs. (6.1) and (6.2), respectively

$$l(\pmb{x_i}) = \min \left(l_{\min} + \eta \frac{f(\pmb{x_i})}{f_{\text{avg}}}, l_{\max} \right) \tag{6.1}$$

$$l(\pmb{x_i}) = l_{\min} + 2\eta \frac{f(\pmb{x_i}) - f_{\min}}{f_{\max} - f_{\min}} \tag{6.2}$$

where $f_{\min}$, $f_{\max}$, and f_{avg} denote the minimum, maximum, and average objective values of the current iteration, respectively; $l_{\max}$ and $l_{\min}$ denote the user-defined maximum and minimum lifetime values, and $\eta = (l_{\max} - l_{\min})/2$.

Similarly, an adaptive population size for GA (APGA) was proposed [3], in which a steady-state GA was employed and the age of the best individual does not grow.

Since the "age" concept, many extended works began to control the population size. For example, the adaptation of population size of subpopulations was implemented for a coevolutionary model [10]. The survival possibility [5] was set for the individuals so that the population size parameter could be replaced. Following the same rules but with the limitation of memory, the GP with tree structures uses the memory limit for the node size to replace the parameter size [20, 21].

However, η in the lifetime mechanism is a newly introduced parameter. Moreover, studies [14] suggested that APGA [3] is not able to fully adapt the population size based on some theoretical analysis and experimental studies. The choice of the lifetime control function becomes a new problem. Therefore, some methods based on estimating or adjusting the population size directly came into view. For example, several populations with different sizes race in the running so that the underdog ones (usually the fitness improvement is used as an indicator) will be terminated gradually. More simply, restarting the algorithm with different sizes as an extension of the evolution strategy can be a parameter control method. To adapt to the search process, some deterministic and dynamic mechanisms have been developed to adjust the population size. The change ratio of the population size was encoded into the genes to get effective self-adaptation [17]. Although the population size parameter is eliminated in some indirect methods, the controllers also incur other parameters, which may become the "new problem."

6.1.1.2 Operator Parameter Control

Among different types of parameters in EAs, operator parameters seem to have a dominant impact on the performance of EAs. Therefore, most of the work focuses on adjusting the parameters of different types of operators, such as mutation, crossover, and selection. Some classic control methods for operator parameters have been introduced in previous chapters, e.g., the control of parameters F and CR in DE in Sect. 5.5.3 and the control of the inertia weight ω in PSO in Sect. 5.4.4.

The initial EP is different from other EAs due to its self-adaptation mechanism of adapting variances (see Sect. 5.2.2). From then on, many algorithms begin to encode significant parameters into chromosomes as shown in Fig. 6.1. The mutation parameters are encoded into the bit-string chromosomes, and the mutation fragments will mutate with other chromosomes, which is not like the classical self-adaptation in ES, where the mutation parameters are sampled by a possibility distribution without the process of mutation. The self-adaptive parameter is treated as a part of the optimization problem along with the problem itself (see Eq. (5.4) in Sect. 5.2).

6.1.1.3 Problem-Related Parameter Control

Owing to the simple implementation, EAs have been used in many kinds of problems. In the following chapters, more types of optimization problems will be

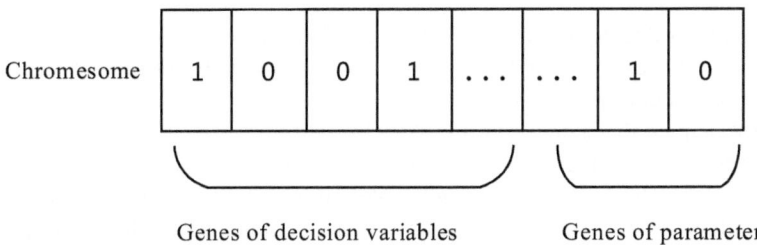

Fig. 6.1 An example of chromosome structure of self-adaptation methods

introduced: multi-objective optimization problems, dynamic optimization problems, constrained optimization problems, etc. In these situations, EAs tend to incur many tailored mechanisms to overcome the corresponding challenges, where the control of problem-relative parameters is an inevitable problem as well. In these chapters, more details will be presented.

6.1.2   Multi-parameter Control

Previous methods focus on the control of specific parameters. Differently, some works control several relevant parameters to obtain higher performance of algorithms. There are two categories of methods to control multi-parameters. One type of method called specific multi-parameter controllers controls more than one parameter, and the other one designs a generic framework to control multiple parameters. The former type of methods can only control specific parameters in a certain algorithm. To address the limitation, some generic parameter controllers were put forward to adjust any numerical parameters in EAs.

Specific multi-parameter controllers are easy to understand, where some unary-parameter controllers are integrated to adjust the values of multiple parameters simultaneously. Here, the following part is mainly about generic parameter control frameworks, which takes the reinforcement learning (RL) paradigm as an example.

With no type constraints of parameters, RL is applied to parameter control with the idea of employing the history decision information. The motivation is to establish an optimal mapping from states sets s to actions sets a. The states reflect certain search processes, and the actions change the parameter setting. State-action values are updated according to the obtained reward. Therefore, the state-action values can estimate how good the action is for a specific state. To optimize the map, many state-action value updating methods can be applied, such as Q-learning [22], SARSA [16], etc. As shown in Eq. (6.3) in the Q-learning algorithm, the state-action value $Q(s, a)$ is updated according to the instantly obtained reward r and the maximum possible state-action value of the next state $\max_{a'} Q(s', a')$

$$Q(s, a) \leftarrow Q(s, a) + \alpha(\gamma \cdot \max_{a'} Q(s', a') + r - Q(s, a)) \qquad (6.3)$$

where γ and α denote reward discount rate and step size.

With the reward obtained in every state transition, the estimated state-action values can be updated.

Eiben defined the state vector (Table 6.1), action vector (Table 6.2), and fitness-related reward function (e.g., the improvement of the best objective value) in the RL framework [6]. Doing exploration and exploitation actions is to update precise state-action values (how good the specific parameter settings are). With the framework, no matter how many parameters or what type the parameters are, if reward, state, and action representation are constructed, EAs with the generic controller will get

Table 6.1 Components of state vector in RL generic controller

Index	State parameter	Range
s_1	Best fitness	0–1
s_2	Mean fitness	0–1
s_3	Standard deviation of the fitness	0–1
s_4	Breeding success number	0–control window
s_5	Average distance from the best	0–100
s_6	Number of evaluations	0–999999
s_7	Fitness growth	0–1
$s_8 - s_{11}$	Previous action vector	0–1

Table 6.2 Components of action vector in RL generic controller

Index	Control parameter	Range
a_1	Population size	3–1000
a_2	Tournament proportion	0–1
a_3	Mutation probability	0–0.06
a_4	Crossover probability	0–1

suitable parameter settings as the evolution process continues. The pseudocode of the generic RL based parameter control is shown in Algorithm 6.1 [6], where ϵ is the probability of taking a random action.

Algorithm 6.1: The training of RL based generic controller [6]

Initialize Q and ϵ;
repeat
    Initialize state s;
    Choose an action a' according to $Q(s, a)$;
    $a = randomize\ a'$ with ϵ possibility;
    repeat
        Take action a, and observe r, s';
        Choose an action a' that optimize the function $Q(s', a')$;
        $a'' = randomize\ a'$ with ϵ possibility;
        Add new training instance Q;
        Use the Q-learning algorithm to update Q-value;
        $s = s'$;
        $a = a''$;
    until s *is the end state*;
until *episode number end*;

In addition to the RL generic parameter controller, other probabilistic rule-driven adaptive models (PRAM) were also proposed [24]. In the algorithm, all parameters were assigned initial values, and a series of greedy rules were devised to automatically adapt the value of parameters. Simultaneously, the evolution process is divided into many pair periods. In the first period, parameters are randomly chosen with the improvement in fitness recorded. In the second period, the parameter selection ratio is proportional to the recorded scores. When coming into another pair period, the initial values are updated toward the best performing one. Based on PRAM, time series prediction was utilized to predict the most promising parameter values for the next periods [1]. With more historical information taken into account, the changing direction of parameters can be predicted more accurately. Following the trend of prediction, more and more similar machine learning algorithms are hybrid with EAs, to predict the performance of different parameter settings.

6.1.3   Discussions

According to the aforementioned works, the adaptive methods tend to sample around the parameter setting range and update the sampling range according to the corresponding recent performance [12]. Therefore, the parameter settings are just the "winner" in a certain interval, but not in the entire search range. Existing value-adjusting rules often lack exploration among the parameter settings, because the optimal parameter setting itself is a complex optimization problem.

For multi-parameter controllers, the expectation is to control uncertain types and the number of parameters in a unified way. However, there is no guide on the composition of parameters. Blindly grouped parameters probably result in a suboptimal result. Comprehensive analysis can be investigated to help users choose the parameters to be controlled.

Generally, the parameter control has not been standardized up to now. Although many pieces of literature present competitive results, there are no universal methods that are widely recognized. In the future, the baseline of the parameter controller can be set, so the control methods can be judged in a more convincing mode. Although the research on parameter controllers has been for a long time, it is still one of the most active but a challenging issue in the EC community.

6.2   Policy Control

Generally speaking, all algorithms have inherent biases against characteristics of problems they intend to solve. Since the analysis of Wolpert in the "No Free Lunch" theorem [23], it is not practical for the single-configured algorithm to pursue overwhelming performance in a variety of test suites. During the last two decades, many works have absorbed various optimization techniques in one so that the ensemble algorithm can be controlled to embody different search specialties tailored to various problems. The control of all kinds of policies in the algorithm also becomes a crucial research topic.

6.2.1 Operator Selection Control

Whether an EA can perform well in an optimization problem depends heavily on whether the search strategies used are compatible with features of the problem. Unfortunately, the classical EA paradigm regards the optimization problem as a "black-box" problem, having no consideration for the features of problems, and algorithms can work just with pre-defined objective functions. A common way, which enables an algorithm to be more generative for solving different kinds of problems, is to ensemble different operators to search synchronously. As a result, the control operator selection becomes the key in the search process of EAs.

When it comes to the adaptive operator control in DE, SaDE [15] is a noticeable work in the field. SaDE inspires many optimizers not only about operator parameter adaptation but also about operator selection. SaDE employs four mutation operators and adaptively choose one with a probability, which is learned from an accumulated success over a time window. The details can be found in Sect. 5.5.3.

Following the similar idea, a self-learning PSO (SLPSO) [13] was proposed to adjust the search policy online during the evolution process. But different from SaDE, the operator selection is implemented at the individual level, i.e., each operator will have different probabilities for different individuals. Compared with the initial PSO, SLPSO incorporated four velocity update equations to serve different objectives to enhance the search ability.

$$\text{exploitation: } v_{i,d} = \omega \cdot v_{i,d} + \eta \cdot (pbest_{i,d} - x_{i,d}) \cdot \mathcal{U}(0,1) \tag{6.4}$$

$$\text{jumping out: } x_{i,d} = x_{i,d} + v_{\text{avg},d} \cdot \mathcal{N}(0,1) \tag{6.5}$$

$$\text{exploration: } v_{i,d} = \omega \cdot v_{i,d} + \eta \cdot (pbest_{rand,d} - x_{i,d}) \cdot \mathcal{U}(0,1) \tag{6.6}$$

$$\text{convergence: } v_{i,d} = \omega \cdot v_{i,d} + \eta \cdot (abest_d - x_{i,d}) \cdot \mathcal{U}(0,1) \tag{6.7}$$

where $pbest_{rand}$ is the $pbest$ of a random particle; v_{avg} is the average velocity of the current swarm; and $abest$ is the best solution constructed based on all historical $gbest$.

The reward value of particle k by operator i at time t ($r_i^k(t)$) is defined by three components: selection probability $\rho_i^k(t)$, fitness improvement $p_i^k(t)$, and success rate $s_i^k(t)$

$$r_i^k(t) = \frac{p_i^k(t)}{\sum_{j=1}^4 p_j^k(t)}\alpha + s_i^k(1-\alpha) + c_i^k \rho_i^k(t) \tag{6.8}$$

where s_i^k is the success ratio of operator i for particle k since the last selection ratio update; $\alpha = \mathcal{U}(0, 1)$; c_i^k is a discount factor for operator i of particle k with the aim to decrease the dominance of the current "optimal" operator, which is defined as follows:

$$c_i^k = \begin{cases} 0.9, & \text{if } s_i^k = 0 \text{ and } \rho_i^k(t) = \max_{j=1}^R (\rho_j^k(t)) \\ 1, & \text{otherwise} \end{cases} \qquad (6.9)$$

where $\rho_i^k(t)$ is the selection ratio of operator i for particle k at the current iteration, which is updated as follows:

$$\rho_i^k(t+1) = \frac{r_i^k(t)}{\sum_{j=1}^4 r_j^k(t)}(1 - 4 * \gamma) + \gamma \qquad (6.10)$$

where $\gamma = 0.01$ is the minimum selection ratio for each operator.

Due to the diversity of search policy, SLPSO can cover different kinds of problems, which brings about better performance on a set of test suites. Furthermore, optimal operators are adaptively selected as the algorithm runs, and the selection policy is adjusted when different optimization problems are applied.

More simplified, the self-adaptation mechanism can be regarded as a versatile framework to control all kinds of parameters, no matter how many parameters or what types of parameters. According to an analogical method of parameter control, similar algorithms are easy to implement to select among different operators. In addition, selecting an operator based on the reward received through the evolving process is to obtain a more efficient search in the next period. Therefore, most of the selection control can be implemented by one of the following techniques: probability matching, adaptive pursuit, and multi-armed bandit algorithms. As stated in [18]: probability matching is concerned with matching the probabilities of operators to the corresponding rewards; adaptive pursuit aims at augmenting the effective ones but impairing the ones in poor conditions recently. The most commonly used probability matching is the one also used in SaDE as shown in Eq. (5.29).

Generally, with some possibility distributed to poor operators, the probability matching mechanism can hardly learn an optimal policy. The adaptive pursuit mechanism is easier to cause "dominant operator effects," where the best operator so far continues to obtain a high selection ratio due to the "greedy" reaction to the recent trends. As a result, only the operator with the maximum estimation value can serve the algorithm.

To learn an optimal policy, some theoretical frameworks have been derived from game theory and reinforcement learning, but they did not gain deserved attention. In nature, "greedy" or "more evenly" is an issue about the trade-off between exploration and exploitation, which is also a fundamental matter in EC. A more detailed discussion can be found in Chap. 7.

6.2.2 Hyper-heuristics

Generally, policy control involves all configurations in EAs: selection and adaptation of algorithms, reproduction operators, niching methods, surrogate models, etc. The setting of an algorithm including all parts can be seen as a hyper-heuristics problem, where the search space does not directly relate to the solution space but the space of search methods or heuristics.

The early idea of hyper-heuristics can be traced back to the 1960s [8], but the term now gradually evolved into the means of automatically designing heuristics. Compared to traditional methods that focus on the search of solution spaces directly, hyper-heuristics operate on the space of heuristics, trying to discover or generate appropriate heuristics for various problems. Given a problem, hyper-heuristics have the advantage to receive "cheaper" and "more precise" solutions than conventional heuristics [4]. With different search spaces, hyper-heuristics usually can be divided into two classes: heuristic selection, which chooses from the existing heuristics, and heuristic generation, which creates heuristics from the existing components. Due to the ensemble capability of versatile heuristic algorithms, hyper-heuristics can handle diverse problems, especially combinatorial optimization problems, which will be presented in detail in Chap. 15. In this section, a heuristic selection-based hyper-heuristic will be discussed.

Given low-level heuristics with different features, hyper-heuristics are aimed at selecting optimal algorithms at lower expenses of "trial and error." In [19], a predefined global optimization algorithm set, which includes SA, GA, PSO, and ACO, is constructed. Diversity detection operators and improvement detection operators determine which heuristic is to be used to find better solutions. To elaborate, the improvement detection operator is used to check whether the solutions found by a heuristic are improved or not in a time interval. If not, the heuristic will not be chosen in the next period. In addition, the diversity detection operator can check distances between different solutions, and if the distances are less than a threshold, all heuristics could be activated. When the algorithm was applied to practical scheduling tasks, the results showed a significant reduction in makespan.

6.2.3 Discussions

To sum up, for EAs with a control mechanism, a general framework skeleton can be shown as Fig. 6.2. The control mechanism aims to adjust the algorithm

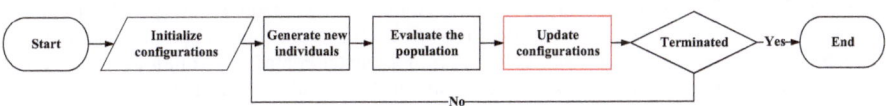

Fig. 6.2 Framework of EAs with parameter control mechanisms

ingredients to adapt to the evolving process, problems with changing dynamics, and problems with different characteristics. Even though many algorithms with versatile inspirations have been proposed by researchers, their specialties were scarcely analyzed. Ensemble framework as a capacity can extract various merits of algorithms into one, where the capability of dealing with unknown and complex problems will be significantly enhanced. Hence, the multi-functional ensemble paradigm deserves additional research to improve the efficiency of optimization algorithms.

In the initial control method, designed mechanisms such as rule-based policies were to create a prompt reaction to the feedback. Therefore, a huge number of database in history have not been explored/utilized effectively. Learning-based algorithms can be promising methods to guide the evolving process, such as parameter tuning being replaced by RL control systems. Moreover, learning-based policies are capable of adjusting the search preference. To some extent, it is also another form of balance on exploration and exploitation, and learning-based policies can provide new insights in designing EAs to cover more complex optimization problems.

References

1. Aleti, A., Moser, I.: Predictive parameter control. In: Proceedings of the 13th Annual Conference on Genetic and Evolutionary Computation, pp. 561–568 (2011)
2. Arabas, J., Michalewicz, Z., Mulawka, J.: GAVaPS—a genetic algorithm with varying population size. In: Proceedings of the First IEEE Conference on Evolutionary Computation, pp. 73–78 (1994)
3. Bäck, T., Eiben, A.E., van der Vaart, N.A.: An empirical study on GAs "without parameters". In: International Conference on Parallel Problem Solving from Nature, pp. 315–324 (2000)
4. Burke, E.K., Hyde, M., Kendall, G., Woodward, J.: A genetic programming hyper-heuristic approach for evolving 2-d strip packing heuristics. IEEE Trans. Evol. Comput. **14**(6), 942–958 (2010)
5. Cook, J.E., Tauritz, D.R.: An exploration into dynamic population sizing. In: Proceedings of the 12th Annual Conference on Genetic and Evolutionary Computation, pp. 807–814 (2010)
6. Eiben, A., Horvath, M., Kowalczyk, W., Schut, M.C.: Reinforcement learning for online control of evolutionary algorithms. In: International Workshop on Engineering Self-Organising Applications, pp. 151–160 (2006)
7. Eiben, Á.E., Hinterding, R., Michalewicz, Z.: Parameter control in evolutionary algorithms. IEEE Trans. Evol. Comput. **3**(2), 124–141 (1999)
8. Fisher, H.: Probabilistic learning combinations of local job-shop scheduling rules, pp. 225–251. Prentice-Hall, London (1963)
9. Goldberg, D.E., Sastry, K., Latoza, T.: On the supply of building blocks. In: Proceedings of the 3rd Annual Conference on Genetic and Evolutionary Computation, pp. 336–342 (2001)
10. Iorio, A., Li, X.: Parameter control within a co-operative co-evolutionary genetic algorithm. In: International Conference on Parallel Problem Solving from Nature, pp. 247–256 (2002)
11. Karafotias, G., Hoogendoorn, M., Eiben, Á.E.: Parameter control in evolutionary algorithms: Trends and challenges. IEEE Trans. Evol. Comput. **19**(2), 167–187 (2014)
12. Kramer, O.: Evolutionary self-adaptation: a survey of operators and strategy parameters. Evol. Intell. **3**(2), 51–65 (2010)

13. Li, C., Yang, S., Nguyen, T.T.: A self-learning particle swarm optimizer for global optimization problems. IEEE Trans. Systems Man Cybern. **42**(3), 627–646 (2011)
14. Lobo, F.G., Lima, C.F.: Revisiting evolutionary algorithms with on-the-fly population size adjustment. In: Proceedings of the 8th Annual Conference on Genetic and Evolutionary Computation, pp. 1241–1248 (2006)
15. Qin, A.K., Huang, V.L., Suganthan, P.N.: Differential evolution algorithm with strategy adaptation for global numerical optimization. IEEE Trans. Evol. Comput. **13**(2), 398–417 (2008)
16. Sutton, R.S., Barto, A.G.: Reinforcement Learning: An Introduction. MIT Press, Cambridge (2018)
17. Teo, J.: Exploring dynamic self-adaptive populations in differential evolution. Soft Comput. **10**(8), 673–686 (2006)
18. Thierens, D.: Adaptive strategies for operator allocation. In: Parameter Setting in Evolutionary Algorithms, pp. 77–90. Springer, Berlin (2007)
19. Tsai, C.W., Huang, W.C., Chiang, M.H., Chiang, M.C., Yang, C.S.: A hyper-heuristic scheduling algorithm for cloud. IEEE Trans. Cloud Comput. **2**(2), 236–250 (2014)
20. Wagner, N., Michalewicz, Z.: Genetic programming with efficient population control for financial time series prediction. In: 2001 Genetic and Evolutionary Computation Conference Late Breaking Papers, pp. 458–462 (2001)
21. Wagner, N., Michalewicz, Z.: Parameter adaptation for GP forecasting applications. In: Parameter Setting in Evolutionary Algorithms, pp. 295–309 (2007)
22. Watkins, C.J.C.H.: Learning from delayed rewards. Ph.D. Thesis, King's College (1989)
23. Wolpert, D.H., Macready, W.G.: No free lunch theorems for optimization. IEEE Trans. Evol. Comput. **1**(1), 67–82 (1997)
24. Wong, Y.Y., Lee, K.H., Leung, K.S., Ho, C.W.: A novel approach in parameter adaptation and diversity maintenance for genetic algorithms. Soft Comput. **7**(8), 506–515 (2003)

Chapter 7
Exploitation Versus Exploration

Abstract This chapter will first introduce the concepts of exploration and exploitation by reviewing several typical optimization algorithms. Next, methods of enhancing exploration or exploitation in EAs will be discussed. The final part will introduce existing methods of how to balance evolutionary exploration and exploitation.

7.1 Introduction

Exploitation refers to search behaviors around the best point found so far in a basin of attraction of an optimum with the aim of locating the optimum quickly. Exploration, on the contrary, refers to searching behaviors in areas rarely visited with the aim of discovering the unexplored basin of attractions of new optima. For example, for a 1-D function maximization problem shown in Fig. 7.1, suppose that we have already evaluated several solutions denoted by black points, and gray points are some potential solutions we will evaluate in the next step. Among these gray points, the diamond ones belong to exploitative search points, and the square ones belong to exploratory search points.

"Exploration and exploitation are the two cornerstones of problem solving by search." [2] Exploitation focuses on speeding up the process of finding the local optimum, while exploration aims at avoiding missing the global optima. In exploitative searching, we can efficiently make improvement with the help of neighboring sampling points, but also never getting a chance to escape the local optimum. Exploration without doubt can solve this problem, but it is also hard for it to find better solutions since there is little neighboring information.

Since exploration and exploitation are complementary to each other, all stochastic search algorithms are in a dilemma of whether to explore more or exploit more. Algorithms with only exploitation make them quickly find an optimum but at risk of trapping into local optima, especially for complex fitness landscapes. Algorithms with only exploration enable them to have the ability of finding the global optimum, but at the cost of high computing time, which is normally unaffordable for problems with large search spaces. Heuristic optimization algorithms are typical stochastic

C. Li et al., *Intelligent Optimization*,
https://doi.org/10.1007/978-981-97-3286-9_7

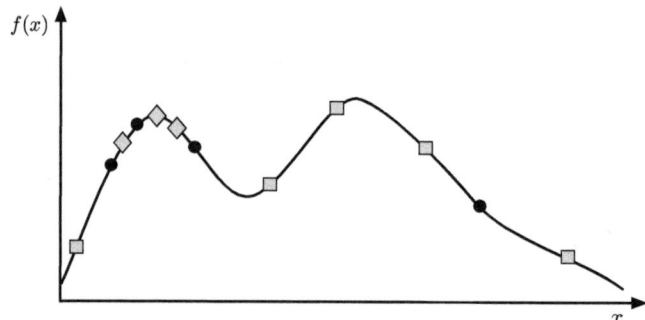

Fig. 7.1 Exploration and exploitation in a 1-D function maximization problem

search algorithms, so their performance is significantly determined by how they deal with the relationship between exploitation and exploration in the search process.

7.2 Exploitation and Exploration Methods

Different types of optimization algorithms have different tendencies to exploitation and exploration. Traditional numerical methods and DFS methods normally consider only exploitation, while in the meta-heuristics methods, exploitation cooperates with exploration with quite different implementation.

7.2.1 Iterative Methods

Most of the numerical optimization methods proposed earlier are purely exploitable algorithms based on the line search, especially the gradient method and Newton method (see Sect. 3.1). These two algorithms use the gradient information or the second derivative of the objective function to iteratively find better neighboring solutions. Figure 7.2 simulates the optimization process of this type of method on a univariate function, where the star denotes the global optimal solution, the solid point denotes the initial solution, and the hollow points denote the iteratively improved solutions. Since the initial solution is located inside the attraction basin of one local optimum shown in Fig. 7.2 on the left, this type of exploitative algorithm will inevitably converge toward a local optimum.

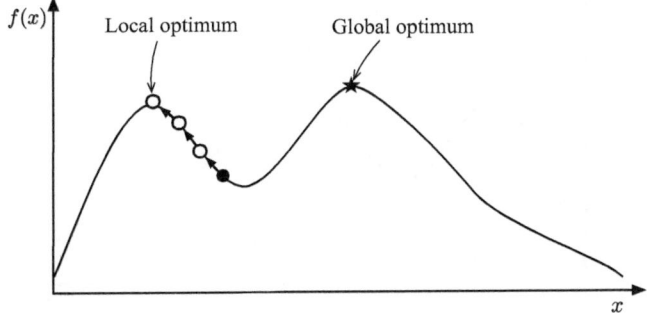

Fig. 7.2 Search process of the iterative method on an 1-D maximization problem

7.2.2   Single-Solution Meta-heuristics

Since iterative methods only have the ability to exploit but do not have the ability to explore, some meta-heuristic optimization algorithms, such as the stochastic hill climbing (see Sect. 3.3.1) and the simulated annealing (see Sect. 3.3.2), were proposed to address this issue. The core point is to add some randomness to the solution update mechanism. The new solution does not necessarily have to be better than the previous one, making it much more possible for the algorithm to explore regions far away from the initial point.

However, to ensure the convergence of the algorithm, the randomness in the solution update mechanism must be reduced with the optimization process. So, the algorithm will gradually lose its exploration ability and finally degenerate into a purely exploitative algorithm. Only by carefully controlling the degeneration speed can the algorithm achieve satisfying performance. Figure 7.3 simulates an ideal optimization process of the simulated annealing with a suitable mutation step size and annealing function on the same problem in Fig. 7.2.

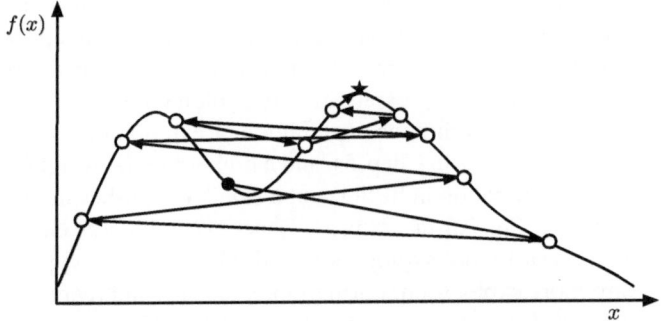

Fig. 7.3 Search process of the simulated annealing on an 1-D maximization problem

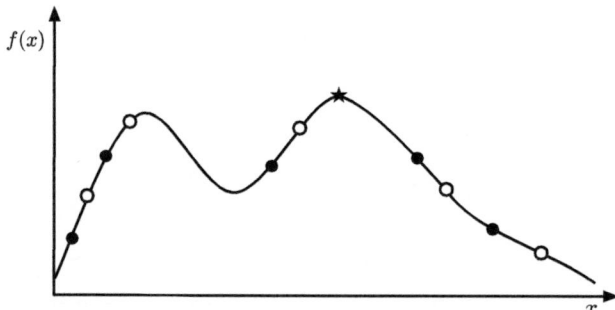

Fig. 7.4 Search process of a simple (5, 5)-EA on an 1-D maximization problem

7.2.3 Population-Based Meta-heuristics

The single-solution meta-heuristic described above retains only one solution throughout the iteration, so the exploration area is completely determined by the current location of the only one solution. To conquer this problem, population-based meta-heuristics represented by variant EAs maintain multiple solutions throughout the evolutionary process. These solutions usually distribute over different areas of the search space at the beginning, which assign EAs a much stronger exploration ability than single-solution meta-heuristics. For example, in Fig. 7.4, if the five solid points mark the distribution of the initial population, then the five hollow points show a very common distribution of the first-generation offspring. It means that EAs can easily achieve the same exploration effect as the ideal simulated annealing.

Furthermore, if we compare the solution update mechanism of the simulated annealing with the canonical survivor selection mechanisms (such as *comma selection* and *plus selection* in Sect. 5.6.1) of EAs, we will find that the randomness in survivor selection is much smaller. It makes the EA possess a much stronger exploitation ability even in the early stage of optimization, when the simulated annealing behaves much more like a random search.

For EAs, all components can have an effect on exploration and exploitation depending on the tendency of search shrinking or not. For example, decreasing the size of the mutation step from a large value or a small value normally results in the transformation of the search from exploration to exploitation. A crossover operator can also generate offspring within the exploitation or exploration zones, such as the crossover of DE (see Fig. 5.11 in Chap. 5), and a higher inertial weight ω pushes the search toward more exploration in SPSO, and a smaller ω pushes the search toward more exploitation (see Fig. 5.8 in Chap. 5). Generally, a higher selection pressure pushes the search toward more exploitation, and a lower selection pressure pushes the search toward more exploration. Furthermore, the size and representation of the population can contribute to exploration and exploitation [1].

7.3 Enhancing Exploration and Exploitation

Although most canonical EAs already have relatively strong exploration and exploitation abilities, various optimization problems that are constantly emerging will always raise higher requirements. For example, some tasks require a final solution with higher precision, and some tasks want a higher optimization speed. In these cases, we need to enhance the exploitation ability of the algorithm. On the other hand, if the fitness landscape of the problem is very deceptive, then the algorithm needs to be more capable of exploration.

7.3.1 Exploration Enhancement Methods

Intervened reproduction, intervened competition, irrigation, and multi-population methods are commonly used to enhance exploration.

Intervened Reproduction
Most typical EAs have parameters in their reproduction process. We list some of them in Table 7.1. By adjusting these parameters, we can control the search range around the parental individuals. For example, if we increase F and Cr in DE, the offspring will be generated far away from the parents.

The input of the reproduction model determines the output. So, by selecting parents that are farther apart, we can generate offspring in a much wider area. For example, if we restrict neighboring individuals, such as p_2, p_3, p_4, and p_5 shown in Fig. 7.5, from being selected simultaneously, then p_2 will be more likely to mate with distant individuals such as p_1, p_7, p_{10}, etc.; thus, the offspring will be more likely to be far away from the attraction basin where p_2 is. This can be achieved by clustering methods.

Intervened Competition
The survivor selection determines the distribution of parents of the next generation, thus indirectly influencing the distribution of the offspring. To achieve this effect, we can make competitions only within the neighborhood of each individual. After adding the *restricted competition*, p_{10} shown in Fig. 7.5 will not be eliminated by p_2, making it much more possible to survive in the offspring selection. Detailed methods include crowding, clearing, and speciation (see the details in Sect. 8.2).

Table 7.1 Reproduction parameters in some typical EAs

Algorithm name	Reproduction parameters
GA	Crossover probability, mutation probability
DE	Mutation factor F, crossover probability Cr
PSO	Inertia weight W, accelerators C_1 C_2

Fig. 7.5 Individuals in the search space of a 2-D problem, where dash lines denote the contour lines of fitness value and points with gray hollow denote the individuals

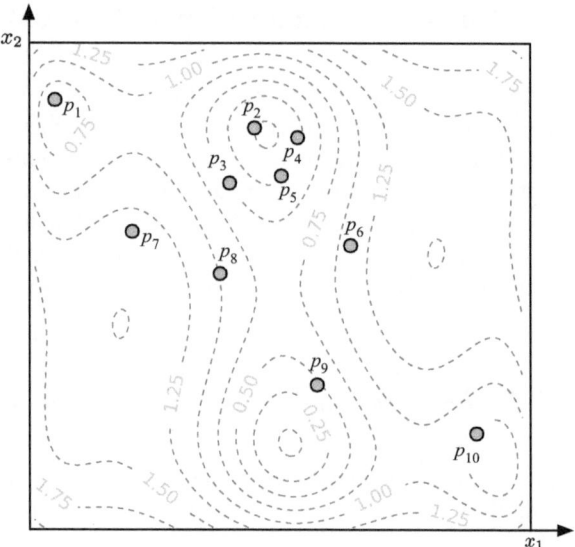

Most EAs are fitness-driven search algorithms, which means that after decreasing the fitness values of the areas around the points frequently visited, the population will naturally go toward other local optima whose fitness values have not been punished yet. Detailed methods include sharing, stretching, and deflation (see the details in Sect. 8.2).

Irrigation

Random immigration is an early and simple method to enhance exploration. We can replace one or more individuals with randomly initialized individuals. These immigrants are totally independent of the distribution of the current population, which makes them much more likely to be generated in areas rarely searched than ordinary offspring.

The archive solution is kind of like history immigration. Since individuals from the earlier stage of the evolution are always located in the midway of the convergence path, we can store some representative ones and make them participate in the reproduction as well as the current population.

Multi-population

Since EAs are stochastic search algorithms, different initial populations may convergence to different points. We can make several populations search simultaneously and occasionally interact with each other. The populations can even come from different EAs.

7.3.2   Exploitation Enhancement Methods

Intervened reproduction and memetic algorithms are used to enhance the exploitation for local optima.

Intervened Reproduction
Since reproduction parameters greatly determine the size of the mutation step, we can decrease the mutation step by adjusting these parameters, thus enhancing the exploitation. However, it should be noted that a too small mutation step size always leads to the premature convergence.

Parental individuals that are similar to each other will also produce similar offspring. So, by choosing neighboring individuals as parents, we can generate offspring gathered around the parents, thus accelerating the exploitation. A typical example is the local informed particle optimizer, in which the global best particle is replaced with the local best particle (see Sect. 5.4.4).

Memetic Algorithm
Memetic algorithm [6] is the combination of population-based EA and single-solution optimizers. In most memetic algorithms, the population is used to enhance diversity, while single-solution optimizers, such as hill climbing or gradient descending, are used to enhance local search. These kinds of methods are quite efficient for some combinatorial optimization problems, on which most reproduction methods proposed in EA field cannot generate valid offspring.

7.4   Balancing Exploration and Exploitation

Reviewing the above methods, we can find that most of the methods that strengthen the exploitation will weaken the exploration and vice versa. For example, the exploration will definitely decrease if we increase the exploitation by adjusting the parameters of the reproduction operator. Therefore, like all other stochastic heuristics, EAs also need to handle the trade-offs between exploration and exploitation.

7.4.1   Explicit Differentiation Methods

Individual Differentiation
Individuals are assigned different tasks. Some individuals only focus on exploitation, while others only do the exploration. There are many approaches to achieve the differentiation, for example, locally informed particle and the globally informed particle in PSO and individuals with large F and small F in DE. These methods are particularly useful in complex problems, such as dynamic environments (see Sect. 11.2.2).

Evolution Stage Differentiation
The evolution stage can be divided into different parts, and each part has different tendencies toward exploitation or exploration. For example, the evolution stage can be divided into three parts based on the maximum evaluation numbers, where the exploration is encouraged in the first stage, while the exploitation is preferred in the last stage.

7.4.2 Population Diversity-Driven Methods

Threshold Value Control Methods
Due to disadvantages of objective-driven search of EAs, each individual ignores the objective value and has a novelty value, which indicates its contribution to the population diversity, to encourage exploration [4]. After manually setting a threshold value, we can add any individuals whose novelty values are larger than the threshold value into the archive solutions or restricted individuals whose novelty values are smaller than the threshold value from reproduction.

Adaptive Parameter Control Methods
As introduced above, the control of parameters can enhance either exploitation or exploration. Some researchers make the parameters dependent on the population-diversity based on their experimental experience. For example, if population diversity decreases or becomes stagnated, parameters will be automatically adjusted to enhance the exploration.

Multi-objective Methods
Pareto dominance relationships are commonly used in multi-objective evolutionary optimization (see Chap. 9) to balance the multiple optimization objectives. Therefore, it is natural to use the Pareto dominance relationship in the survivor selection by adding helper objectives [3] to balance diversity and fitness of the population.

7.4.3 Non-overlapping Multi-population Methods

In these methods, initial individuals are usually grouped into different sub-populations based on their distance relationship. Each sub-population is expected to search for one local optimum. So their search areas are required to be non-overlapping. Since each individual can only interact with individuals of the same sub-population, non-overlapping multi-population methods (e.g., the clustering PSO [5] for locating and tracking multiple optima) are kind of like the mixture of inbreeding mating and restricted competition.

Overlapping multi-population methods introduced before are usually used to increase population diversity. It sacrifices many additional evaluations to only enhance exploration ability. However, non-overlapping multi-population methods enhance not only exploitation but also exploration, if we allow some interactions between sub-populations.

7.4.4 Space Partitioning-Based Methods

In order to learn unexplored search areas, search history can be clustered to recognize promising areas to explore. The crucial point is how to partition the search space. There are several ways, such as drawing grid lines or projection lines in objective spaces, using binary space tree or k-d space tree in continuous search spaces, using self-organizing maps in combinatorial search spaces, etc. After space partitioning, frequently visited areas can be identified by sub-spaces rather than crowded individuals. Thus, following searches in these areas will be prohibited or inhibited.

In space partitioning-based methods, the exploitation is achieved by normal survivor selection. However, its exploration differs it from traditional methods. These methods, compared to fitness modification methods (sharing, stretching, etc.), are much closer to non-visiting or tabu search.

7.5 Discussions

Exploration and exploitation are two fundamental concepts. The trade-offs between them significantly influence the performance of EAs. Although many studies have been conducted on this topic, much more attention and effort are still needed to provide users and researchers with deeper insights. Specifically, several issues require further attention.

Measurements of Exploration and Exploitation
The delicate balance between exploration and exploitation is crucial for effective decision-making. Within the EC community, this balance remains an open question. To address it, we must first identify the distinct behaviors associated with exploration and exploitation.

Existing approaches often rely on indirect methods, particularly diversity measurements. However, the relationship between population diversity and the equilibrium between exploration and exploitation remains unclear. Notably, a highly diverse population does not necessarily guarantee robust global search exploration capabilities.

Control of Exploration and Exploitation
Many factors contribute to the balance of exploration and exploitation, such as selection, reproduction operators, population size, and even the representation. Therefore, when and how to control the rate of exploration and exploitation must be comprehensively investigated.

Identification of Promising Areas

To avoid local optima and accelerate the search for the global optimum, i.e., to effectively balance exploration and exploitation, we must learn promising areas that contain relatively good optima through the exploration and exploitation behaviors of the whole search process. If we can achieve this, exploration would be guided. One possible way is to build learning models to predict the degree of promising areas based on search history, and this can be achieved based on space partitioning-based methods equipped with learning models to predict the improvement of objective value of the population for the future search.

References

1. Črepinšek, M., Liu, S.H., Mernik, M.: Exploration and exploitation in evolutionary algorithms: a survey. ACM Comput. Surv. **45**(3), 1–33 (2013)
2. Eiben, A.E., Schippers, C.A.: On evolutionary exploration and exploitation. Fundamenta Informaticae **35**(1), 35–50 (1998)
3. Jensen, M.T.: Helper-objectives: using multi-objective evolutionary algorithms for single-objective optimisation. J. Math. Model. Algorithms **3**(4), 323–347 (2004)
4. Lehman, J., Stanley, K.O.: Abandoning objectives: evolution through the search for novelty alone. Evol. Comput. **19**(2), 189–223 (2011)
5. Li, C., Nguyen, T.T., Yang, M., Mavrovouniotis, M., Yang, S.: An adaptive multipopulation framework for locating and tracking multiple optima. IEEE Trans. Evol. Comput. **20**(4), 590–605 (2015)
6. Moscato, P.: On evolution, search, optimization, genetic algorithms and martial arts: towards memetic algorithms. Technical report, California Institute of Technology, Pasadena (1989)

Chapter 8
Multimodal Optimization

Abstract This chapter will first introduce the definition of multimodal optimization. Next, most representative evolutionary multimodal optimization algorithms, which are also known as niching methods, will be introduced. Future challenges will be discussed in the final part.

8.1 Introduction

In most optimization problems, only the best possible solution needs to be found. However, some problems may have more than one global optima, and the decision-maker tends to locate all the global optima. Figure 8.1 shows a simple example.

In addition, in some real-world applications, the single best solution is far enough away from the decision-maker due to physical or cost constraints. For example, it may be challenging to deploy the solution in practice, the price of changing it in the future may be too high, and its objective value may be too sensitive to the precision of decision variables. In these cases, some sub-optimal but also preferable solutions will become very important. Besides, multiple different candidate solutions may help the decision-maker better understand the inner structure of these black-box optimization problems.

If a problem seems to have any one of these above characteristics, the decision-maker will tend to locate all global optima or good local optima as many as possible. These problems are called multimodal optimization problems (MMOPs).

To solve MMOPs, we need to restart traditional single-solution heuristics such as hill climbing and simulated annealing many times, each time from different initial points, trying to explore the basins of attraction of different local optima. On the contrary, population-based evolutionary algorithms have natural superiority over them. Efforts made by EAs to avoid stagnation keep the population relatively diverse. Since individuals located in different optima will both have high fitness values, in the ideal situation, the population will converge toward several points.

However, most traditional EAs are designed to solve global optimization problems (GOPs), which means that if there exists a very little height difference between peaks (which refer to different local optima in the fitness landscape), all individuals

© China University of Geosciences Press 2024
C. Li et al., *Intelligent Optimization*,
https://doi.org/10.1007/978-981-97-3286-9_8

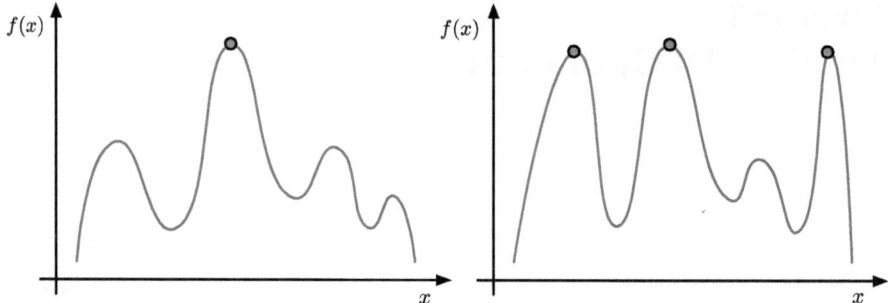

Fig. 8.1 Landscapes of two 1-D multimodal functions, where gray points indicate solutions that need to be found

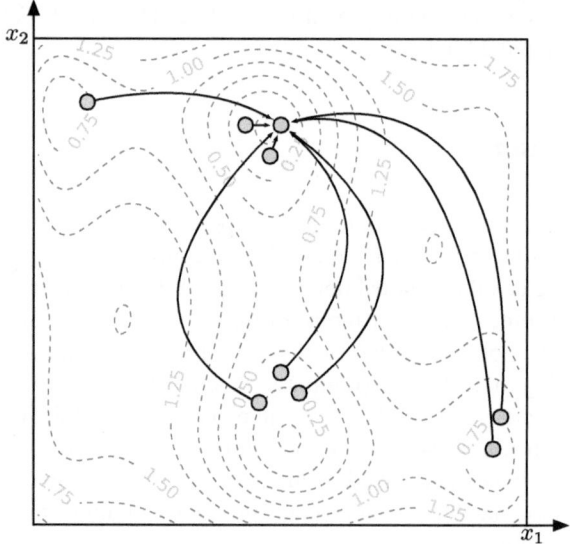

Fig. 8.2 A typical example of the converging trend of population from global optimization EAs when minimizing a 2-D function with two global optima and two local optima, where dash lines denote the contour lines of fitness value

will inevitably converge toward the best one in the final stage of evolution. Even if different peaks have the same height, since they may have different local fitness landscape structures and thus having different search difficulties, there will still exist fitness differences between the best individuals of different peaks.

Therefore, for these global optimization EAs, which rarely limit interactions between individuals, the population distribution that covers different local optima will only persist in the early stage and thus will lose its ability to locate these local optima with enough high precision (as shown in Fig. 8.2). In the face of this problem, we can still find answers from the process of natural evolution.

8.2 Niching Methods for Traditional EAs

To prevent convergence to a single point, EAs borrowed a concept called niching (shown in Fig. 8.3), which originates in the ecosystem. In most biological environments, there are many niches. Competitions only occur between creatures that live in the same niche by dividing resources among them, as shown in Fig. 8.2. Similarly, suppose that the best individual can only affect individuals of the same niche, the population can naturally converge toward different optimal solutions. So the problem now is how to achieve niching in EAs?

8.2.1 Fitness Sharing

To answer this question, Goldberg and Richardson in [4] assumed the decision space as an environment where individuals in the same niche should compete with each other for limited resources. The more individuals in a niche, the fewer chances they have to survive. Each individual has a niche count c that indicates the number of individuals less than a given distance (usually denoted as niche radius r) from it. And the final fitness value of each individual will be divided by their niche counts. Figure 8.4 shows the modified fitness values of four individuals in a niche.

$$c(\mathbf{p}_i) = |\{ \mathbf{p}_k \mid \forall k \in [1:N], \|\mathbf{x}_{\mathbf{p}_i} - \mathbf{x}_{\mathbf{p}_k}\| < r\}| \tag{8.1}$$

$$f'(\mathbf{p}_i) = \frac{f(\mathbf{p}_i)}{c(\mathbf{p}_i)} \tag{8.2}$$

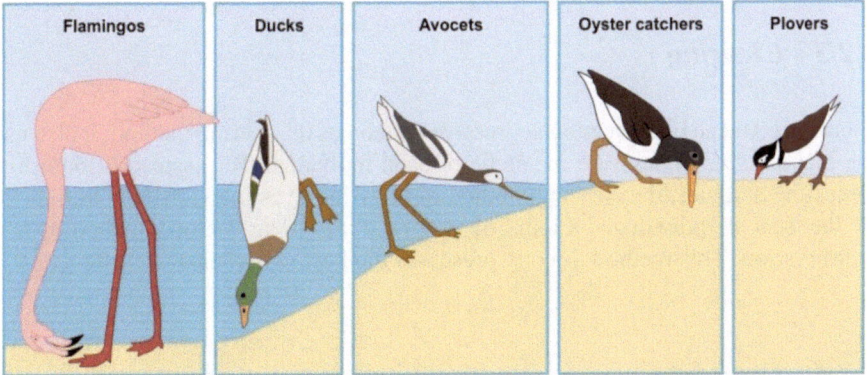

Fig. 8.3 Species inhabit different niches to avoid competition (this figure is from BioNinja's blog)

Fig. 8.4 Fitness sharing

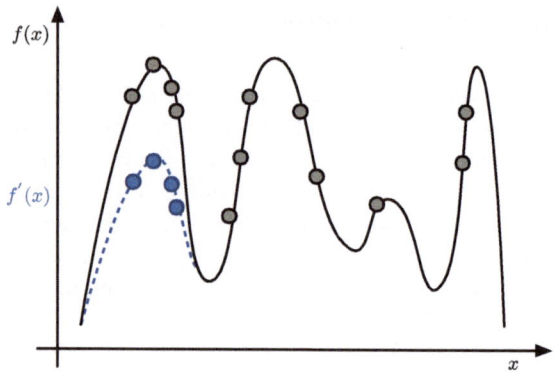

However, the niche radius is difficult to set. Therefore, Yin in [14] used the adaptive Macqueen's K-means clustering algorithm to group individuals into different niches.

8.2.2 Restricted Tournament Selection

Harik in his paper [5] modified the traditional tournament selection to a niching version, namely, restricted tournament selection (RTS). In the tournament selection, w parents are randomly selected from the population. The parent with a higher fitness value will be selected with a larger probability. However, in RTS, the closest parent to the offspring will be selected. So in the survivor selection process, an individual will tend to compete with individuals close to it.

8.2.3 Clearing

Clearing [10] can be seen as an aggressive version of the sharing method. It also uses the niche radius to describe the niche of each individual. But instead of punishing their fitness values, the clearing method directly removes individuals within niches of the best k individuals. Li in [6] proposed a similar method named species conservation. This method directly preserves the best individuals in each niche.

8.2.4 Crowding

Crowding was firstly proposed by De Jong [1] to maintain population diversity and later modified by Mahfoud[9] for multimodal optimization. After two offspring are

Fig. 8.5 Paring of two parents and two offspring

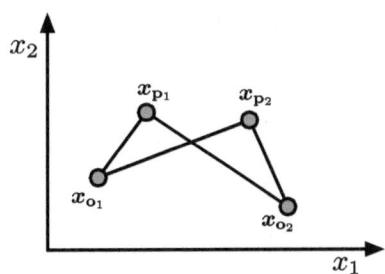

produced by two parents, these four individuals will be paired in a manner that makes the sum of distance smaller. For example, in Fig. 8.5, $\mathbf{p}_1$ and $\mathbf{p}_2$ are parents, and $\mathbf{o}_1$ and $\mathbf{o}_2$ are offspring. The sum of $\|x_{\mathbf{p}_1} - x_{\mathbf{o}_1}\|$ and $\|x_{\mathbf{p}_2} - x_{\mathbf{o}_2}\|$ is less than the sum of $\|x_{\mathbf{p}_1} - x_{\mathbf{o}_2}\|$ and $\|x_{\mathbf{p}_2} - x_{\mathbf{o}_1}\|$. So $\mathbf{o}_1$ will compete with $\mathbf{p}_1$, and $\mathbf{o}_2$ will compete with $\mathbf{p}_2$.

8.3 Niching Methods of Emerging EAs

Since those above niching methods are integrated into traditional EAs, such as simple GA and canonical ES, their niching techniques are restricted by the paradigm of these EAs. With the emergence of some later-proposed EAs such as PSO and DE, there comes a variety of unique niching methods that rely on the specific algorithm.

8.3.1 DE with Neighbor Mutation

As introduced in Sect. 5.6, mutation plays the primary role in the DE reproduction process. The final survivor selection is between the base vector x_i and the target vector v_i. In DE/rand/1, the target vector is generated by

$$v_i = x_{r_1} + F(x_{r_2} - x_{r_3}) \tag{8.3}$$

To achieve niching, Epitropakis proposed DE/nrand/1 [3], where x_{r_1} is replaced with the nearest neighbor of the base vector.

$$v_i = x_{nearest} + F(x_{r_2} - x_{r_3}) \tag{8.4}$$

Qu et al. proposed a neighbor mutation strategy of DE [13], which selects all donor vectors (e.g., x_{r_1}, x_{r_2}, and x_{r_3} in Fig. 8.6) among the m-th neighborhood niche of its base vector.

Fig. 8.6 Mutation process of
DE/rand/1

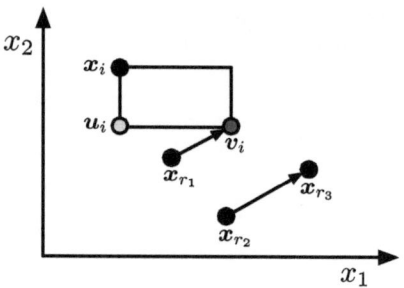

8.3.2 PSO with Restricted Communication Topology

As introduced in Sect. 5.4, each particle will be propelled toward the stochastic average of its personal best location *pbest* and the best location of the swarm *gbest*. So if *gbest* is restricted in the neighborhood of the particle, different niches will naturally form. Li in [8] defined the neighborhood using a ring topology, where each particle's neighbor consists only of its previous and subsequent particles according to their indexes in the swarm. In addition, a Euclidean distance-based niching PSO, namely, LIPS [12], defined *gbest* as the nearest neighbor best of the particle. A fitness Euclidean-distance ratio-based PSO named FER-PSO [7] used the fitness Euclidean-distance ratio to replace *gbest* with the nearest-and-fittest neighbor best of the particle.

8.4 Other Evolutionary Niching Methods

In recent years, with the development of the research on maintaining population diversity, there are also some general niching methods.

8.4.1 Multi-population

Niches in multi-population methods are non-overlapping and evolve independently. This is different from the clustering-based GA [14] we introduced in the fitness sharing. The latter conducts reproduction across different niches, so the niching behavior is achieved by the fitness punishment. However, the multi-population methods limit both the reproduction and selection inside each niche. A typical example is the NEA2 proposed by Preuss in [11], where a nearest-better clustering method is applied in the beginning to separate individuals into several subpopulations, and each population evolves separately using CMA-ES until reaches convergence.

8.4.2 Multi-objective Selection

Deb and Saha in [2] converted the single-objective multimodal problem into a bi-objective optimization problem

$$\min/\max \ f_1 = f'(\mathbf{p})$$
$$\max \ f_2 = nichingEffect(\mathbf{p})$$

where f_1 is the original objective function and f_2 is the niching effect contributed by the individual $\mathbf{p}$, e.g., the difference between $\mathbf{p}$ and all other individuals or the negative number of $\mathbf{p}$'s niche count.

8.5 Challenges

Niching based methods employ the idea of *divide and conquer*. Although many algorithms have been proposed, more effort needs to be paid attention to. In fact, to solve MMOPs, we need to well address the challenging issues of exploitation and exploration discussed in Chap. 7. Identification and formation of niches are key to the performance of niching methods, which is equivalent to identification of promising areas of complex multimodal problems during the search. For niching methods, specifically, the following issues should be addressed.

Search Efficiency
Some niching methods achieve good performance simply by suppressing the convergence of population, even if the problem is simple and monotonic. However, since the optimization problem is black box, an ideal niching method should be able to learn whether the problem is multimodal or not. It should not only cover all optimal solutions but also locate it as fast as possible.

Niching Structure
Some earlier niching methods need a user-defined niche radius that is difficult to set properly. Some latter proposed methods replace it with the number of niches, which is easier to set but still has a great influence on the niching performance. Therefore, more elegant structures, which is able to depict complex shapes of niche, are needed to design.

Measuring Performance
The existing artificial benchmark functions and metrics concentrate too much on finding "peaks of the same height." But as we introduced at the beginning of this chapter, multiple close and robust sub-optima may be much better than multiple global optima that far away from each other. We should not regard MMOP as a sub-area of GOP or DOP.

References

1. De Jong, K.A.: An Analysis of the Behavior of a Class of Genetic Adaptive Systems. University of Michigan, Michigan (1975)
2. Deb, K., Saha, A.: Finding multiple solutions for multimodal optimization problems using a multi-objective evolutionary approach. In: Proceedings of the 12th Annual Conference on Genetic and Evolutionary Computation, pp. 447–454. ACM, New York (2010)
3. Epitropakis, M.G., Plagianakos, V.P., Vrahatis, M.N.: Finding multiple global optima exploiting differential evolution's niching capability. In: Proceedings of the 2011 IEEE Symposium on Differential Evolution, pp. 1–8. Springer, Paris (2011)
4. Goldberg, D.E., Richardson, J.: Genetic algorithms with sharing for multimodal function optimization. In: Proceedings of the Second International Conference on Genetic Algorithms on Genetic Algorithms and Their Application, pp. 41–49. Lawrence Erlbaum, Hillsdale (1987)
5. Harik, G.R.: Finding multimodal solutions using restricted tournament selection. In: Proceedings of the 6th International Conference on Genetic Algorithms, pp. 24–31. Morgan Kaufmann, San Francisco (1995)
6. Li, J.P., Balazs, M.E., Parks, G.T., Clarkson, P.J.: A species conserving genetic algorithm for multimodal function optimization. Evol. Comput. **10**(3), 207–234 (2002)
7. Li, X.: A multimodal particle swarm optimizer based on fitness Euclidean-distance ratio. In: Proceedings of the 9th Annual Conference on Genetic and Evolutionary Computation, pp. 78–85. ACM, New York (2007)
8. Li, X.: Niching without niching parameters: Particle swarm optimization using a ring topology. IEEE Trans. Evol. Comput. **14**(1), 150–169 (2010)
9. Mahfoud, S.W.: Niching methods for genetic algorithms. Ph.D. Thesis, University of Illinois Urbana-Champaign, Urbana (1995)
10. Pétrowski, A.: A clearing procedure as a niching method for genetic algorithms. In: Proceedings of 1996 International Conference on Evolutionary Computation, pp. 798–803. IEEE Press, Piscataway (1996)
11. Preuss, M.: Niching the cma-es via nearest-better clustering. In: Proceedings of the 12th Annual Conference Companion on Genetic and Evolutionary Computation, pp. 1711–1718 (2010)
12. Qu, B.Y., Suganthan, P.N., Das, S.: A distance-based locally informed particle swarm model for multimodal optimization. IEEE Trans. Evol. Comput. **17**(3), 387–402 (2013)
13. Qu, B.Y., Suganthan, P.N., Liang, J.J.: Differential evolution with neighborhood mutation for multimodal optimization. IEEE Trans. Evol. Comput. **16**(5), 601–614 (2012)
14. Yin, X., Germay, N.: A fast genetic algorithm with sharing scheme using cluster analysis methods in multimodal function optimization. In: Proceedings of the International Conference on Artificial Neural Nets and Genetic Algorithms, pp. 450–457. Springer, Innsbruck (1993)

Part IV
Advanced Topics and Applications

This part provides several advanced applications of evolutionary computation algorithms, including multi-objective optimization, constrained optimization, dynamic optimization, robust optimization, large-scale optimization, and expensive optimization. In this part, three real-world applications are also introduced, i.e., antenna design, vehicle routing, and contamination source identification in water distribution systems.

Chapter 9
Multi-objective Optimization

Abstract In recent decades, evolutionary multi-objective optimization has attracted a growing interest due to the fact that many real-world applications are multi-objective optimization problems (MOPs). This chapter introduces basic concepts regarding multi-objective optimization, then several popular multi-objective optimization evolutionary algorithms (MOEAs) are described, and performance measures and visualization of Pareto front are also introduced.

9.1 Introduction

In the real world, MOPs widely exist in various fields such as economic management, engineering practice, and scientific research. Here, take the decision to buy a vehicle as an example, as shown in Fig. 9.1. The horizontal axis is the cost, and the vertical axis is the comfort of the vehicle. Usually, the most comfort with the smallest cost is the most desired goal of decision-makers, but they are two contradictory objectives, i.e., the more comfort it has, the more cost it will take. Therefore, what the decision-maker can do is try to find the one that is highly cost-effective. This kind of problem belongs to the multi-objective optimization problem.

Based on the above example, MOPs have the following characteristics: ① objectives conflict with each other, and improving the performance of one objective will cause the performance of one or more other objectives to decline; ② there is no solution to make all objectives optimal; and ③ there is a set of solutions with trade-offs between the objectives. A formal description of MOPs can be seen in Eq. (1.1).

9.1.1 Basic Concepts

Several basic concepts of multi-objective optimization are discussed below.

Fig. 9.1 Decision: buying a
vehicle

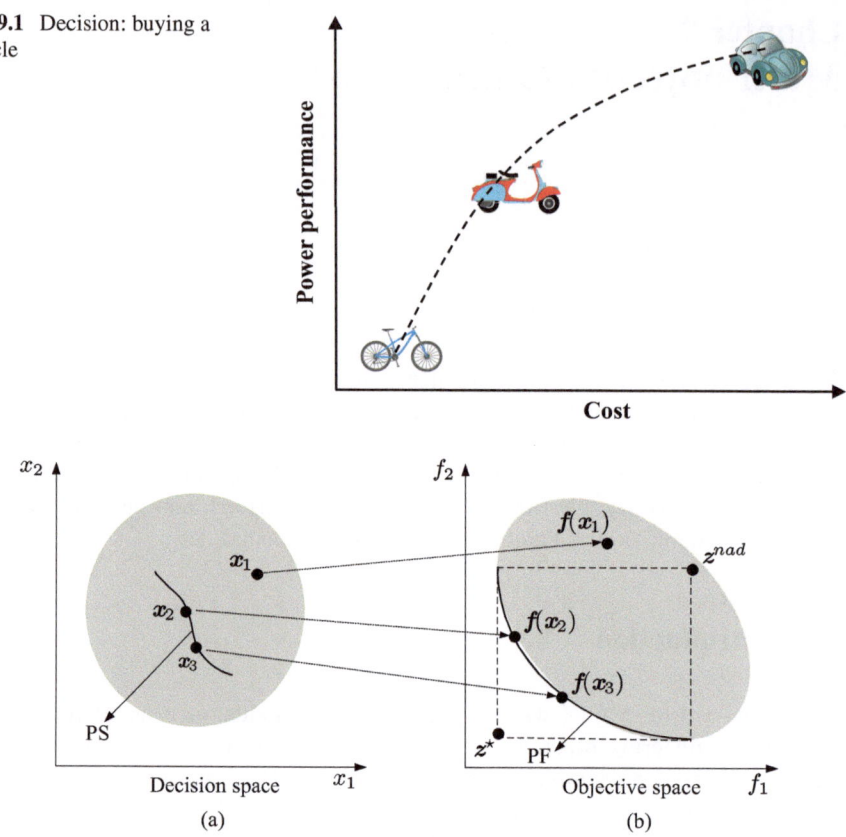

Fig. 9.2 The decision space and the objective space for a two-dimensional bi-objective minimization problem, where regions in gray are feasible regions

Definition 1 (Feasible solution) If x is distributed in the feasible region $\mathbb{X}_f$ in the decision space $\mathbb{X}$, i.e., $\forall x \in \mathbb{X}_f \subseteq \mathbb{X}$, then x is a feasible solution, e.g., x_1 in the region in gray (feasible region) in Fig. 9.2.

Definition 2 (Pareto dominance) For two feasible solutions x_1 and x_2, the necessary and sufficient condition that x_1 dominates x_2 is

$$\forall i = 1, 2, \cdots, M, f_i(x_1) \leq f_i(x_2) \wedge \exists j = 1, 2, \cdots, M, f_j(x_1) < f_j(x_2) \tag{9.1}$$

written as $x_1 \prec x_2$; if

$$\exists i = 1, 2, \cdots, M, f_i(x_1) < f_i(x_2) \wedge \exists j = 1, 2, \cdots, M, f_j(x_1) > f_j(x_2) \tag{9.2}$$

then x_1 and x_2 are non-dominated; x_1 weakly dominates x_2 if $\forall i = 1, 2, \cdots, M, f_i(x_1) \le f_i(x_2)$ (written as $x_1 \preceq x_2$). In Fig. 9.2, $x_2 \prec x_1, x_3 \prec x_1$, x_2, and x_3 are non-dominated.

Definition 3 (Pareto optimal solution) A solution x^* is called Pareto optimal solution (also termed as non-inferior, efficient, or admissible solution) if and only if it satisfies the following conditions:

$$\nexists x \in \mathbb{X}_f : x \prec x^* \tag{9.3}$$

where $\mathbb{X}_f$ is feasible solution sets.

Definition 4 (Pareto optimal set (POS or PS)) The set of all non-dominated solutions in the decision space, i.e., the set of x^*, such as the line in Fig. 9.2a.

Definition 5 (Pareto optimal front (POF or PF)) The mapping of PS in objective space is called Pareto optimal front, such as the curve in Fig. 9.2b.

Definition 6 (Ideal point z^* and nadir point z^{nad}) z^* is the point that has the best value on each objective with the set of PF, and z^{nad} is the point that makes all objectives worst with the set of PF.

9.1.2 Properties of PF

The characteristics of the PF of MOPs have an important effect on the performance of an MOEA, common characteristics include continuous or discontinuous or discrete, convex or concave, unimodal or multi-peak, or deceptive or non-deceptive, which are shown in Fig. 9.3. For academic research, these characteristics are simulated using benchmark problems. Commonly used test suites include the ZDT [25], DTLZ [10], GLT [13], MOEA_F [19], WFG [15], etc.

Among the test suites mentioned above, ZDT, GLT, and MOEA_F have the following characteristics: ① They are composed of two or three objective functions, and their Pareto front are known, and ② the dimension of decision variable is arbitrary and flexible to set. A notable characteristic of MOEA_F is that the shape of the PS is complicated. The DTLZ and WFG test suites are easily scalable to any number of objectives and to control the shape of the PF.

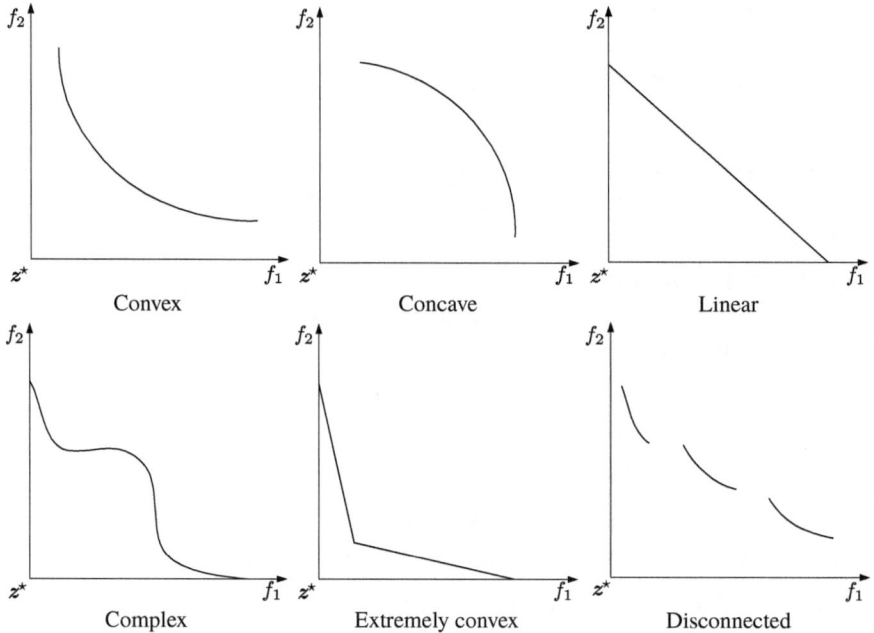

Fig. 9.3 Properties of PF

9.2   Multi-objective Evolutionary Algorithms

With the development of research on multi-objective evolutionary optimization, three kinds of classical algorithm frameworks of multi-objective evolutionary algorithm have been formed, which are domination-based, indicator-based, and decomposition-based methods.

9.2.1   *Domination-Based Algorithms*

Regarding dominance-based MOEA, it mainly uses the Pareto dominance relationship between individuals and the degree of crowding to select individuals to maintain diversity. There are many early research on the framework of this kind of algorithms, for example, strength Pareto EA (SPEA2) [27], Pareto archived ES (PAES) [18], Pareto envelope-based selection algorithm (PESA) [7], non-dominated sorting GA (NSGA-II) [9], etc. Among them, the most popular algorithm is the NSGA-II that was proposed by Deb et al. The working mechanism of NSGA-II is illustrated in Fig. 9.4.

The framework of NSGA-II is shown in Algorithm 9.1. An initial population $\mathbb{P}_0$ with the size of N is randomly initialized at the beginning. On this basis, a

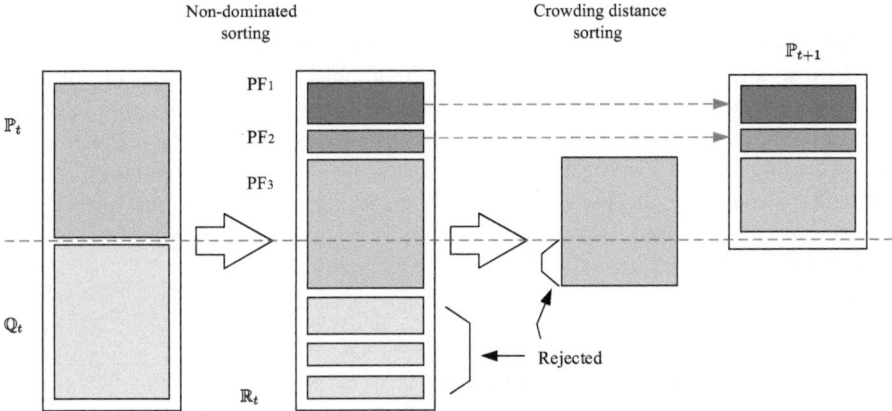

Fig. 9.4   Illustration of the working mechanism of NSGA-II

new population $\mathbb{Q}_0$ with the same size is generated by the crossover and mutation operation. As shown in Fig. 9.4, non-dominated sorting on $\mathbb{R}_{t+1} = \mathbb{P}_t + \mathbb{Q}_t$ is performed, and then individuals are selected from the front rank to the next rank in turn, until the rank of sets, which overflows the population size, is selected and the crowding distance of each individual in the set is calculated, where the crowding distance of an individual in the objective space is the size of the largest cuboid containing the individual without including any other individual of the population. Finally, the remaining individuals are selected according to the crowding distance.

Algorithm 9.1: The work flow of NSGA-II

Initialize population $\mathbb{P}_0 = \{x^1, \cdots, x^N\}$, $t = 0$;
while *termination criterion is not fulfilled* **do**

   $\mathbb{Q}_t = make\text{-}new\text{-}pop\ (\mathbb{P}_t)$;
   $\mathbb{R}_t = \mathbb{P}_t \cup \mathbb{Q}_t$;
   $\mathbb{F} = nondominated\text{-}sort\,(\mathbb{R}_t)$;
   // generate all sorting subsets $\mathbb{F} = \{\mathbb{F}_1, \mathbb{F}_2, \cdots\}$
   $\mathbb{P}_{t+1} = \varnothing$ and $i = 1$;
   while $|\mathbb{P}_{t+1}| < N$ **do**

      $\mathbb{P}_{t+1} = \mathbb{P}_{t+1} \cup \mathbb{F}_i$;
      $i = i + 1$;

   $Crowding\text{-}distance\text{-}assignment\ (\mathbb{F}_i)$;
   // compute the crowding distance of individuals in $\mathbb{F}_i$ set
   $Descending\text{-}Sort\ (\mathbb{F}_i)$;
   // sort $\mathbb{F}_i$ based on crowding distance in descending order
   $\mathbb{P}_{t+1} = \mathbb{P}_{t+1} \cup \mathbb{F}_i\,[1 : (N - |\mathbb{P}_{t+1}|)]$;
   $t = t + 1$;

Domination-based MOEAs, particularly NSGA-II, have been successfully applied in many applications. However, some shortcomings in this kind of MOEAs need to be addressed: ① When the number of objectives increases, the number of non-dominated individuals increases exponentially; accordingly, the selection pressure decreases exponentially, leading to slow evolution progress. ② Although the crowding distance indicator is used, the diversity of the population is still difficult to maintain. In view of these shortcomings, there are many improved algorithms, for example, ϵ-ranking evolutionary multi-objective optimization [11] and grid-based EA [23].

9.2.2   Indicator-Based Algorithms

Indicator-based MOEAs guide the search direction of algorithms through evaluation indicators. The indicator-based EA (IBEA) proposed by Zitzler et al. [26] introduces performance indicators into the selection strategies of multi-objective evolutionary algorithms, such as the hypervolume indicator (see Sect. 9.3.2): the higher the value of the hypervolume, the better the quality of the solution set. An advantage is that no additional diversity preservation mechanism needs to be considered.

The framework of IBEA is shown in Algorithm 9.2. The IBEA first defines a fitness value in terms of a binary performance measure/indicator and then directly uses the fitness value, which is defined as follows, in the selection process:

$$F(x) = \sum_{x' \in \mathbb{P}_{t+1} \setminus \{x\}} -e^{-I(\{x'\},\{x\})/k} \tag{9.4}$$

where $k > 0$ is a scaling factor and I is a binary quality indicator used to compare the quality of two solution sets relatively of each other. The binary additive ϵ-indicator $I_{\epsilon+}$ [29] is the minimum ϵ such that for any solution x_2 in $\mathbb{B}$, there is at least one solution x_1 in $\mathbb{A}$ that is not worse with a factor of ϵ in all objectives.

$$I_{\epsilon+}(\mathbb{A}, \mathbb{B}) = \min\{\epsilon \mid \forall x_2 \in \mathbb{B} \; \exists x_1 \in \mathbb{A} : f_i(x_1) - \epsilon \le f_i(x_2)$$
$$\forall i \in \{1, \cdots, M\}\} \tag{9.5}$$

The individual, with the smallest fitness value, is removed from the current population, then generates an offspring, and calculates the fitness value until termination conditions are met. Generally, the IBEA can be combined with arbitrary indicators.

This kind of MOEA has the advantage of simple structure and only needs to consider the performance indicators used by the algorithm, but the calculation of the general performance indicators is very time-consuming and will affect the performance of the algorithm. To solve this computational problem, the gradient information of I_H is used to guide population evolution [12]. A hypervolume esti-

---

Algorithm 9.2: IBEA

Initialize population $\mathbb{P}_0 = \{x^1, \ldots, x^N\}$, parameter k, and $t = 0$
while *termination criterion is not fulfilled* **do**
$\quad$ $\mathbb{Q}_t = $ *make-new-pop* $(\mathbb{P}_t)$;
$\quad$ $\mathbb{P}_{t+1} = \mathbb{P}_t \cup \mathbb{Q}_t$;
$\quad$ Assign fitness for each $x \in \mathbb{P}_{t+1}$ by Eq. (9.4);
$\quad$ **while** $|\mathbb{P}_{t+1}| > N$ **do**
$\quad\quad$ Choose an individual x^* with $F(x^*) \leq F(x)$ for all $x \in \mathbb{P}_{t+1}$;
$\quad\quad$ Remove x^* from $\mathbb{P}_{t+1}$;
$\quad\quad$ Update the fitness values of the remaining individuals by
$\quad\quad$ $F(x) = F(x) + e^{-I(\{x^*\},\{x\})/k}$;
$\quad$ $t = t + 1$;

---

mation method [3], which uses the estimation method to calculate the hypervolume index, was proposed. The algorithm can compromise the accuracy of the estimation and avoid high computing resources. Generally, the accuracy of the estimation of the indicator value and the computing resources needed are contradictory with each other, and there is no optimal situation for both.

9.2.3 Decomposition-Based Algorithms

A number of papers have applied the idea of decomposition to MOEA, but it has not gained much attention until a multi-objective optimization algorithm based on decomposition (MOEA/D) [24] was proposed in 2007, in which traditional mathematical programming methods were integrated. The main idea is to decompose a MOP into multiple single-objective subproblems and then optimize these subproblems at the same time to obtain a set of Pareto optimal solutions in a single run.

The framework of MOEA/D is shown in Algorithm 9.3. First, a set of weight vectors $(\lambda^1, \lambda^2, \cdots, \lambda^N)$ are generated according to the weight vector generation method in [8]. Next, decompose the problem into a set of single-objective subproblems by decomposition approaches, e.g., the weighted sum (WS) approach, the Tchebycheff (TCH) approach, and penalty-based boundary intersection (PBI) approach. Each subproblem is assigned an individual, and all individuals constitute the current population $\mathbb{P}$ and then repeatedly generate new individuals by choosing two parent individuals from the neighborhood of each weight vector and update subproblems by these new individuals until the termination condition is satisfied.

The decomposition method includes WS, TCH, and PBI, which are shown in Fig. 9.5. They all use a set of weight vectors $(\lambda^1, \lambda^2, \ldots, \lambda^N)$. For any weight vector $\lambda^i = (\lambda^i_1, \lambda^i_2, \ldots, \lambda^i_M)^T$, it meets $\sum_{j=1}^{M} \lambda^i_j = 1$, $i = 1, \ldots, N$. Figure 9.5 displays the optimization process of the three decomposition methods. The WS

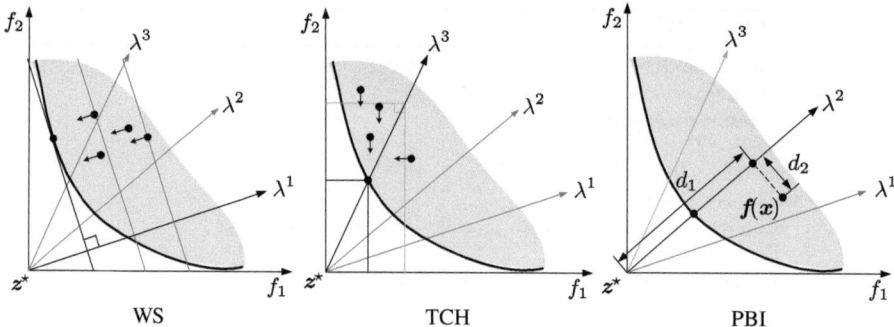

Fig. 9.5 Decomposition methods

Algorithm 9.3: MOEA/D

Initialize population $\mathbb{P}_0 = \left\{ x^1, \cdots, x^N \right\}$;
Set reference point $z = (\min_{i=1}^{N}(x_1^i), \cdots, \min_{i=1}^{N}(x_M^i))$;
Set $\mathbb{A} = \emptyset$;   // output
foreach *weight vector* λ^i **do**
    Get its T neighborhood $\lambda_1^i, \cdots, \lambda_T^i$ based on the Euclidean distance;
    Set $\mathbb{N}_i = \{i_1, \cdots, i_T\}$;

while *stop criteria are not satisfied* **do**
    foreach $i \in \{1, \cdots, N\}$ **do**
        Randomly select two individuals from $\mathbb{N}_i = \{i_1, \cdots, i_T\}$ and generate a new
        individual x through reproduction operators ;
        Apply a problem-specific repair/improvement heuristic on x to produce x' ;
        foreach $j = 1, \cdots, M$ **do** if $z_j > f_j(x')$, then set $z_j = f_j(x')$ **foreach** $j \in \mathbb{N}_i$
        do if $g^{te}(x'|\lambda^j) \leq g^{te}(x^j|\lambda^j)$, then set $x^j = x'$ Remove all vectors dominated by
        $f(x')$ from $\mathbb{A}$;
        Add $f(x')$ to $\mathbb{A}$ if no vectors in $\mathbb{A}$ dominate $f(x')$;

---

method mainly considers MOPs with convex PF (see Fig. 9.3), and the scalar optimization problem is

$$\min g^{ws}(x|\lambda) = \sum_{j=1}^{M} \lambda_j f_j(x) \tag{9.6}$$

The optimal solution of Eq. (9.6) is a Pareto optimal solution of the MOP. For the TCH method, the scalar optimization problem is

$$\min g^{te}(x|\lambda, z^*) = \max_{1 \leq j \leq M} \left\{ \lambda_j \left| f_j(x) - z_j^* \right| \right\} \tag{9.7}$$

For the PBI method, the scalar optimization problem is

$$\min g^{pbi}(x|\lambda, z^*) = d_1 + \theta d_2 \tag{9.8}$$

where $d_1 = \frac{(f(x)-z^*)^\mathrm{T}\lambda}{\|\lambda\|}$, $d_2 = \left\| f(x) - z^* - \frac{d_1}{\|\lambda\|}\lambda \right\|$, and θ is the penalty parameter.

9.3   Performance Evaluation

When solving MOPs with MOEAs, a set of Pareto optimal solutions can be obtained. There are three evaluation indexes for the quality of the set of solutions, namely, convergence, extensity, and uniformity, which are shown in Fig. 9.6. Generally, extensity and uniformity are collectively referred to as diversity.

Convergence measures the closeness of the set of solutions obtained by MOEAs to the true PF of the problem. Extensity measures the coverage of the set of solutions obtained by MOEAs over the space of the true PF. Uniformity means that the set of solutions obtained by MOEAs is evenly distributed.

There are many performance metrics regarding the multi-objective optimization. Here we list several metrics that are widely used for evaluating the performance of MOEAs.

9.3.1   C Indicator

C Indicator [28] measures the dominance relationships of two solution sets. For any two solution sets $\mathbb{A}$ and $\mathbb{B}$, $C(\mathbb{A}, \mathbb{B})$ is the proportion of solutions in $\mathbb{B}$ that is at least

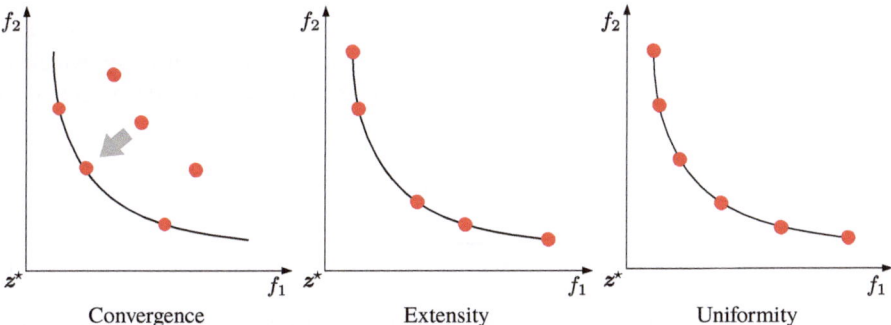

Fig. 9.6 Three evaluation indexes for the quality of a solution set

dominated by one of the solutions in $\mathbb{A}$, which is defined by

$$C(\mathbb{A}, \mathbb{B}) = \frac{|\{b \in \mathbb{B} \mid \exists\, a \in \mathbb{A} : a \preceq b\}|}{|\mathbb{B}|} \tag{9.9}$$

where the range of $C(\mathbb{A}, \mathbb{B})$ is $[0, 1]$; $C(\mathbb{A}, \mathbb{B}) = 0$ indicates that no solution in $\mathbb{B}$ is dominated by any solution of $\mathbb{A}$; and $C(\mathbb{A}, \mathbb{B}) = 1$ indicates that all solutions in $\mathbb{B}$ are dominated by solutions in $\mathbb{A}$. Note that when comparing the quality of $\mathbb{A}$ and $\mathbb{B}$, both $C(\mathbb{A}, \mathbb{B})$ and $C(\mathbb{B}, \mathbb{A})$ need to be considered due to $C(\mathbb{A}, \mathbb{B}) \neq C(\mathbb{B}, \mathbb{A})$ generally.

9.3.2   Generational Distance

Generational distance (GD) [22] measures the average distance from the set $\mathbb{P}$ to a set of reference points of the PF. Let $\mathbb{P} = \{p_1, p_2, \cdots, p_N\}$ in the objective space and $\mathbb{Z}$ be a given population and a reference point set of the PF, respectively. The GD to the PF is defined by

$$\mathrm{GD}(\mathbb{P}) = \frac{1}{N} \sqrt{\sum_{i=1}^{N} d(p_i, \mathbb{Z})^2} \tag{9.10}$$

where $d(p_i, \mathbb{Z})$ is the Euclidean distance between p_i and its nearest point in $\mathbb{Z}$. From the equation, the metric of GD can only reflect the convergence of a set of solutions to the PF.

9.3.3   Maximum Spread

Maximum spread (MS) [25] is used to measure extensity of set usually, which is defined as the sum of the maximum range of a solution set on each objective as follows:

$$\mathrm{MS}(\mathbb{P}) = \sqrt{\sum_{j=1}^{M} \max_{p,q \in \mathbb{P}} (p_j - q_j)^2} \tag{9.11}$$

where p_j and q_j are two points that have the largest distance on the j-th objective. The higher the MS value is, the better extensity the point set has. When $M = 2$, the MS of a solution set is the Euclidean distance of its two outer solutions. According to the definition of MS, it only considers the extreme solutions of the set and cannot reflect uniformity. Moreover, since it does not involve the convergence of

set, solutions far from the PF usually contribute a lot to MS, which can easily lead
to incorrect assessments.

9.3.4 Spacing

Spacing [21] is defined as the variation of the distance between solutions in
a solution set, used to measure uniformity usually. Given a solution set $\mathbb{P} =
\{p_1, p_2, \cdots, p_N\}$ in the objective space, the spacing value of $\mathbb{P}$ is calculated by

$$S(\mathbb{P}) = \sqrt{\frac{1}{N-1} \sum_{i=1}^{N} (\overline{d} - d(p_i, \mathbb{P}/p_i))^2} \tag{9.12}$$

where $\overline{d}$ is the mean of all $d(p_i, \mathbb{P}/p_i)$ and $d(p_i, \mathbb{P}/p_i)$ denotes the Manhattan distance of p_i to the set $\mathbb{P}/p_i$, which is calculated by $d(p_i, \mathbb{P}/p_i) =
\min\limits_{p \in \mathbb{P}/p_i} \sum_{j=1}^{M} \left| p_i^j - p^j \right|$, where p_i^j is the jth objective of p_i.

The lower the spacing value, the better the uniformity of $\mathbb{P}$, e.g., $S(\mathbb{P}) = 0$ means
all solutions of the set $\mathbb{P}$ are distributed equidistantly on the basis of Manhattan
distance. However, the spacing indicator only measures the uniformity of a set of
solutions.

9.3.5 Inverted Generational Distance

Inverted generational distance (IGD) [6], which is the inverted version of DG,
measures the average distance from a set of reference points $\mathbb{Z}$ of the PF to the
approximation set $\mathbb{P}$. The calculation is given below, which is shown in Fig. 9.7.

$$\text{IGD}(\mathbb{P}, \mathbb{Z}) = \frac{1}{|\mathbb{Z}|} \sum_{z \in \mathbb{Z}} d(z, \mathbb{P}) \tag{9.13}$$

where $d(z, \mathbb{P})$ is the Euclidean distance between the reference point z and its nearest
individual in $\mathbb{P}$.

IGD could measure both the diversity and convergence of $\mathbb{P}$ since the reference
point set is normally benchmark dataset provided with the benchmark problem, i.e.,
the smaller the IGD value, the better performance of the population $\mathbb{P}$.

Fig. 9.7 Calculation of IGD

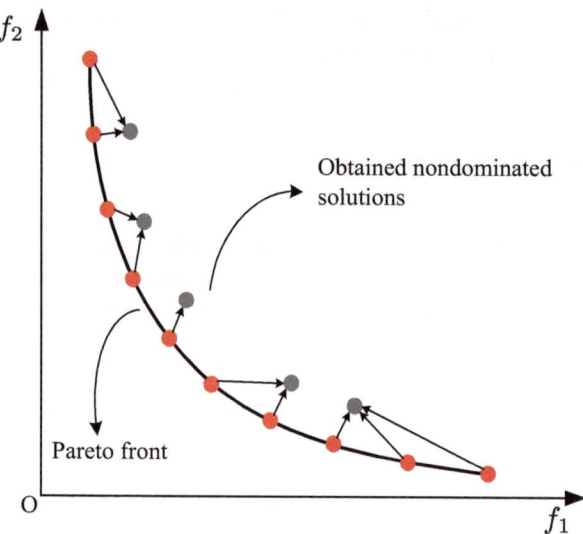

9.3.6 *Hypervolume*

Hypervolume (HV) [28] calculates the volume covered by the set of solutions obtained in the objective space. Let $z^r = \left(z_1^r, \cdots, z_M^r\right)^{\mathrm{T}}$ be a reference point of the objective space, and it is dominated by all Pareto-optimal solutions. Suppose $\mathbb{P}$ is the obtained solution set by an MOEA. Then, the Hypervolume indicator I_H of $\mathbb{P}$ with regard to z^r is the volume of the region dominated by $\mathbb{P}$ and bounded by z^r. It can be defined as below, which is shown in Fig. 9.8.

$$I_H(\mathbb{P}) = \ell\left(\bigcup_{p \in \mathbb{P}} [p, z^r]\right) \tag{9.14}$$

where $\ell(\cdot)$ is the Lebesgue measure.

Alternatively, we can interpret Eq. (9.14) as

$$I_H(\mathbb{P}) = \ell\left(\bigcup_{p \in \mathbb{P}} \{x \mid p \preceq x \preceq z^r\}\right) \tag{9.15}$$

Obviously, hypervolume indicator strictly obeys the dominant principle. If individual x_1 dominates individual x_2, hypervolume indicator value of x_1 must be greater than x_2. The higher the hypervolume, the better the approximation.

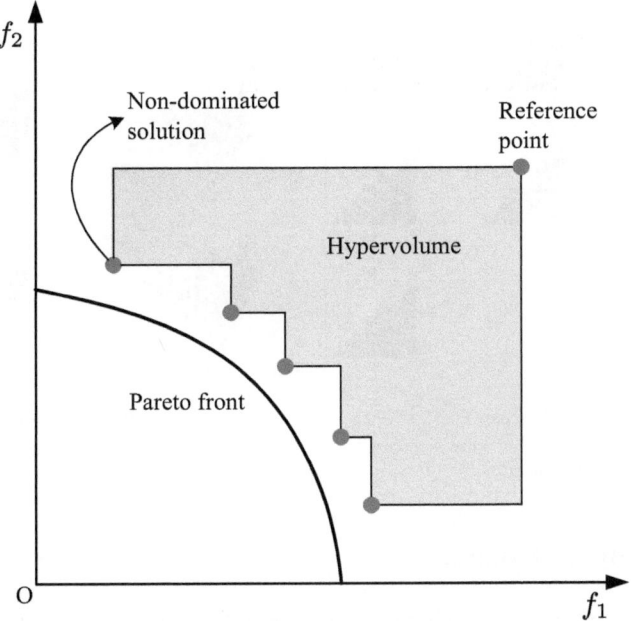

Fig. 9.8 Calculation of the hypervolume

9.4   Visualization in the Objective Space

In multi-objective optimization, to better observe the evolutionary process of the population in the objective space at each generation, it is hoped that the data in the population can be visualized in the form of images or tables. Obviously, in a multi-objective problem with two or three objectives, individuals in the population can be drawn directly in 2-D or 3-D coordinates, and then their distribution can be observed. However, for many-objective optimization problems (MaOPs) with more than three objectives (i.e., $M > 3$), it is difficult to express all objective values of individuals in a coordinate system. Therefore, the development of visualization tools for high-dimensional data has gradually attracted the attention of researchers. This section describes several existing high-dimensional data visualization methods, which are categorized into three classes.

9.4.1   Visualization Using Original Values

The solution set can be visualized directly by the following three means.

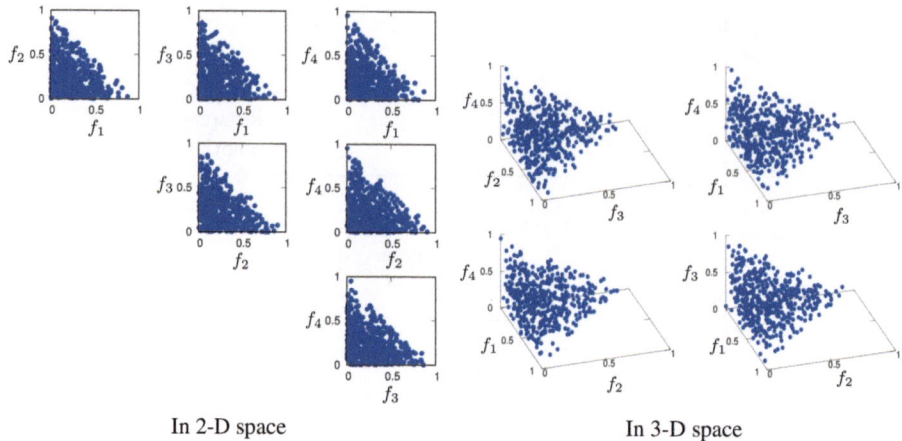

In 2-D space   In 3-D space

Fig. 9.9 Scatterplot matrix

9.4.1.1 Scatterplot Matrix

A scatterplot matrix is a matrix of all possible combinations of two or three objectives, including a scatterplot in 2-D and 3-D spaces. As shown in the right graphs in Fig. 9.9, a three-dimensional visualization is used for a solution set with four objectives. This visualization method is fast, simple, and robust. However, it cannot visualize some characteristics of the solution set in high-dimension spaces, e.g., the Pareto dominance relation, the shape, and knee region of the PF. In addition, for solution sets with M objectives, there are totally $\frac{m(m-1)}{2}$ combinations in a 2-D space. This will be a huge challenge for decision-makers to have an overview of the relationships among visualized solutions.

9.4.1.2 Parallel Coordinate

The difference between perpendicular coordinates and parallel coordinates [17] is the use of parallel axes to represent the dimension of the dataset. In other words, a vertical line expresses the projection of each dimension, and the maximum and minimum values of each dimension are mapped to the upper and lower boundaries on these vertical lines.

As shown in Fig. 9.10, a polyline made up of $M-1$ line connecting M dimension values represents an M-dimensional point. Because of easy implementation and an overview is provided of the Pareto domination relationships for a solution set, it is one of the most frequently used visualization tools for analyzing the solution distribution for MaOPs. However, different orders of the index of objectives will create quite different clustering results for the same solution set, which makes it hard to interpret different characteristics of the solution set, especially in the case of a large size of the solution set.

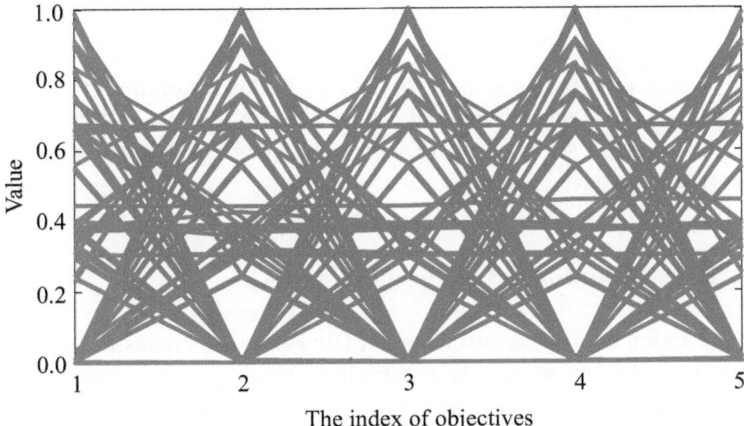

Fig. 9.10 Parallel coordinate

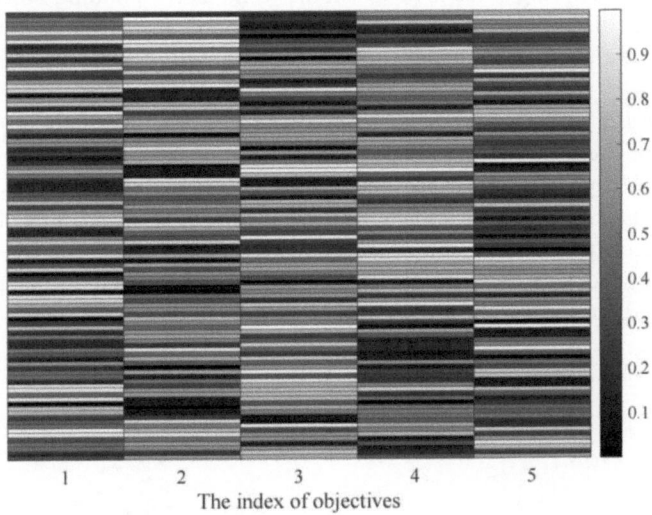

Fig. 9.11 Heat maps

9.4.1.3   Heat Maps

Instead of mapping solutions into parallel axes, heat maps [20] use an array of cells with different gray colors to represent multi-dimensional data, where the color of each cell is set based on the data value. As shown in Fig. 9.11, a brighter color usually represents a higher objective value and vice versa. Similar to parallel coordinates, heat maps also suffer from the same issue when analyzing the relationships for a solution set.

9.4.2 Visualization Using Transformed Values

It is hard to have an overview analysis of the distribution of solutions by means of the direct visualization of the original data, especially on MaOPs with a large dataset. To alleviate this issue, several visualization techniques based on transformed values of the original data have been proposed.

9.4.2.1 Radial Coordinate Visualization

The idea of radial coordinate visualization (RadVis) [14] comes from physics. For a M objective MOP, M objectives are distributed evenly on the circumference of a unit circle, as shown in Fig. 9.12 for a four-objective MOP. Each objective vector is fixed with M springs, and these springs are connected to the anchors. Therefore, the position of the objective vector is the point where the spring forces are at equilibrium. The spring force is proportional to the corresponding objective value. The closer the distance to f_i, the higher the value of f_i. For example, if the solution has the same value on each objective, then the point will be exactly in the center of the circle.

Although RadVis is able to provide an overview of the distribution of a solution set, it cannot preserve the Pareto dominance relationship, the convergence to the PF. To address this issue, a 3D version of RadVis, namely, three-dimensional radial coordinate visualization (3D-RadVis) [16], was proposed, where a third dimension was introduced to visualize the shape and convergence of a solution set. It was reported that the relative position of solutions, shape of the PF, distribution of solutions, and convergence to the PF can be preserved.

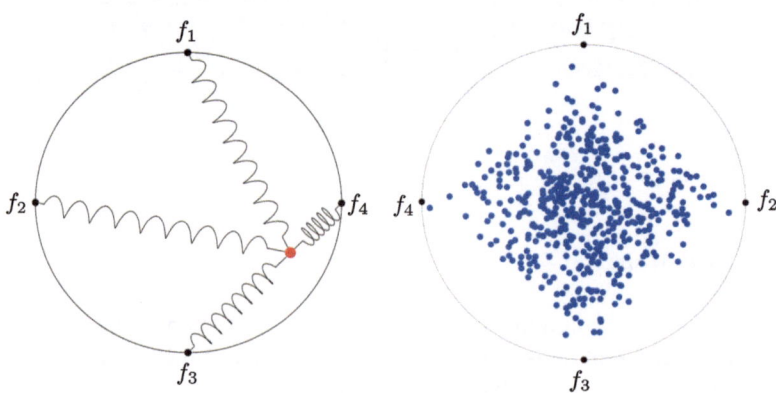

Fig. 9.12 Radial coordinate visualization

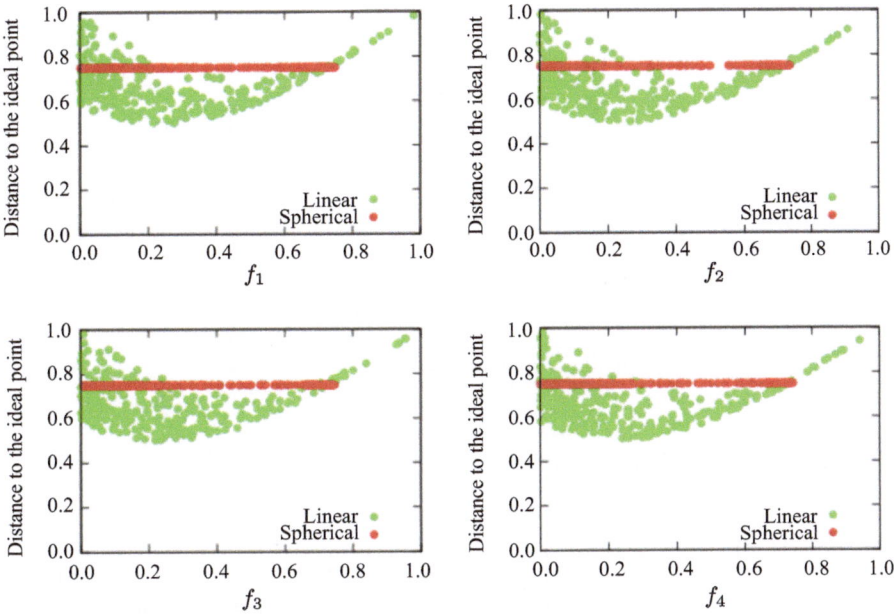

Fig. 9.13 Level diagrams

9.4.2.2 Level Diagrams

If there are M objectives, the level diagrams [4] include M graphs, the abscissa of each graph represents the value on the objective, and the ordinate represents the nearest distance to the ideal point, which can be measured by a specific norm of normalized objectives, as shown in Fig. 9.13. It was reported that some characteristics of the PF can be analyzed, such as discontinuities, closeness to the ideal point, and ranges of attainable values.

9.4.2.3 Hyper-radial Visualization

Hyper-radial visualization [5] is similar to the level diagram method. Objective vectors keep distance from ideal objective vectors (their hyper-radius) but keep distance from two subsets of objectives, respectively. Visualizing the two bases can keep the shape of the approximation set well, and the distribution of quantities can be correctly expressed as a linear basis rather than a spherical basis, as shown in Fig. 9.14.

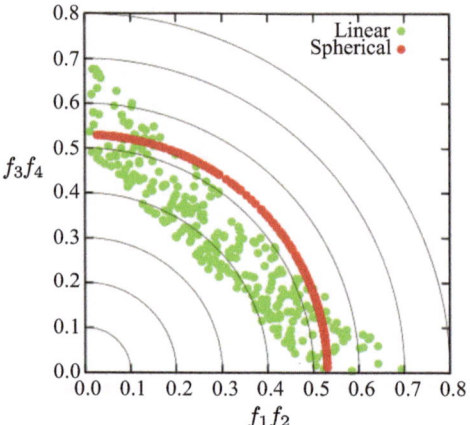

Fig. 9.14 Hyper-radial visualization

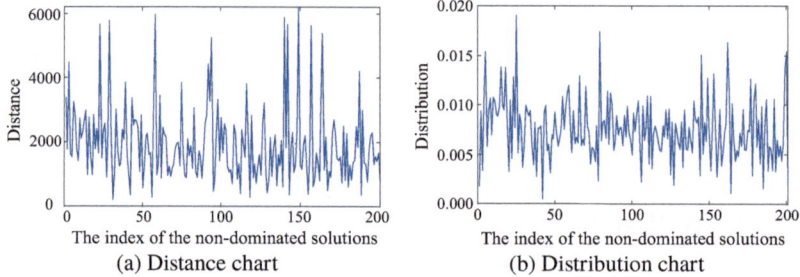

(a) Distance chart   (b) Distribution chart

Fig. 9.15 Distance and distribution charts

9.4.3   *Visualizing the Distribution Relation of Solution*

Transforming high-dimensional data into a lower-dimensional space to show some particular characteristics is another possible way to analyze the solution set, even though some information will be lost after the dimension reduction.

9.4.3.1   Distance and Distribution Charts

Distance and distribution charts [2] plot the solution according to their distance to the Pareto front (i.e., the distance to the nearest solution of the PF) and distance to other solutions (one kind of measurements for diversity), i.e., it includes two figures, the distance chart and the distribution chart, respectively, which are shown in Fig. 9.15. This approach is able to reflect the coverage of the non-dominated solutions in the objective space regardless of the number of objectives. The distance

Fig. 9.16 Hyper-space
Diagonal Counting

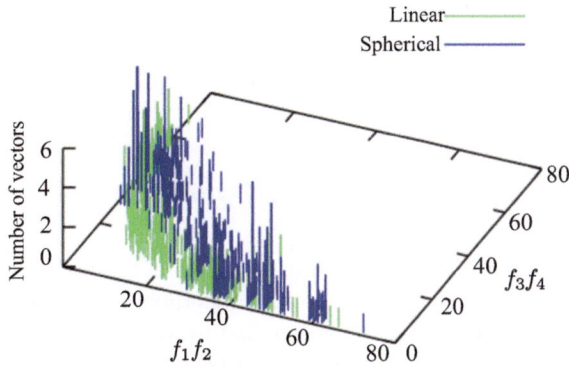

chart is straightforward and easy to analyze, but the distribution chart needs to be carefully analyzed due to the disconnection of the PF.

9.4.3.2 Hyper-space Diagonal Counting

Hyper-space diagonal counting [1] approach works as follows: First, each objective is divided into a predefined number of bins. The bins of a pair of objectives are then counted using hyper-space diagonal counting, producing indices for this pair of objectives. These indices are plotted on two axes (one for each pair of objectives), while the third axis is used to plot the number of vectors of the approximation set that fall in the same set of indices. Take a four-dimensional space as an example, which is shown in Fig. 9.16.

The plot better captures the distribution of vectors than the shape of the approximation sets. In addition, this method does not maintain the dominance relations between objective vectors.

9.5 Challenges

Over the past several decades, many MOEAs, test problems, performance metrics, and visualization techniques have been proposed, but there are still many challenges to be addressed in the future.

9.5.1 Many-objective Test Problems

Benchmark problems are very important for the development of algorithms, and several test suites have been developed for academic research. However, it is necessary to develop a wide variety of test problems, especially the problems that are able to reflect features of real-world applications. In addition, it is also needed to analyze the gap between existing test problems and real-world applications.

9.5.2   Many-Objective Optimization Algorithms

In the real world, characteristics of MaOPs are more complicated. Current MOEAs have good performance in solving problems with two and three objectives. However, when these MOEAs are applied to MaOPs, they will encounter many difficulties. For example, ① in dominance-based MOEAs, the reduction of selection pressure slows down the evolution progress due to the loss of selection pressure and significantly affects the performance of the algorithms; ② in indicator-based MOEAs, increasing the number of objectives will increase the amount of computation; ③ in decomposition-based MOEAs, it is difficult to determine the criteria for generating uniform weight vectors for MaOPs. Some scholars have proposed adaptive generation methods of weight vectors, but there are still many issues to be solved. Although various improved algorithms have been proposed to solve these problems, there are still many limitations, so it is very necessary to develop new many-objective optimization algorithms.

9.5.3   Performance Evaluation Methods

The evaluation of a solution set for a many-objective problem is an important research issue. These well-known and frequently used performance indicators, such as IGD and I_H described above, are not always applicable to many-objective optimization problems. When IGD is calculated, the uniformity of selected reference points affects the calculation results, and optimal distributions of solutions are not always intuitive. Similarly, the optimal distribution for I_H depends on the shapes of the Pareto front. So it is very necessary to develop new evaluation performance.

9.5.4   Visualization of Many-Objective Optimization

Although many visualization techniques have been developed, such as the direct and indirect visualization techniques introduced above, there still a lack of methods that are capable of visualizing all characteristics of the solution set through the evolutionary process, e.g., the shape of the PF, the Pareto dominance relationships between solutions, the distribution of solutions, and the convergence trend to the PF. Therefore, more powerful visualization techniques are needed.

References

1. Agrawal, G., Bloebaum, C., Lewis, K.: Intuitive design selection using visualized n-dimensional Pareto frontier. In: 46th AIAA/ASME/ASCE/AHS/ASC Structures, Structural Dynamics and Materials Conference, pp. 1813–1826. American Institute of Aeronautics and Astronautics, Reston (2005)
2. Ang, K.H., Chong, G., Li, Y.: Visualization technique for analyzing non-dominated set comparison. In: Proceedings of the 4th Asia-Pacific Conference on Simulated Evolution and Learning, pp. 36–40. Nanyang Technical University, Orchid Country Club, Singapore (2002)
3. Bader, J., Zitzler, E.: Hype: an algorithm for fast hypervolume-based many-objective optimization. Evol. Comput. **19**(1), 45–76 (2011)
4. Blasco, X., Herrero, J.M., Sanchis, J., Martínez, M.: A new graphical visualization of n-dimensional Pareto front for decision-making in multiobjective optimization. Inf. Sci. **178**(20), 3908–3924 (2008)
5. Chiu, P.W., Bloebaum, C.: Hyper-radial visualization (HRV) for decision-making in multi-objective optimization. In: 46th AIAA Aerospace Sciences Meeting and Exhibit, p. 908. American Institute of Aeronautics and Astronautics, Reston (2008)
6. Coello, C.A.C., Sierra, M.R.: A study of the parallelization of a coevolutionary multi-objective evolutionary algorithm. In: Mexican International Conference on Artificial Intelligence, pp. 688–697. Springer, Heidelberg (2004)
7. Corne, D.W., Knowles, J.D., Oates, M.J.: The Pareto envelope-based selection algorithm for multiobjective optimization. In: International Conference on Parallel Problem Solving from Nature, pp. 839–848. Springer, Berlin (2000)
8. Das, I., Dennis, J.E.: Normal-boundary intersection: a new method for generating the Pareto surface in nonlinear multicriteria optimization problems. SIAM J. Optim. **8**(3), 631–657 (1998)
9. Deb, K., Pratap, A., Agarwal, S., Meyarivan, T.: A fast and elitist multiobjective genetic algorithm: NSGA-II. IEEE Trans. Evol. Comput. **6**(2), 182–197 (2002)
10. Deb, K., Thiele, L., Laumanns, M., Zitzler, E.: Scalable test problems for evolutionary multiobjective optimization. In: Abraham, A., Jain, L., Goldberg, R. (eds.) Evolutionary Multiobjective Optimization, pp. 105–145. Springer, London (2005)
11. Deb, K., Mohan, M., Mishra, S.: Towards a quick computation of well-spread Pareto-optimal solutions. In: Evolutionary Multi-criterion Optimization, pp. 222–236. Springer, Berlin (2003)
12. Emmerich, M., Beume, N., Naujoks, B.: An EMO algorithm using the hypervolume measure as selection criterion. In: International Conference on Evolutionary Multi-criterion Optimization, pp. 62–76. Springer, Berlin (2005)
13. Gu, F., Liu, H.L., Tan, K.C.: A multiobjective evolutionary algorithm using dynamic weight design method. Int. J. Innov. Comput. Inf. Control **8**(5(B)), 3677–3688 (2012)
14. Hoffman, P., Grinstein, G., Marx, K., Grosse, I., Stanley, E.: DNA visual and analytic data mining. In: IEEE Conference on Visualization, pp. 437–441. IEEE Computer Society Press, Los Alamitos (1997)
15. Huband, S., Barone, L., While, L., Hingston, P.: A scalable multi-objective test problem toolkit. In: International Conference on Evolutionary Multi-criterion Optimization, pp. 280–295. Springer, Berlin (2005)
16. Ibrahim, A., Rahnamayan, S., Martin, M.V., Deb, K.: 3D-RadVis: visualization of Pareto front in many-objective optimization. In: IEEE Congress on Evolutionary Computation, pp. 736–745. IEEE, Vancouver (2016)
17. Inselberg, A.: Parallel Coordinates: Visual Multidimensional Geometry and Its Applications. Springer, New York (2009)
18. Knowles, J.D., Corne, D.W.: Approximating the nondominated front using the Pareto archived evolution strategy. Evol. Comput. **8**(2), 149–172 (2000)
19. Li, H., Zhang, Q.: Multiobjective optimization problems with complicated Pareto sets, MOEA/D and NSGA-II. IEEE Trans. Evol. Comput. **13**(2), 284–302 (2009)

20. Pryke, A., Mostaghim, S., Nazemi, A.: Heatmap visualization of population based multi objective algorithms. In: International Conference on Evolutionary Multi-criterion Optimization, pp. 361–375. Springer, Berlin (2007)
21. Scott, J.R.: Fault tolerant design using single and multicriteria genetic algorithm optimization. Master's thesis, Massachusetts Institute of Technology, Cambridge (1995)
22. Van Veldhuizen, D.A., Lamont, G.B.: Evolutionary computation and convergence to a Pareto front. In: Late Breaking Papers at the Genetic Programming 1998 Conference, pp. 221–228. Stanford University Bookstore, California (1998)
23. Yang, S., Li, M., Liu, X., Zheng, J.: A grid based evolutionary algorithm for many objective optimization. IEEE Trans. Evol. Comput. **17**(5), 721–736 (2013)
24. Zhang, Q., Li, H.: MOEA/D: a multiobjective evolutionary algorithm based on decomposition. IEEE Trans. Evol. Comput. **11**(6), 712–731 (2007)
25. Zitzler, E., Deb, K., Thiele, L.: Comparison of multiobjective evolutionary algorithms: empirical results. Evol. Comput. **8**(2), 173–195 (2000)
26. Zitzler, E., Künzli, S.: Indicator-based selection in multiobjective search. In: International Conference on Parallel Problem Solving from Nature, pp. 832–842. Springer, Berlin (2004)
27. Zitzler, E., Laumanns, M., Thiele, L.: SPEA2: improving the strength Pareto evolutionary algorithm. Technical report, Eidgenössische Technische Hochschule Zürich, Zurich (2001)
28. Zitzler, E., Thiele, L.: Multiobjective optimization using evolutionary algorithms – a comparative case study. In: International Conference on Parallel Problem Solving from Nature, pp. 292–301. Springer, Berlin (1998)
29. Zitzler, E., Thiele, L., Laumanns, M., Fonseca, C.M., Da Fonseca, V.G.: Performance assessment of multiobjective optimizers: an analysis and review. IEEE Trans. Evol. Comput. **7**(2), 117–132 (2003)

Chapter 10
Constrained Optimization

Abstract In this chapter, we will introduce the concept of constrained optimization problems (COPs), commonly used constraint-handling techniques based on EAs, and future research on constrained optimization.

10.1 Introduction

Most real-world optimization problems in the fields of science and engineering are subject to constraints. These problems can be classified as COPs. The presence of constraints divides the search space into many discrete feasible and infeasible regions, which makes it hard to solve. A COP can be defined as the minimization of an objective function subject to a set of constraints. It can be described in Eq. (1.1), where $M = 1$, and the vectors $g(x) \leq 0$ and $h(x) = 0$ represent inequality and equality constraints, respectively. To handle an equality constraint $h(x) = 0$, it is typically transformed into an inequality constraint, i.e., $|h(x)| - \delta \leq 0$, where δ is a sufficiently small constant.

Unlike unconstrained optimization, which aims at seeking the global optimum of a problem, the difficulty of solving COPs normally comes from constraints. Therefore, constrained optimization normally focuses on finding a feasible solution, that is, how to deal with constraints. In fact, many real-world optimization problems may not even have an objective function, such as the antenna design problem introduced in Sect. 15.1, which has only performance and geometric constraints.

To find a feasible solution, the concept of the degree of constraint violation of a solution is introduced, which reflects the distance of a solution to the feasible space. It is defined as follows:

$$\phi(x) = \sum_{j=1}^{N} \max(0, g_j(x)) + \sum_{k=1}^{P} \max(0, h_k(x) - \delta) \tag{10.1}$$

where δ is a small constant close to zero. The feasible space is the set of all solutions that satisfy all constraints.

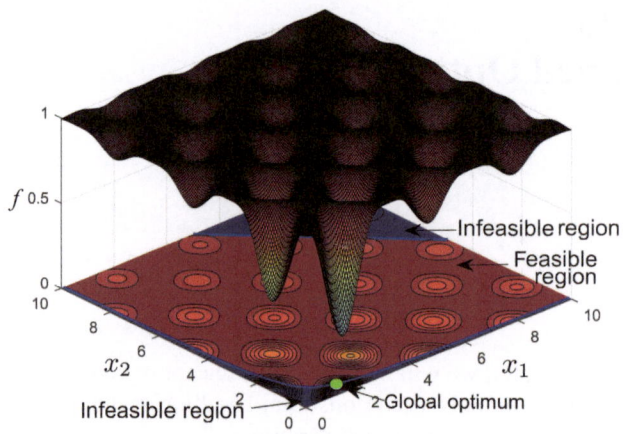

Fig. 10.1 The fitness landscape of a COP

Now, we consider the following COP

$$
\min \quad f(x) = 1 - \left| \frac{\cos^4 x_1 + \cos^4 x_2 - 2\cos^2 x_1 \cos^2 x_2}{\sqrt{x_1^2 + x_2^2}} \right|
$$

$$
\text{s. t.} \quad g_1(x) = 0.75 - x_1 x_2 \leq 0 \tag{10.2}
$$

$$
g_2(x) = x_1 + x_2 - 15 \leq 0
$$

where $0 \leq x_1, x_2 \leq 10$

Figure 10.1 plots the fitness landscape of Eq. (10.2), where the blue areas are two separate infeasible regions, points located in infeasible regions are called infeasible solutions; otherwise, they are feasible solutions. The green point in Fig. 10.1 is the global optimum which is situated precisely on the boundary between the feasible and infeasible regions. It is obvious that the fitness landscape exhibits multimodality, making it more challenging for an algorithm to locate the global optimum.

Traditional mathematical programming methods for solving COPs are usually gradient-based search approaches, including sequential quadratic programming, Lagrangian method, projection gradient method, interior point method, exterior point method, and so on. However, traditional methods have certain limitations when handling nonlinear COPs:

(1) They need a feasible and promising initial solution.
(2) They typically rely on the gradient information of the optimization problem.
(3) They can indeed encounter significant difficulties when faced with problems with one of the characteristics of non-differentiability, disconnection, and lack of explicit mathematical expression.
(4) They often get trapped in local optima.

In the last two decades, EAs have attracted growing interest and have been widely used to solve COPs, leading to the development of various constraint-handling techniques for EAs, namely, constrained optimization evolutionary algorithms (COEAs). Compared to the traditional mathematical methods, EAs have the following advantages:

(1) They are much more easier to implement.
(2) Population-based EAs, in particular, are well-suited for multimodal optimization scenarios.
(3) They are robust to various characteristics of the problem.
(4) They do not need the explicit mathematical expression of the problem.

10.2   Constraint-Handling Techniques

Constraint-handling EAs can be categorized into four types based on the type of constraint-handling techniques they utilize: penalty function methods, methods that optimize objectives and constraints separately, methods based on multi-objective optimization techniques, and methods based on an ensemble of constraint-handling techniques.

Figure 10.2 illustrates the objective and constraint violation space of a COP. For any solution x, the objective and constraint violation space can be divided into four regions: A, B, C, and D. It is evident that the solutions in region A have better objective values and smaller constraint violations compared to solution x; thus, region A is the most desirable region for a COP. On the contrary, the solutions in region D have worse objective values and larger constraint violations than solution x, and thus region D does not contain any useful information. Region

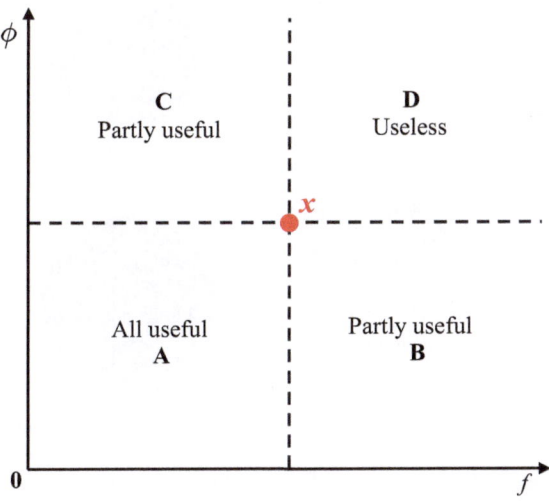

Fig. 10.2 The objective and constraint violation space of a COP

B and Region C are somewhat useful; the former contains solutions with smaller constraint violations than x but generally poorer objective values, while the latter contains solutions with better objective values than x but typically larger constraint violations. The primary distinction among different constraint-handling techniques lies in how they use the useful information found in regions B and C.

10.2.1   Penalty Function

As shown in Fig. 10.3, the penalty function method can be roughly divided into three categories: static penalty function, dynamic penalty function, and adaptive penalty function.

10.2.1.1   Static Penalty Function

Penalty function converts a COP to an unconstrained one by adding product of the degree of constraint violation and a nonnegative predefined penalty factor λ to the objective [1], which is the simplest and the earliest constraint-handling technique

$$F(x) = f(x) + \lambda \phi(x) \qquad (10.3)$$

However, setting a suitable penalty factor λ in advance is a complicated task, since it is usually problem-dependent. If the penalty is set too small, the algorithm might be unable to reach feasible regions. Conversely, if the penalty is set too

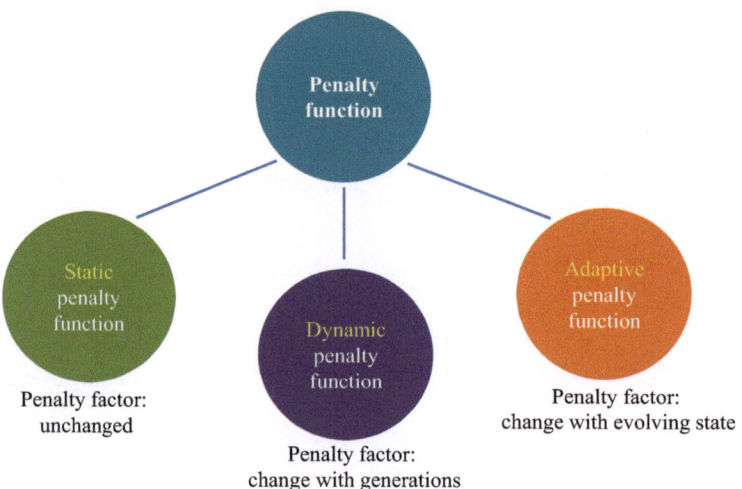

Fig. 10.3   The category of penalty function method

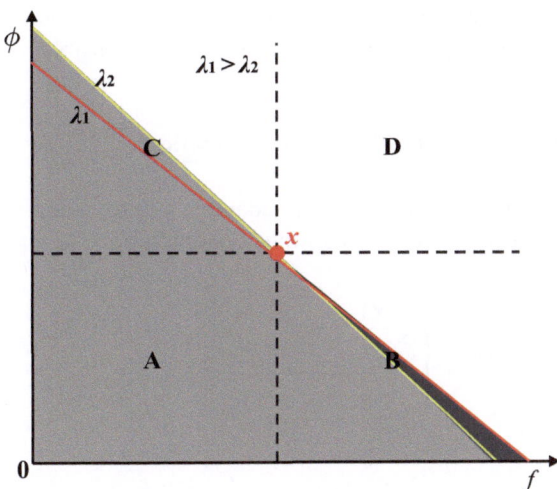

Fig. 10.4 The search bias of penalty function

large, the algorithm might converge to feasible regions too quickly to locate optimal solutions situated on the boundary between feasible and infeasible regions.

The search bias of the penalty function is demonstrated in Fig. 10.4. As shown in Fig. 10.4, solutions below the diagonal are considered superior to solution x. The lower diagonal area comprises region A, part of region B, and part of region C. It is worth noting that the larger the penalty factor, the larger the region C becomes, which contains more constraint information, while region B shrinks and contains less objective information, and vice versa. Clearly, one of the primary challenges with the penalty function approach is determining an appropriate value for λ throughout the search process.

10.2.1.2 Dynamic Penalty Function

Initially, the penalty function method usually employed a constant as a penalty factor. However, it has been recognized that different optimization stages of the same optimization problem may require different penalty factors. Consequently, several dynamic adjustment methods for the penalty factor have been proposed. A nonstationary penalty function [9] which defines the fitness function as

$$F(x) = f(x) + \lambda(t)\phi(x) \tag{10.4}$$

where $\lambda(t)$ represents the dynamic penalty factor which changes according to the number of generations.

During the early stage of evolution, a smaller $\lambda(t)$ will make the population diverge in the search space due to the small value of constraint, which enhances the population diversity. In other words, the algorithm pays more attention to the objective than constraints. Conversely, in the later stage of evolution, a larger $\lambda(t)$ pushes the population to converge to the feasible region.

10.2.1.3   Adaptive Penalty Function

The penalty factor in the adaptive penalty function is adjusted according to the feedback information of the population. In [14], the fitness function is defined as

$$
F(x) = \begin{cases} \hat{f}(x) & \text{for feasible solution} \\ \hat{\phi}(x) & \text{no feasible solution} \\ \sqrt{\hat{f}(x)^2 + \hat{\phi}(x)^2}[(1 - r_f)\hat{\phi}(x) + r_f \hat{f}(x)] & \text{otherwise} \end{cases}
$$

(10.5)

where $\hat{f}(x)$ and $\hat{\phi}(x)$ denote the normalized objective value and constraint violation value, respectively; r_f represents the feasible ratio which equals to the number of feasible solutions of the current population divided by the population size.

From Eq. (10.5), we can see that if there is no feasible solution in the current population, the objective function value of the population will be totally ignored, and all solutions will be given a fitness value solely based on their constraint violation to find more feasible solutions. If the solutions of the current population are all feasible, then solutions are compared based on their normalized objective value to improve the objective value within the feasible region. If there are both feasible and infeasible solutions, the fitness function will be adjusted based on the feasible ratio of the population to select solutions with both low objective values and low constraint violations.

10.2.2   Separation of Objectives and Constraints

The combination of objective and constraint by adding the penalty factor simplified the problem when comparing solutions. However, adjusting the value of the penalty factor can be challenging. Therefore, researchers try to solve COPs by optimizing the objectives and constraints separately using the following approaches.

10.2.2.1   Feasibility Rule

Feasibility rule [3] is another efficient constraint-handling technique that does not require users to define parameters in advance, and it is easy to implement. In feasibility rule, one solution x_a is considered superior to x_b, if they meet any of the following three conditions:

Fig. 10.5 The search bias of feasibility rule

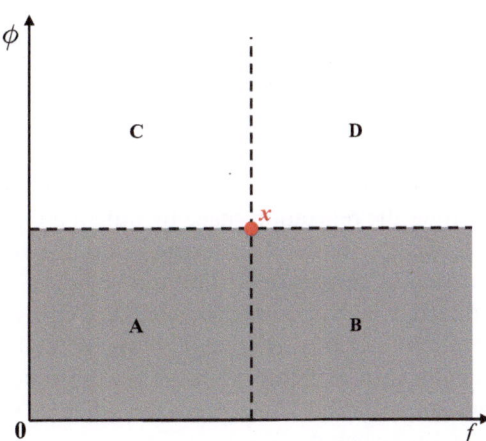

(1) x_a is feasible and x_b is infeasible.
(2) Both x_a and x_b are feasible, but the objective value of x_a is better than that of x_b.
(3) Both x_a and x_b are infeasible, but the degree of constraint violation of x_a is smaller than that of x_b.

Feasibility rule is typically employed as the selection strategy when comparing a parent solution with a child one [5, 6]. In this way, the comparison between any two solutions can be easily implemented.

Nevertheless, it is well-known that the main weakness of the feasibility rule lies in its preference for feasible solutions. Some infeasible solutions that carry important information may be ignored during the selection process, which may cause an EA trapped in local optima, particularly when the global optimum is situated on the boundary of the feasible region or when the feasible regions are disjointed from each other. Figure 10.5 shows the search biases of the feasibility rule approach. EA prefers solutions in regions A and B compared to x while ignoring solutions in region C.

To alleviate the greediness of the feasibility rule for constraints, a replacement mechanism [16] which incorporates the promising objective function information into the feasibility rule and a cluster-replacement-based operator [18] was integrated into the feasibility rule to alleviate the greediness.

10.2.2.2 ε-Constraint Methods

Due to the fact that some infeasible solutions also contain important objective information, ε-constraint method [13] offers a partial remedy to the feasibility rule's limitations by introducing a relaxation factor, which is shown in Eq. (10.6).

$$(f(x_a), \phi(x_a)) <_\varepsilon (f(x_b), \phi(x_b)) \tag{10.6}$$

$$\equiv \begin{cases} f(\boldsymbol{x}_a) < f(\boldsymbol{x}_b) & \text{if } \phi(\boldsymbol{x}_a), \phi(\boldsymbol{x}_b) \leq \varepsilon \\ f(\boldsymbol{x}_a) < f(\boldsymbol{x}_b) & \text{else if } \phi(\boldsymbol{x}_a) = \phi(\boldsymbol{x}_b) \\ \phi(\boldsymbol{x}_a) < \phi(\boldsymbol{x}_b) & \text{otherwise} \end{cases} \qquad (10.7)$$

ε-constraint method first relaxes the constraint boundaries and then gradually narrows the relaxation space by utilizing the ε parameter. By dynamically controlling the epsilon level ε, certain infeasible solutions can be conditionally accepted as feasible solutions, so that the objective information of these solutions can be utilized, which is beneficial to tackle COPs with small feasible regions. The epsilon level ε is a critical parameter. In the special case where $\varepsilon = 0$, ε-constraint method is equivalent to feasibility rule. As illustrated in Fig. 10.6, the candidate solutions in regions A and C are preferred when $\phi(\boldsymbol{x}) < \varepsilon$, whereas regions A and B are preferred when $\phi(\boldsymbol{x}) \geq \varepsilon$.

10.2.2.3   Stochastic Ranking

Both the feasibility rule and the ε-constraint method prioritize the degree of constraint violation over the objectives. In contrast, stochastic ranking [11] aims to strike a balance between the degree of constraint violation and the objectives. In the stochastic ranking, a probability parameter ρ is introduced to determine whether the comparison between two solutions is determined by the objective value or the degree of constraint violation. The comparison will be based on the objective value if a random value is less than ρ; otherwise, the comparison is based on the degree of constraint violation. When $\rho = 0$, stochastic ranking is also equivalent to the feasibility rule. As shown in Fig. 10.7, solutions in regions A and B are preferred

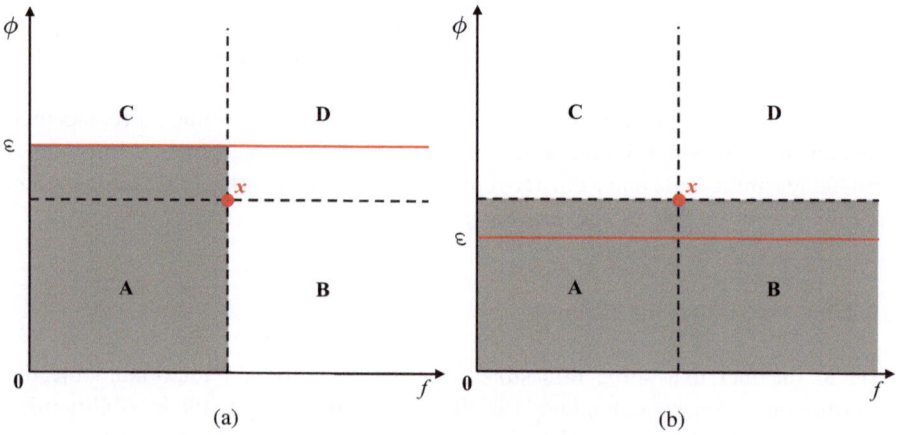

Fig. 10.6 The search bias of the ε-constraint method. (**a**) $\phi(\boldsymbol{x}) < \varepsilon$. (**b**) $\phi(\boldsymbol{x}) \geq \varepsilon$

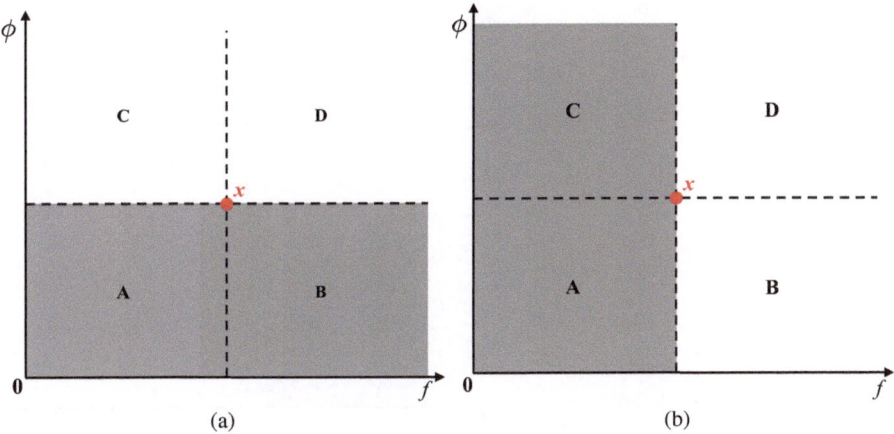

Fig. 10.7 The search bias of stochastic ranking. (**a**) $rand \geq \rho$. (**b**) $rand < \rho$

with a probability $1 - \rho$, while solutions in regions A and C are preferred with a probability of ρ.

Evidently, determining the appropriate values for the probability parameter ρ in stochastic ranking and the epsilon level ε in the ε-constraint method significantly influences the performance of a constraint-handling EA.

10.2.2.4 Multi-objective Methods

In recent years, significant efforts have been devoted to the field of MOEAs (MOEAs) (see Chap. 9), offering a fresh perspective for addressing COPs. The key to solving COPs is how to balance the objective function and constraints. In multi-objective-based approaches, the degree of constraint violation $\phi(x)$ is treated as an additional objective

$$F(x) = (f(x), \phi(x)) \tag{10.8}$$

or each constraint violation is treated as a single objective

$$F(x) = (f(x), G_1(x), \cdots, G_k(x)) \tag{10.9}$$

where $G_i(x)$ is the degree of constraint violation of x with respect to the i-th constraint, i.e., $G_i(x) = \max(0, g_i(x))$ for $i = 1, \cdots, N$ and $G_{N+j}(x) = \max(0, h_j(x) - \delta)$ for $j = 1, \cdots, P$.

By transforming the constraints into additional objectives, population diversity can be maintained while optimizing objective function and constraints simultaneously. A survey on the utilization of MOEAs for single-objective COPs can be found

Fig. 10.8 The search bias of multi-objective-based constraint-handling technique

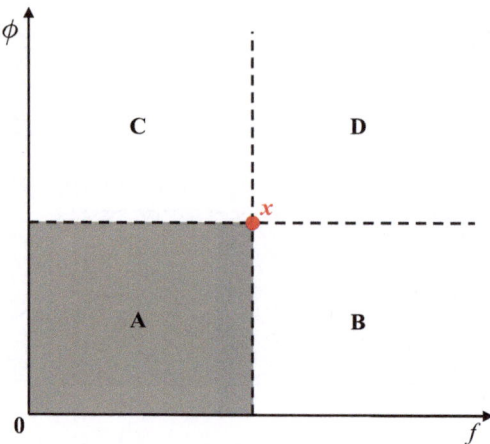

in [12]. As shown in Fig. 10.8, only region A is preferred for multi-objective-based constraint-handling techniques.

A constrained multi-objective DE (CMODE) [15], which combines the multi-objective optimization with DE (see Sect. 5.6), was proposed to deal with COPs. In CMODE, a novel infeasible solution replacement mechanism based on multi-objective optimization was proposed, with the purpose of guiding the population toward promising solutions and feasible regions simultaneously. The approach adopts the Pareto dominance to compare solutions.

A self-adaptive selection method based on the multi-objective technique [7] has been introduced, which intends to exploit both informative infeasible and feasible solutions. A bi-objective evolutionary method with the penalty function [2] has been adopted to tackle COPs. In this approach, the penalty parameters are dynamically tuned. The coevolutionary dual-population DE (DPDE) [4] divided the population into two subpopulations based on the feasibility of the solutions. In DPDE, the COP is treated as a bi-objective optimization problem. Meanwhile, the dual subpopulations communicate with each other through an information-sharing strategy.

10.2.3   Ensemble Methods

The penalty function methods are easy to implement, but they are hard to adjust the penalty coefficient λ; the feasibility rule methods are free of the penalty factor and easy to implement but tend to fall into a local optimum; the ε-constraint method has a stronger global search capability by introducing the information of objective function, but the setting of ε is critical; the stochastic ranking methods also introduce the information of objective function, but the setting of ρ is critical; the multi-objective methods are able to maintain the population diversity, but the decrease of selection pressure is difficult to handle.

To make use of the advantages of different constraint-handling approaches, the ensemble of constraint-handling techniques (ECHT) [10] has been proposed to deal with COPs. Four constraint-handling approaches were integrated into the ECHT framework: feasibility rule, self-adaptive penalty function, ε-constraint method, and stochastic ranking.

In ECHT, each constraint-handling method operates with its own population, and every function call made by each population, associated with a specific constraint-handling technique, is efficiently utilized. This approach aims to make use of the advantages of different constraint-handling techniques in different optimization. However, it is important to note that ensemble methods typically lead to increased design complexity and computational burden.

10.3 Challenges and Future Directions

Although a number of constraint-handling EAs have been proposed in recent years, there are still several unresolved open issues that require attention.

How to Use Infeasible Solutions Efficiently

Since the optima of most COPs are located on the boundary of the feasible region, it is advantageous to incorporate both feasible and infeasible solutions [8]. Moreover, the presence of infeasible solutions helps algorithms move between disjoint feasible regions. Consequently, exploitation of solutions at the edge of feasible regions becomes much easier, and the optimal solution may be found along a more direct and efficient path by traversing the infeasible part of the search space [17]. However, keeping infeasible solutions can expand the search space and potentially lead to a waste of resources as an EA explores unpromising solutions. Thus, how to make use of infeasible solutions efficiently is also an important research topic.

Expensive Constrained Optimization

Many engineering optimization problems require time-consuming computer simulations or physical experiments to evaluate candidate solutions. Traditional EAs are not suitable for such scenarios, since the large number of function evaluations is unaffordable. A common approach is surrogate models, where part of expensive fitness evaluations is replaced by computationally cheap approximate models, often referred to as surrogates or meta-models (see Chap. 14). How to find the optimum within small feasible regions under the limited computation budget is also a challenging topic.

Constrained Multi-/Many-objective Optimization

Multi-/ many-objective optimization problems are hot topics in the field of EC, resulting in many MOEAs. Nevertheless, most of them are designed for unconstrained optimization problems. For constrained multi-objective optimization problems (CMOPs), the difficulty is how to balance multiple conflicting objectives while satisfying constraints. Apparently, solving this kind of problem is difficult, since the constraint-handling technique developed for single-objective optimization cannot be easily integrated into a MOEA.

Dynamic Constrained Optimization

Dynamic COPs in which the objective function and/or constraints change over time are encountered in many real-world problems (see Sect. 11.5). However, most existing studies on constraint handling only focus on stationary COPs. The challenges for dynamic constrained optimization might involve the following three aspects:

(1) Constraint dynamics may cause changes in the geometric structure of feasible regions.
(2) Objective function dynamics may lead to the global optimum to switch from one feasible area to another.
(3) In problems with fixed objective functions and dynamic constraints, changing infeasible regions might expose a new global optimum.

References

1. Coello, C.A.C.: Theoretical and numerical constraint-handling techniques used with evolutionary algorithms: a survey of the state of the art. Comput. Methods Appl. Mech. Eng. **191**(11), 1245–1287 (2002)
2. Datta, R., Deb, K.: Individual penalty based constraint handling using a hybrid bi-objective and penalty function approach. In: IEEE Congress on Evolutionary Computation, pp. 2720–2727 (2013)
3. Deb, K.: An efficient constraint handling method for genetic algorithms. Comput. Methods Appl. Mech. Eng. **186**(2), 311–338 (2000)
4. Gao, W.F., Yen, G.G., Liu, S.Y.: A dual-population differential evolution with coevolution for constrained optimization. IEEE Trans. Cybern. **45**(5), 1108–1121 (2014)
5. Hamza, N.M., Essam, D.L., Sarker, R.A.: Differential evolution with a constraint consensus mutation for solving optimization problems. In: 2014 IEEE Congress on Evolutionary Computation (CEC), pp. 991–997. IEEE (2014)
6. Hamza, N.M., Essam, D.L., Sarker, R.A.: Constraint consensus mutation-based differential evolution for constrained optimization. IEEE Trans. Evol. Comput. **20**(3), 447–459 (2015)
7. Jiao, L., Li, L., Shang, R., Liu, F., Stolkin, R.: A novel selection evolutionary strategy for constrained optimization. Inf. Sci. **239**, 122–141 (2013)
8. Jiao, R., Zeng, S., Li, C.: A feasible-ratio control technique for constrained optimization. Inf. Sci. **502**, 201–217 (2019)
9. Joines, J., Houck, C.: On the use of non-stationary penalty functions to solve non-linear constrained optimization problems with gas. In: IEEE International Conference on Evolutionary Computation, pp. 579–584 (1994)
10. Mallipeddi, R., Suganthan, P.N.: Ensemble of constraint handling techniques. IEEE Trans. Evol. Comput. **14**(4), 561–579 (2010)
11. Runarsson, T.P., Yao, X.: Stochastic ranking for constrained evolutionary optimization. IEEE Trans. Evol. Comput. **4**(3), 284–294 (2000)

12. Segura, C., Coello, C.A.C., Miranda, G., León, C.: Using multi-objective evolutionary algorithms for single-objective constrained and unconstrained optimization. Ann. Oper. Res. **240**(1), 217–250 (2016)
13. Takahama, T., Sakai, S.: Constrained optimization by the ε constrained differential evolution with gradient-based mutation and feasible elites. In: IEEE Conference on Evolutionary Computation, pp. 1–8 (2006)
14. Tessema, B., Yen, G.G.: An adaptive penalty formulation for constrained evolutionary optimization. IEEE Trans. Syst. Man Cybern. **39**(3), 565–578 (2009)
15. Wang, Y., Cai, Z.: Combining multiobjective optimization with differential evolution to solve constrained optimization problems. IEEE Trans. Evol. Comput. **16**(1), 117–134 (2012)
16. Wang, Y., Wang, B.C., Li, H.X., Yen, G.G.: Incorporating objective function information into the feasibility rule for constrained evolutionary optimization. IEEE Trans. Cybern. **46**(12), 2938–2952 (2015)
17. While, L., Hingston, P.: Usefulness of infeasible solutions in evolutionary search: an empirical and mathematical study. In: IEEE Congress on Evolutionary Computation, pp. 1363–1370 (2013)
18. Xu, B., Chen, X., Tao, L.: Differential evolution with adaptive trial vector generation strategy and cluster-replacement-based feasibility rule for constrained optimization. Inf. Sci. **435**, 240–262 (2018)

Chapter 11
Dynamic Optimization

Abstract Dynamic optimization problems (DOPs) have attracted more and more interest in recent years. On the one hand, there are numerous real-world optimization problems that exhibit dynamic characteristics. On the other hand, compared with traditional static optimization problems, DOPs are inherently challenging and difficult to solve. This chapter mainly introduces some dynamic handling methods, formulations of dynamic optimization problems, and some measure indicators for performance comparison. Finally, challenges for future research are discussed for future work.

11.1 Introduction

An optimization problem, in which the fitness landscape changes over time, is referred to as a dynamic optimization problem. Dynamic characteristics are widespread in real-world optimization problems. For instance, when someone plans to drive from city A to city B, the optimal route's travel time can be influenced by various dynamic factors, including transportation conditions, weather, and speed constraints. These dynamic factors may change over time for different reasons. Dynamic factors will change the structure of the optimization problems, making the problems more challenging to solve.

We take the minimization as an example. A DOP can be described as below

$$\min \boldsymbol{f}(\boldsymbol{x}, t) = \{f_i(\boldsymbol{x}_i^f(t), \boldsymbol{\phi}_i^f(t)), i = 1, 2, \cdots, m(t)\}$$

$$\text{s. t.} \boldsymbol{g}(\boldsymbol{x}, t) = \{g_i(\boldsymbol{x}_j^g(t), \boldsymbol{\phi}_j^g(t)), j = 1, 2, \cdots, n(t)\} \le \boldsymbol{0} \qquad (11.1)$$

$$\boldsymbol{h}(\boldsymbol{x}, t) = \{h_i(\boldsymbol{x}_k^h(t), \boldsymbol{\phi}_k^h(t)), k = 1, 2, \cdots, p(t)\} = \boldsymbol{0}$$

where $m(t)$ is the number of objectives to be optimized at time t; $\boldsymbol{g}$ are inequality constraints functions; $\boldsymbol{h}$ are equality constraints functions; and $\boldsymbol{\phi}_i^f$, $\boldsymbol{\phi}_j^g$, and $\boldsymbol{\phi}_k^h$ are control parameters of the objectives, inequality constraints, and equality constraints, respectively. A change can be caused by any variable or control parameters.

© China University of Geosciences Press 2024
C. Li et al., *Intelligent Optimization*,
https://doi.org/10.1007/978-981-97-3286-9_11

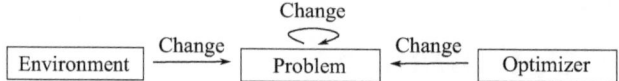

Fig. 11.1 Source of changes

11.1.1 Basics of Changes

The problem considered in this chapter refers to a system that runs in a certain environment. The running of the system is controlled by an optimizer to maximize the performance of the system. In this subsection, we will introduce the fundamental principles of change.

What to Change

Change may come from the internal structure of the problem, the environment, or the optimizer, as shown in Fig. 11.1. Note that the concept of environmental change is not the term of problem change.

Environmental changes refer to changes in the environment where the system operates, e.g., changes in pressure, light, humidity, temperature, and so on. Changes in these factors can cause changes in objectives and/or constraints. Consequently, the feasibility and optimality of solutions are also affected, thus impacting the system's performance.

The decisions made by the optimizer also influence the problem. In some real-world problems with time-linkage properties, such as vehicle scheduling and routing, decisions aimed at maximizing present performance may have implications for future performance. Another example is in a game, where players always choose the best strategy, while it is the worst one for their opponents.

Most of the changes come from the internal structure of the problem. A change normally will take effect on the fitness landscape of DOPs, including physical changes in the solution and/or objective space, e.g., the shape of the PS or PF, the basins of attraction of optima in the fitness landscape, and the feasible/infeasible region.

In particular, we list the following two cases.

(1) The number of constraints

- If only boundary constraints for decision variables exist, the problem is called dynamic non-constrained optimization problem.
- If there are inequality and/or equality constraints, the problem is a dynamic COPs, constraint-handling strategies should be taken into consideration when designing algorithms.

(2) The number of objectives
- If only one objective needs to be optimized, the problem is called a dynamic single-objective optimization problem.
- If more than one objective needs to be optimized, the problem is called a dynamic multi-/many-objective optimization problem. Therefore, strategies for solving objective conflicts should be considered.

When to Change
The frequency of changes refers to the frequency with which changes occur. Generally, it is controlled by a time pattern, which is typically derived from real-world applications, such as continuous, discontinuous, triggered by cases, or even a random time pattern.

How to Change
Depending on the source of changes, the way to change the problems can be divided into three categories: environmental-related changes, decision-related changes, and state-related changes (internal changes).

11.1.2   Characteristics of Changes

Components of dynamic problems will change over time. Here, we list several features that are widespread in real-world applications.

Change Frequency
Change frequency is a basic attribute when a problem changes. It denotes how often a problem changes and how long a static environment lasts. A high change frequency means that problems change rapidly, and algorithms have limited time to search for optima, and vice versa. Change frequency is an important parameter to evaluate the convergence speed of an algorithm. If an algorithm can achieve rapid convergence in a high-frequency change environment, it indicates that the algorithm can quickly adapt to a new environment.

Change Severity
Change severity is also a basic feature of DOPs. It denotes the degree of difference of the problem before and after a change. For example, in artificial benchmarks, the distance that the optima move after a change can serve as a measure of the severity of the change of the optima. It is obvious that severity is strongly associated with algorithm performance. If the change severity is small, new optima are close to old optima, making it easy for algorithms to discover new optima and vice versa.

Change Mode
Change mode describes the regularity of changes, which is typically controlled by a pattern of change, e.g., whether changes are periodical, random, or chaotic [20]. Some real-world scenarios exhibit these change modes; for instance, traffic flow on a street may follow a weekly pattern. The change mode reflects the regularity

of a system, which algorithms can learn. But for problems without obvious change patterns, their changes are unpredictable, posing a significant challenge to algorithmic search efficiency.

Detectability
When solving a DOP, an algorithm can respond to changes quickly if it can detect the changes. Therefore, change detection is a widely used method for DOPs. These algorithms assume that changes are easy to detect, that is, algorithms can detect changes with small computational resources. However, in reality, detectability is also a characteristic of change, and sometimes changes are not easily detectable. For example, changes may occur only in a portion of the solution space, or there may exist noise that hinders the accuracy of change detection. Therefore, algorithms may encounter significant obstacles when detection strategies fail to detect changes.

Dynamic Number of Objectives and/or Variables
The number of objectives and/or variables may also change over time. This depends on the current production objectives, decision-making preferences, or even government requirements. When the number of objectives or variables decreases, the problem becomes easier, because the objective space or solution space, respectively, shrinks. However, when the search space expands exponentially with an increase in variables, the search efficiency of an algorithm sharply declines, because it becomes challenging for the algorithm to find the global optimum with limited computational resources. Moreover, when the number of objectives increases, it introduces significant challenges for an algorithm to find the Pareto optimal front (PF) due to the lack of selection pressure. Additionally, this may lead to the degeneration of the PF in multi-objective optimization problems.

Modality Change
Modality (see Sect. 2.2.1) is a common feature for both static problems and DOPs. The unimodal problem is an easy problem, but it represents only a small scenario in real-world applications. Most optimization problems are multimodal, which require algorithms to simultaneously track as many optima as possible, while dynamic multi-objective optimization problems may result in disconnected PS, requiring algorithms to search different promising regions simultaneously.

11.2   Dynamism Handling Methods

Compared to methods for static optimization problems, DOP methods need not only to locate optima but also to prepare for changes, so that optima can be quickly tracked after each change. It requires algorithms to take more considerations into account. This section provides a simple introduction to these considerations.

11.2.1 Difficulties

Before designing an algorithm for DOPs, it is necessary to analyze the key challenges for the design of an efficient algorithm.

Diversity Loss

The balance between exploration and exploitation is always one of the most important issues for EAs (see Chap. 7). Exploration is particularly important for EAs solving DOPs, because it keeps the population having good exploration capability for unexplored search space and hence enables EAs to find new moving optima.

Therefore, we need special strategies to maintain population diversity. In multi-objective scenarios, population diversity in both objective space and decision space needs to be maintained, because the mapping from PS to PF is typically not linear. Population diversity in the objective space offers the optimizer a variety of non-dominated solutions for multi-objective DOPs, while diversity in the decision space acts as a primary driving force for evolution for single-objective DOPs. Therefore, algorithms designed for different types of DOPs should incorporate different diversity maintenance strategies.

Hard-to-Detect Changes

The detection of changes is an important mechanism for most algorithms for DOPs. Reevaluating existing solutions and checking for differences in objective values is a common way to detect changes. This method is effective for detecting changes at the global level, where the objective values of all solutions in the solution space change simultaneously. However, it may not be effective if the selected solutions for reevaluation are located in a static local search space [21]. Prediction methods based on population change cannot also guarantee successful detection. Inappropriate detection methods can cause failures when changes occur and fail to respond to changes or even trigger false alarms.

Unknown Change Time

A worse situation is that the moment of change is presumed to be unknown; in other words, a change could potentially occur at any point during the runtime. Therefore, we need to detect changes from time to time. As a result, most computational resources are wasted to detect changes, and limited resources are available for the search process. How to balance computational resources between detection and search is a challenging issue.

Outdated Memory

One potential issue for most EAs in dynamic environments is that the objective values of solutions will be outdated due to changes in the problem. This means that the objective values may be incorrect and should be updated. Otherwise, decisions made with incorrect information may mislead the search.

The Loss of Tracking

When changes occur, the current optima covered by existing solutions may shift to somewhere else, and new optima may also appear in the area where no solutions are distributed. Only when the population can find the basins of shifted optima, the population will have a chance to track the shifted optima in a short time; otherwise, it may miss the shifted optima.

11.2.2 Dynamism Handling Strategies

As mentioned above, DOP algorithms need to improve convergence speed and maintain population diversity at the same time. Some dynamism handling strategies have been designed to achieve this purpose.

Diversity Enhancement

This approach increases diversity when change is detected. Generally, detection is performed by reevaluating the best solution or specific "sentinel" solutions. Whenever a change is detected, the population will adopt a certain policy to increase diversity. Some typical diversity enhancement strategies will be introduced here.

The simplest approach is to re-initialize the population from scratch after each change so that diversity can be increased at the beginning of each new problem [1].

Additionally, increasing the probability of mutation also has a good effect on handling changes. A hypermutation technique was used as an adaptive operator in GA [6], where the GA employs a higher mutation probability when the time-averaged best-of-generation performance decreases after a change; otherwise, the GA employs a low mutation rate.

Introducing a number of new random individuals into the population is also an effective method. Deb et al. introduced some random or mutated solutions into the population to track the changing PF based on the NSGA-II [7].

Another way is to perform long-distance mutations when a change occurs. Tinos and Yang used a self-adaptive value q to regulate the distribution of q-Gaussian mutations. When a change occurs, the value of q will increase, allowing individuals to make long jumps to escape local optima. Subsequently, after the change, the value decreases to enhance exploitation for improved optima [29].

Diversity enhancement is indeed an effective approach to managing changes. Nevertheless, it relies on change detection techniques, which, as mentioned above, may fail in various situations. Moreover, these methods increase diversity without taking the problem's inherent structure into account, potentially leading to a wasteful allocation of computational resources.

Diversity Maintenance

Diversity maintenance is another strategy based on increasing the diversity of the population. Differently from diversity enhancement, diversity maintenance does not rely on changes detection by maintaining diversity throughout the optimization process. The population has many strategies to maintain its diversity, which will be described later.

Grefenstette introduced random individuals at the beginning of each generation [14] and replaced an equal number of the worst solutions in the population.

Toffolo et al. incorporated diversity indicators as additional objectives to be simultaneously optimized with the original objective [30]. Common diversity indicators include individual age, population entropy, and distance between individuals in decision space, among others. Another well-known diversity maintenance approach is multi-charged PSO, proposed by Blackwell et al. [3]. This method follows the repulsion mechanism observed among particles in physics, ensuring that the entire population remains diverse within a defined range to preserve diversity.

During the whole running time, the population always has good diversity, which means it has a good chance to explore the whole space to react to changes. Therefore, this method can bring the following advantages:

(1) Suitable for solving problems with severe changes, especially in the case where optima shift significantly from their original positions, i.e., abrupt changes in the fitness landscape.
(2) Suitable for solving dynamic problems with a low change frequency that provides more time for these algorithms to converge.

However, these methods also have some drawbacks:

(1) It pays more attention to diversity, which slows down the convergence speed.
(2) It is not effective for small changes, since the global level of diversity maintenance does not help the tracking of optima, which are close to the old optima.

Memory-Based Methods

Memory-based methods reuse information from previous generations, which may be useful to handle the current change. Usually, the best solutions in every generation are stored in memory. When a change occurs, algorithms employ the most closely related solutions from memory as the starting point for the next population. These methods can significantly improve search efficiency in situations where changes are small or periodic.

There are two kinds of memory: implicit and explicit memory. Implicit memory employs a redundant encoding scheme, such as the diploid genomes in [12]. In this scheme, a pair of chromosomes is used, with one chromosome representing the solution and the other storing additional information used to control diversity within the population. An explicit memory uses an extra storage with explicit rules for storing and retrieving information. Ramsey et al. [27] were the first to use a case-based memory to store high-quality solutions as well as their associated environmental information. Similarly, Branke devised a memory population in which distinct individuals were chosen to serve as memory storage [4]. Yang abstracted allele genes as environmental information and integrated them with individuals to serve as a memory, facilitating the generation of new populations in novel environments [34].

Memory-based methods are highly effective when dealing with changes that exhibit periodic or recurrent patterns. When optima relocate to regions containing memorized solutions, the population can be initialized around these memorized solutions and converge to new optimums rapidly. However, they also have some drawbacks:

(1) Algorithms heavily rely on the memory. If optima moves to regions without any memorized solutions, this strategy becomes ineffective.
(2) They rely on change detection. They will not work due to the failure of change detection.
(3) Redundant information requires more storage space and retrieval time.

Prediction-Based Methods

Prediction-based methods are designed to improve their efficiency with the prediction model. When a change is detected, it is inefficient to introduce individuals randomly across the whole search space. In some cases, changes may imply some patterns, which are predictable. Algorithms could learn from historical data and predict the change pattern, e.g., the trajectory of the global optimum. Thus, algorithms could initialize the population in the predicted promising regions to where the global optimum might move.

Hatzakis and Wallace [15] introduced a dynamic queuing multi-objective optimizer that uses historical data to forecast the future change of problems. They proposed a feedforward prediction scheme to record the current optima's positions and then used the data to train a prediction model for estimating the location of new global optima. In [35], a population-based prediction strategy was presented to predict the location of the entire population. They stored the central position and the manifold of the Pareto set in each environment and used a linear autoregression model to predict the central position and Gaussian disturbance to predict the shape of the Pareto set's manifold. Unlike the former methods, the approach presented in [23] utilizes the Kalman filter to predict the location of the Pareto set. The Kalman filter method employs a feedback modification scheme to reduce prediction errors.

Prediction-based approaches can improve search efficiency regardless of the severity of a change, especially when the change pattern is regular. However, they also come with certain drawbacks:

(1) Insufficient Training Data: During the initial stages of the search process, it may be challenging to obtain an adequate amount of data for model training. This can lead to significant prediction errors. A misguided prediction can mislead the population toward unpromising regions. Moreover, when there is a lack of data around the new optima, predictions may fail to predict the locations of the new optima.
(2) Difficulty in Data Labeling: For example, different operators generate solutions with different numbers of solutions. In an algorithm integrated with many operators, it is a complex task to label a solution with other solutions. Incorrect labeling can substantially impact the model accuracy.
(3) Limited Applicability to Regular Changes: If the changes are stochastic or lack a regular pattern, then prediction methods do not work.

Multi-populations Methods

The multi-population method is a form of divide-and-conquer strategy. Its basic idea is to perform simultaneous searches in distinct subregions to track multiple optima at the same time. Each subpopulation exclusively explores its designated subregion. This approach can be viewed as a combination of diversity maintenance and diversity enhancement.

Compared to single-population methods, multipopulation methods have several advantages. Firstly, population diversity can be maintained at a global level, since populations are distributed in different subregions in the fitness landscape. Second, it can locate and track multiple optima simultaneously, which is very helpful to locate and track the movement of target optima. Third, any strategy designed for the single-population method can be easily extended to a multi-population version. However, the management of populations in multi-population methods presents several challenging issues [19]. These include task allocation for each subpopulation, determining the number of populations, specifying the number of individuals in each subpopulation, controlling the merging or splitting of populations to prevent overlap or overcrowding in the search process, and defining the search areas for each subpopulation. Moreover, all of these factors should adapt to changes.

To deal with these issues, many algorithms have been proposed. Oppacher and Wineberge [26] employed multiple small populations for searching and a single large population to track changing peaks. Branke [5] used a large main population to search for optima and several small populations to track changes for each optimum. Another approach, for example, introduced by Ursem [32] integrated both the searching and tracking process within each subpopulation. Once a subpopulation discovers a new optimum, it is split into two subpopulations to ensure that each subpopulation exclusively tracks one optimum. To adapt to changes in the number of optima, Li [18] proposed a feedback learning method to control the number of populations. In this method, an evolutionary status is defined as the current number of populations. The predicted number of populations is estimated based on a statistical value derived from all historical adaptations with the current evolutionary status and a feedback value based on the difference in the number of populations between two adaptations.

To solve DMOPs, Greeff and Engelbrecht [13] proposed a vector-evaluated PSO, in which each swarm focuses on solving a single objective function, and share knowledge with each other using a ring or random topology. Goh and Tan [11] proposed a dynamic competitive-cooperative coevolutionary algorithm that enables adaptive problem decomposition, incorporating stochastic competitors to track moving optima.

11.3   Benchmark Problems and Metrics for Dynamic Single-Objective Optimization

This section introduced several popular benchmark generators in continuous space and discussed various performance measures.

11.3.1   Classical Problem Generators

Artificial test suites, which simulate real-world applications, are used to evaluate algorithm performance. These suites often offer flexible configurations to adjust problem difficulty. Generally speaking, the fitness landscape of most problems consists of peaks, making parameters such as peak height, width, and location critical for shaping the fitness landscape. Here are some classical generators used to create such dynamic landscapes.

DF Generator
Morisson and De Jong [22] designed the DF1 in two steps. First, they designed some basic functions that have complex morphology, as shown in Fig. 11.2. Then they introduced dynamics to these functions. These functions can be extended to any number of dimensions, and their peak locations are artificially controlled. An example in two-dimensional space is shown below:

$$f(x, y) = \max_{i=1,\cdots,N} [H_i - R_i \cdot \sqrt{(x - X_i)^2 + (y - Y_i)^2}] \tag{11.2}$$

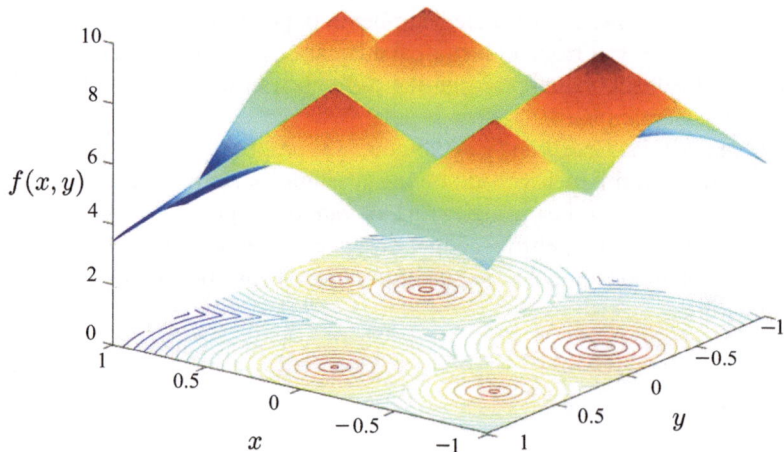

Fig. 11.2 The landscape of the DF1 generator

where N denotes the number of the peaks and (X_i, Y_i) is the location of the i-th peaks in two-dimensional space, H_i and R_i are the height and slope of the i-th peak. The height, location, and slope can vary with time according to user-defined rules.

Moving Peak Benchmark

Interestingly, Branke et al. [4] proposed a method similar to DF1 at the same conference. In their approach, the dynamic fitness landscape is generated by specifying the location, height, and width of peaks. This method is straightforward to describe and analyze, and the problem is scalable and tunable. It not only mirrors the complexity of real-world applications but also provides researchers with fresh perspectives on algorithm behaviors.

The moving peak benchmark (MPB) generator is defined as below:

$$F(\boldsymbol{x}, t) = \max_{i=1,\cdots,N} \frac{H_i(t)}{1 + W_i(t) \sum_{j=1}^{D}(x_j - X_j^i(t))^2} \tag{11.3}$$

where H_i and W_i denote the height and the width of the i-th peak, respectively; X_j^i is the location of the j-th dimension of the i-th peak.

A method similar to the above format is defined in Eq. (11.4), and it combines all the values of each peak.

$$F(\boldsymbol{x}, t) = \sum_{i=1}^{N} \frac{H_i(t)}{1 + W_i(t) \sum_{j=1}^{D}(x_j - X_j^i(t))^2} \tag{11.4}$$

And the landscapes of these two methods are shown in Fig. 11.3.
The height, width, and location of each peak vary with time as follows

$$H_i(t) = H_i(t-1) + 7 \cdot \sigma$$
$$W_i(t) = W_i(t-1) + 0.01 \cdot \sigma \tag{11.5}$$
$$X_i(t) = X_i(t-1) + v$$

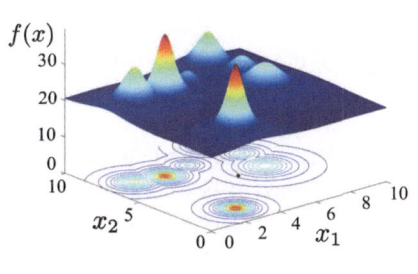

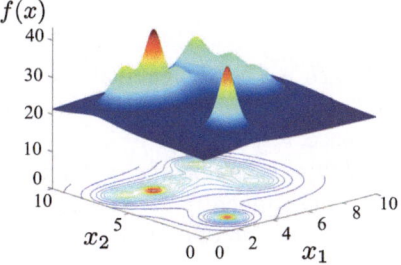

(a) the max function   (b) the superposition function

Fig. 11.3 The MPB landscapes defined by Eqs. (11.3) **(a)** and (11.4) **(b)**

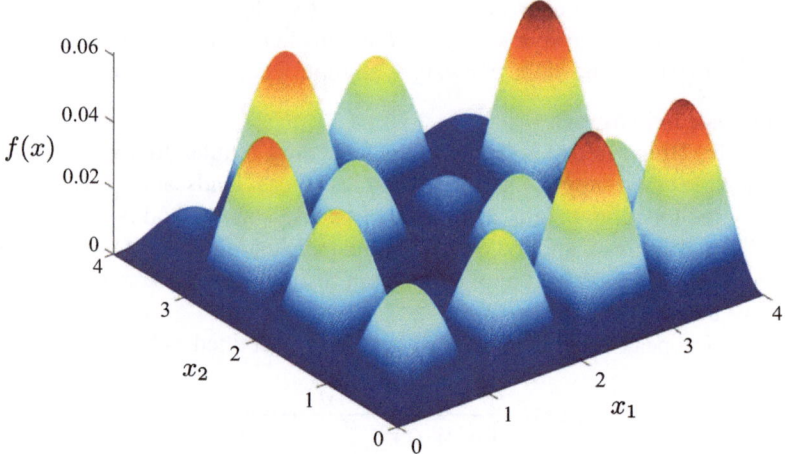

Fig. 11.4 Dynamic height of the peaks

where σ is a random number generated from a standard normal distribution $\sigma \in N(0, 1)$; $\boldsymbol{v}$ consists of a random direction and a user-defined step size, representing the change of the location of each optimum.

Divide and Design
Krzysztof et al. [31] introduced a generator that divides the search space into uniform and disjoint subspaces and then established a unimodal function in each subspace. The overall objective function is then constructed by aggregating these subfunctions.

Figure 11.4 illustrates the fitness landscape generated using this generator, where each subspace has a peak with a user-defined height. In each subspace, the objective function is defined as:

$$f_i(\boldsymbol{x}) = H_i \cdot \prod_{j=1}^{D} (l_j - x_j)(x_j - u_j) \tag{11.6}$$

where H_i is a coefficient that is related to the height of the peak in the i-th subspace and l_j and u_j are the lower and upper boundary of the j-th dimension of the i-th subspace, respectively.

In this generator, the width of each peak remains constant, but it offers the flexibility to modify the height and location of each peak. It is important to note that the location of each peak can only vary within its respective subspace. A peak matrix can be used to adjust the height of all peaks according to a specific pattern. For example, the matrix in Eq. (11.7) is the height of the peaks in Fig. 11.4.

$$H = \begin{bmatrix} 0.50 & 0.60 & 0.92 & 0.92 \\ 0.65 & 0.11 & 0.50 & 0.51 \\ 0.80 & 0.52 & 0.29 & 0.97 \\ 0.23 & 0.84 & 0.65 & 0.20 \end{bmatrix} \tag{11.7}$$

Generalized Dynamic Benchmark Generator
Li et al. [20] proposed a generalized dynamic benchmark generator (GDBG) for creating dynamic changes by tuning the system control parameters with a deviation $\Delta\phi$:

$$f(x, \phi, t + 1) = f(x, \phi \oplus \Delta\phi, t) \tag{11.8}$$

Based on this idea, GDBG offers six different change patterns for $\Delta\phi$, including small step changes, large step changes, random changes, chaotic changes, recurrent movements, and recurrent movements with noise.

They designed two benchmark generators in continuous space. One is the dynamic rotation peak benchmark generator (DRPBG), which is defined as follows:

$$f(x, t) = \min_{i=1,\cdots,m} \left(H_i(t) + W_i(t) \cdot \left(\exp \left(\sqrt{\sum_{j=1}^{D} \frac{(x_j - X_j^i(t))^2}{D}} \right) - 1 \right) \right) \tag{11.9}$$

Different from the movement of the MPB, the location of each peak is rotated using a rotation matrix, which is defined as follows:

$$X(t + 1) = X(t) \cdot R(t) \tag{11.10}$$

$$R(t) = R_{1,2}(\theta(t)) \cdot R_{3,4}(\theta(t)) \cdot \cdots \cdot R_{2l-1,2l}(\theta(t)), \ 2l \le D \tag{11.11}$$

where l is the number of dimensions that are randomly selected to rotate. The rotation matrix $R_{ij}(\theta)$ is obtained by rotating the projection of x in the plane $i - j$ with an angle θ from the i-th axis to the j-th axis.

Figure 11.5 shows a landscape with multiple peaks in a two-dimensional search space before and after the rotation.

Another generator named dynamic composition benchmark generator (DCBG) is defined as

$$F(x, t) = \sum_{i=1}^{m} (\omega_i \cdot (f_i'((x - O_i(t) + O_{iold})/\lambda_i \cdot M_i) + H_i(t))) \tag{11.12}$$

Fig. 11.5 Landscape
rotation. (**a**) Original
landscape. (**b**) Rotated
landscape

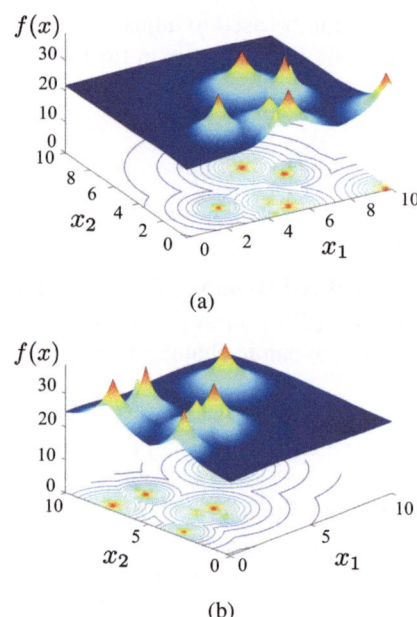

(a)

(b)

where $f_i'(x)$ is the scaled i-th basic function $f_i(x)$; ω_i denotes the weigh of the i-th basic function; $O_i(t)$ denotes the optimum of the f_i at time t; O_{iold} is the original optimum of the $f_i(x)$; M_i is the orthogonal matrix for f_i; and $H_i(t)$ is an additional height added to f_i.

By combining several basic functions, e.g., Sphere function, Rastrigin function, Griewank function, Ackley function, and Weierstrass function, DCBG can generate a very complex fitness landscape with diverse characteristics.

11.3.2 Performance Measures

When evaluating algorithms for solving DOPs, we pay more attention to accuracy and adaptability. Several performance metrics have been introduced for this purpose.

Offline Performance
$\hat{F}_{\text{BOG}}$ records the best value at each generation and calculates an average value over all generations

$$\hat{F}_{\text{BOG}} = \frac{1}{G} \times \sum_{t=1}^{G} F_{\text{BOG}_j} \tag{11.13}$$

where $\hat{F}_{BOG}$ is the mean best-of-generation objective or fitness value, G is the number of generations in a run, and F_{BOG_j} is the best-of-generation objective or fitness value of generation j on a problem.

It is important to note that problems may have different objectives or fitness values after a change. When comparing the performance of different algorithms, it is essential to consider the normalization of the best-of-generation value. Otherwise, the measure indicators may be biased, because algorithms could get the bigger values when the problem's objective or fitness value is at a high level.

Offline Error
E is calculated as the average of the difference between the objective value of the best individual and the optimum value at each generation

$$E = \frac{1}{G} \sum_{i=1}^{G} e_i \qquad (11.14)$$

where e_i is the error at the i-th generation. This measure can be employed when the global optimum is known for every change.

Best-Before-Change Error
This measure is calculated as the average of the difference between the value of the best-found solution and the optimum value at the end of each change period

$$E = \frac{1}{T} \sum_{t=1}^{T} e_t \qquad (11.15)$$

where e_t is the best error just before the t-th change and T is the number of changes.

11.4   Benchmark Problems and Metrics for Dynamic Multi-objective Optimization

Compared to DOPs, DMOPs optimize multiple objectives simultaneously, leading to additional characteristics related to changes in PS and PF. Farina et al. [9] classified DMOPs into four types based on the changes of the PS and PF.

(1) Type I: The PS changes with time, but the PF keeps static.
(2) Type II: Both the PS and PF change with time.
(3) Type III: The PS keeps stable, but the PF changes with time.
(4) Type IV: Both the PS and PF keep static; other factors change with time.

11.4.1   Classical Benchmark Problems

Creating benchmark problems with distinct characteristics aids in assessing an algorithm's capacity to tackle various challenges encountered in real-world scenarios. Compared to MOPs and DOPs, designing a DMOP is more difficult, because it integrates all the features of the other two benchmarks.

Some recognized common features that a DMOP benchmark should have include:

(1) The curvature and the continuity of the PF should change with time, which could affect the uniformity of the solutions on PF.
(2) Dynamic shape and degeneration of the PF.
(3) Dynamic number of objectives and variables.
(4) The shape and continuity of the PS should change with time, and each dimension has its own speed of change.
(5) Dynamic modality, detectability, and predictability.
(6) The type of the DMOP can be changed with time.
(7) Current PF and PS may depend on former PF and PS.
(8) Benchmark problems should resemble features of real-world applications.

Over the last two decades, researchers have developed many dynamic multiobjective benchmark problems by building upon static MOPs. A simple overview is presented below.

FDA Benchmark
Farina et al. designed a DMOP benchmark test suite named FDA [9] based on several existing MOP benchmark problems. The PF and PS of FDA can be designed independently. The FDA test suite comprises five test functions, three of them having two objectives, while the remaining two are scalable to random objectives. The key features of FDA include changes in PS following a trigonometric function and the periodic transformation of the shape of PF from convex to concave. For FDA functions with more than two objectives, when the number of objectives is greater than two, the PF shape may take the form of a hyperplane or hypersphere.

DMOPs Benchmark
Goh et al. designed a benchmark test suite named DMOPs [11] based on FDA. It comprises three functions, and all of them are two-objective functions. DMOP1 is categorized as type III problem, DMOP2 as type II, and DMOP3 as type I. The biggest difference between FDA and DMOP3 is that one of the objectives in DMOP3 is random, introducing greater complexity in finding Pareto solutions.

F Benchmark
Zhou et al. [35] found that the FDA has several drawbacks in the FDA benchmark, including the static range of PS, the linear mode of the change for each dimension of the PS, and the PS shape limited to lines or rectangles. All these designed features make FDA not universal and relatively straightforward to tackle. Thus, they developed a benchmark, namely, the F test suite.

F test suite consists of ten functions, most of which are two-objective functions, and only one is a triple-objective function. Their notable feature is the presence of nonlinear relationships among different variables, resulting in complex PS shapes. Additionally, the F test suite exhibits several other characteristics:

(1) Dividing all variables into two nonoverlapping parts, with each objective optimizing one part.
(2) The range of the PS changes dynamically.
(3) The PS will have a large jump periodically, and so on.

HE Benchmarks

Helbig et al. focused on designing difficult problems, including those with isolated PS or PF, deceptive PS or PF, and complicated PS. Therefore, they designed the HE benchmark test suite, comprising ten functions [16].

Most of the functions have two objectives, and they also adopt variable grouping strategies. Its main drawback is that only the PF changes while the PS remains invariant. Changes only occur in the PF, and the previous PS is still the new PS, and thus algorithms do not need to track the PS.

Recent Benchmarks

In recent years, many researchers have shifted their focus within DMOPs test suites, emphasizing the challenges posed by dynamic characteristics rather than solely by multi-objective characteristics. Algorithms must adapt to dynamic changes, leading to the emergence of new benchmarks with rich dynamic characteristics.

Jiang and Yang designed a new benchmark test suite named JY [17] based on FDA. Its prominent features include the continuous change in the shape of the PF between convex and concave, varying change rates in different dimensions of the PS, and random change in problem types. These features make this benchmark more comprehensive and more challenging to algorithms.

Gee et al. designed a benchmark test suite named GTA [10] that also incorporates dynamic features into the problems. The key features of GTA include:

(1) Dynamic changes in problem modality, such as change from multimodal to unimodal and vice versa.
(2) Dynamic degeneration of the PF, depending on the relationships between all sub-objectives.
(3) Dynamic changes in PF continuity, raising some challenges for algorithms.

11.4.2 Performance Measures

When evaluating the performance of algorithms for DMOPs, both the quality of the distribution of PF and the ability to respond to changes should be considered. As described in Sect. 9.3, certain indicators evaluate the convergence or diversity of the set of solutions for static optimization problems. Here, we will introduce several modified versions of these indicators to make them suitable for evaluating algorithms on DMOPs.

mIGD

As introduced in Sect. 9.3.5, IGD measures both the convergence and diversity of a solution set. In DMOPs, the PF undergoes changes, so we calculate the average IGD over all changes throughout the entire process as follows:

$$m\text{IGD} = \frac{1}{T} \sum_{t \in T} \text{IGD}(\mathbb{Z}_t^\star, \mathbb{P}_t) \qquad (11.16)$$

where $\mathbb{Z}_t^\star$ is the set of solutions in a given approximation PS at each change t; $\mathbb{P}_t$ is the set of solutions obtained at the end of each change t; and T is the number of changes in a run.

Maximum Spread

Maximum spread is a diversity indicator that measures the distance between the boundary solutions in the obtained solution set. The calculation has been introduced in Sect. 9.3.3. Goh et al. proposed a modified version in [11]

$$m\text{MS} = \frac{1}{T} \sum_{t=1}^{T} \sqrt{\frac{1}{M} \sum_{k=1}^{M} \left[\frac{\min[\overline{\mathbb{P}}_{kt}, \overline{\mathbb{Z}}_{kt}^\star] - \max[\underline{\mathbb{P}}_{kt}, \underline{\mathbb{Z}}_{kt}^\star]}{\overline{\mathbb{Z}}_{kt}^\star - \underline{\mathbb{Z}}_{kt}^\star} \right]^2} \qquad (11.17)$$

where $\overline{\mathbb{P}}_k$ and $\underline{\mathbb{P}}_k$ are the maximum and minimum value of the k-th objective of the obtained solution set, respectively; $\overline{\mathbb{Z}}_k^\star$ and $\underline{\mathbb{Z}}_k^\star$ are the maximum and minimum value of the k-th objective in a given approximation PF, respectively.

Spacing Metric

Scott's spacing metric [28] is used to measure the degree of uniformity of the solution set obtained:

$$mS = \frac{1}{T} \sum_{t=1}^{T} S(\mathbb{P}_t) \qquad (11.18)$$

where $\mathbb{P}_t$ is the solution set obtained at the end of each change t and $S(\mathbb{P}_t)$ is the spacing metric defined in Eq. (9.12).

Hypervolume

Hypervolume is a comprehensive indicator that could represent the convergence and diversity of a solution set. The calculation of HV for algorithms for static MOPs has been described in Eq. (9.14), and the average HV value over all changes can also be calculated for DMOPs.

$$mI_H = \frac{1}{T} \sum_{t=1}^{T} I_H(\mathbb{P}_t) \qquad (11.19)$$

11.5 Dynamic Constrained Optimization

Dynamic constrained optimization problems are prevalent in real-world applications. For example, the torque cannot exceed the strength of the object when manufacturing a component. The arrival time of goods must meet the customer's requirements. Therefore, optimizing a dynamic constrained problem holds significant practical importance. The basic dynamic characteristics of DCOPs include three types.

(1) Objectives change, but constraints remain static.
(2) Constraints change, but objectives remain static.
(3) Both objectives and constraints change.

When constraints are considered, algorithms face distinct challenges. For example, the feasible search space may be separated into discontinuous regions, and optima might be located on the boundary of feasible regions. At the same time, dynamic changes may occur on these constraints, making it more challenging for algorithms.

How to adjust constraint-handling techniques and dynamic handling techniques for DCOPs is an important issue. Compared with dynamic unconstrained optimization, there is little research on dynamic constrained optimization due to the difficulties. Nguyen presented a concrete introduction to DCO in his dissertation [24]. To promote the research, Nguyen et al. [25] and Wang et al. [33] designed a single-objective DCOPs benchmark. A dynamic constraint multi-objective optimization test suite [2] was proposed based on [8].

Nowadays, the main methods for solving DCOPs include two parts, constraints-handling techniques and dynamism response strategies. They divide a DCOP into a constrained part and a dynamic part. Dynamic handling strategies introduce diversity when changes are detected. Then a DCOP can be treated as a static COP, and many constraint-handling techniques can be used to solve it. Details of the constraint-handling strategies can be referred to Sect. 10.2.

11.6 Challenges

Dynamic optimization is an emerging research topic in the field of intelligent optimization. Although many scholars have paid much attention to it and made great progress in recent years, there are still many challenges to be overcome. We discuss these challenges in terms of problem modeling, algorithm design, and performance metrics.

(1) To evaluate and improve the performance of algorithms, the design of academic benchmark problems is very important. However, there is still a big gap between academic research and real-world applications. The design of academic benchmark problems often prioritizes the ease of algorithmic performance analysis,

potentially overlooking many essential features of real-world applications. For example, many test suites for DMOPs are made up of composite functions with objectives that exhibit clear relationships, which may not hold in real-world scenarios.

(2) Multi-population methods are very effective in dealing with changes. However, a significant challenge they face is the allocation of resources among these populations. One potential solution is to predict promising regions with a learning model based on all historical data.

(3) As for the performance measures, it is necessary to design effective metrics that can evaluate algorithm performance when the real PF and PS are not known.

(4) Dynamic constrained optimization is also important, and there is a need for both test suites and algorithms in this area.

References

1. Abbass, H.A., Sastry, K., Goldberg, D.E.: Oiling the wheels of change: the role of adaptive automatic problem decomposition in non–stationary environments. Technical report, Illinois Genetic Algorithms Laboratory, University of Illinois Urbana-Champaign, Urbana (2004)
2. Azzouz, R., Bechikh, S., Ben Said, L.: Multi-objective optimization with dynamic constraints and objectives: new challenges for evolutionary algorithms. In: Proceedings of the 2015 Annual Conference on Genetic and Evolutionary Computation, pp. 615–622. Association for Computing Machinery, New York (2015)
3. Blackwell, T., Branke, J.: Multiswarms, exclusion, and anti-convergence in dynamic environments. IEEE Trans. Evol. Comput. 10(4), 459–472 (2006)
4. Branke, J.: Memory enhanced evolutionary algorithms for changing optimization problems. In: Proceedings of the 1999 Congress on Evolutionary Computation, pp. 1875–1882. IEEE, New York (1999)
5. Branke, J., Kaußler, T., Smidt, C., Schmeck, H.: A multi-population approach to dynamic optimization problems. In: Parmee, I. (ed.) Evolutionary Design and Manufacture, pp. 299–307. Springer, London (2000)
6. Cobb, H.G.: An investigation into the use of hypermutation as an adaptive operator in genetic algorithms having continuous, time-dependent nonstationary environments. Technical report, Naval Research Lab, Washington (1990)
7. Deb, K., Karthik, S., et al.: Dynamic multi-objective optimization and decision-making using modified NSGA-II: a case study on hydro-thermal power scheduling. In: Proceedings of the International Conference on Evolutionary Multi-criterion Optimization, pp. 803–817. Springer, Berlin (2007)
8. Deb, K., Pratap, A., Meyarivan, T.: Constrained test problems for multi-objective evolutionary optimization. In: Proceedings of the International Conference on Evolutionary Multi-criterion Optimization, pp. 284–298. Springer, Berlin (2001)
9. Farina, M., Deb, K., Amato, P.: Dynamic multiobjective optimization problems: test cases, approximations, and applications. IEEE Trans. Evol. Comput. 8(5), 425–442 (2004)
10. Gee, S.B., Tan, K.C., Abbass, H.A.: A benchmark test suite for dynamic evolutionary multiobjective optimization. IEEE Trans. Cybern. 47(2), 461–472 (2016)
11. Goh, C.K., Tan, K.C.: A competitive-cooperative coevolutionary paradigm for dynamic multiobjective optimization. IEEE Trans. Evol. Comput. 13(1), 103–127 (2008)

12. Goldberg, D.E., Smith, R.E.: Nonstationary function optimization using genetic algorithms with dominance and diploidy. In: The Second International Conference on Genetic Algorithms, pp. 59–68. L. Erlbaum Associates, Hillsdale (1987)
13. Greeff, M., Engelbrecht, A.P.: Solving dynamic multi-objective problems with vector evaluated particle swarm optimisation. In: IEEE Congress on Evolutionary Computation, pp. 2917–2924. IEEE, New York (2008)
14. Grefenstette, J.J.: Genetic algorithms for changing environments. In: International Conference on Parallel Problem Solving from Nature, pp. 137–144. Elsevier, Netherlands (1992)
15. Hatzakis, I., Wallace, D.: Dynamic multi-objective optimization with evolutionary algorithms: a forward-looking approach. In: The 8th Annual Conference on Genetic and Evolutionary Computation, pp. 1201–1208. Association for Computing Machinery, New York (2006)
16. Helbig, M., Engelbrecht, A.P.: Benchmarks for dynamic multi-objective optimisation algorithms. ACM Comput. Surv. **46**(3), 1–39 (2014)
17. Jiang, S., Yang, S.: Evolutionary dynamic multiobjective optimization: benchmarks and algorithm comparisons. IEEE Trans. Cybern. **47**(1), 198–211 (2016)
18. Li, C., Nguyen, T.T., Yang, M., Mavrovouniotis, M., Yang, S.: An adaptive multipopulation framework for locating and tracking multiple optima. IEEE Trans. Evol. Comput. **20**(4), 590–605 (2015)
19. Li, C., Nguyen, T.T., Yang, M., Yang, S., Zeng, S.: Multi-population methods in unconstrained continuous dynamic environments: the challenges. Inf. Sci. **296**, 95–118 (2015)
20. Li, C., Yang, S.: A generalized approach to construct benchmark problems for dynamic optimization. In: Asia-Pacific Conference on Simulated Evolution and Learning, pp. 391–400. Springer, Berlin (2008)
21. Li, C., Yang, S.: A general framework of multipopulation methods with clustering in undetectable dynamic environments. IEEE Trans. Evol. Comput. **16**(4), 556–577 (2012)
22. Morrison, R.W., De Jong, K.A.: A test problem generator for non-stationary environments. In: Proceedings of the 1999 Congress on Evolutionary Computation, pp. 2047–2053. IEEE, New York (1999)
23. Muruganantham, A., Tan, K.C., Vadakkepat, P.: Evolutionary dynamic multiobjective optimization via Kalman filter prediction. IEEE Trans. Cybern. **46**(12), 2862–2873 (2015)
24. Nguyen, T.T.: Continuous dynamic optimisation using evolutionary algorithms. Ph.D. thesis, University of Birmingham, Birmingham (2011)
25. Nguyen, T.T., Yao, X.: Continuous dynamic constrained optimization-the challenges. IEEE Trans. Evol. Comput. **16**(6), 769–786 (2012)
26. Oppacher, F., Wineberg, M.: The shifting balance genetic algorithm: improving the ga in a dynamic environment. In: Proceedings of the 1st Annual Conference on Genetic and Evolutionary Computation, pp. 504–510. Morgan Kaufmann, San Francisco (1999)
27. Ramsey, C.L., Grefenstette, J.J.: Case-based initialization of genetic algorithms. In: Proceedings of the 5th International Conference on Genetic Algorithms, pp. 84–91. Morgan Kaufmann, San Francisco (1993)
28. Scott, J.R.: Fault tolerant design using single and multicriteria genetic algorithm optimization. Master's thesis, Massachusetts Institute of Technology, Cambridge (1995)
29. Tinós, R., Yang, S.: Evolutionary programming with q-gaussian mutation for dynamic optimization problems. In: IEEE Congress on Evolutionary Computation, pp. 1823–1830. IEEE, New York (2008)
30. Toffolo, A., Benini, E.: Genetic diversity as an objective in multi-objective evolutionary algorithms. Evol. Comput. **11**(2), 151–167 (2003)
31. Trojanowski, K., Michalewicz, Z.: Searching for optima in non-stationary environments. In: Proceedings of the 1999 Congress on Evolutionary Computation, pp. 1843–1850. IEEE, New York (1999)
32. Ursem, R.K.: Multinational ga optimization techniques in dynamics environments. In: Proceedings of Genetic and Evolutionary Computation, pp. 19–26. Morgan Kaufmann, San Francisco (2000)

33. Wang, Y., Yu, J., Yang, S., Jiang, S., Zhao, S.: Evolutionary dynamic constrained optimization: test suite construction and algorithm comparisons. Swarm Evol. Comput. **50**, 100559 (2019)
34. Yang, S.: Associative memory scheme for genetic algorithms in dynamic environments. In: Workshops on Applications of Evolutionary Computation, pp. 788–799. Springer, Berlin (2006)
35. Zhou, A., Jin, Y., Zhang, Q.: A population prediction strategy for evolutionary dynamic multiobjective optimization. IEEE Trans. Cybern. **44**(1), 40–53 (2013)

Chapter 12
Robust Optimization

Abstract The definition of robust optimization originates from the field of engineering design. Due to the limitation of production technology, the production process is full of uncertain factors, such as the change of temperature and humidity in the environment, the production error of the instrument, etc. If the production process is sensitive to these uncertainty factors, many products will be unqualified. Therefore, in practical production, the production process should be robust to these uncertainty factors.

In this chapter, we will introduce the concept of robust optimization, the commonly used robust mathematical models, some robust optimization algorithms, and discussions on robust optimization.

12.1 Introduction

In this section, we will introduce the concept of robust optimization problems from the point of view of the system principle [2]. As shown in Fig. 12.1, in a real-world application, the optimizer provides solutions to the system, which is embedded in the environment, and the system runs with the solutions provided. The performance of the system is evaluated by some means and then goes to the optimizer as feedback on the quality of solutions. During the runtime, uncertainties and dynamics come from the outside and inside of the system.

From the figure, we can see that there are four sources of dynamics and uncertainties.

(1) Uncertainties in variables [4]. In practical production, there are uncertainties in the production equipment due to instrument errors, and thus the product may be a little different from the solution designed by the optimizer. This uncertainty is known as the uncertainty of the decision variables.

(2) Uncertainties from the environment [4]. This uncertainty comes from the environment or operating conditions. The system is always subject to disturbance from the running environment. For example, real-time traffic conditions on the

Fig. 12.1 Sources of
dynamics and uncertainties

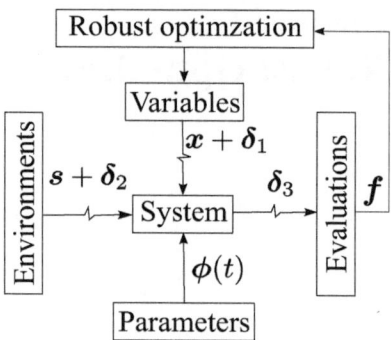

road will be affected by many environmental factors from weather, drivers, and
vehicles.

(3) Uncertainties in the evaluation. In this case, the evaluation of the system
performance is not accurate for three reasons:

- Measuring instruments are not accurate, so uncertainty is introduced into the
evaluation.
- The analytical fitness function is not available, and thus a mathematical
model is built to evaluate the system performance. The mathematical model
is not the same as the real fitness function, and the evaluation is contaminated
with uncertainties.
- The accurate evaluation is quite expensive, and we can only get the cheaper
approximate evaluation by a computing simulation.

(4) Changes in the system. Changes in the system mean that the parameters of the
system change with time, which is also known as dynamic optimization (see
Chap. 11). Due to the practical demand in real-world applications, the solution
is required to be robust over time for the following reasons:

- Change detection is quite complex. Changes may occur at any time, which
causes some types of changes that cannot be guaranteed to be detected, e.g.,
a change may occur during the detection of changes.
- In practice, changing a solution will bring extra costs, so the solution should
be robust over time.
- In some cases, a solution should be maintained as long as its performance
is acceptable. In the scheduling of aircraft takeoff/landing, any changes
to the scheduling would adversely affect the airport services, e.g., making
passengers wait longer.

Considering the four sources of dynamics and uncertainties, robust optimization
is defined as follows

$$f(x + \delta_1, s + \delta_2, \phi(t)) + \delta_3 \tag{12.1}$$

where x is the decision variables; s is the environmental factors; $\phi(t)$ is the control parameters of the system which change with time; and $\delta_1, \delta_2, \delta_3$ are the uncertainties which come from the decision variables, the environment, and the evaluation, respectively.

In this section, we introduce only the dynamics and uncertainties in the objective evaluation. Actually, the dynamics and uncertainties can also be introduced into the constraints, and the feasible area will become fragmented and time-varying, which makes the problem more difficult to solve.

12.2 Robust Optimization Algorithms

The definitions of robustness are quite different for different types of uncertainties. Therefore, there are different mathematical models to measure the robustness of the four types of uncertainties described above.

12.2.1 Uncertainties in Decision Variables

Noise is introduced into decision variables, which is very common in industrial production. For example, due to the inaccurate measurement of the length of screws of a certain product in manufacturing, there is a certain error, which will impact the performance, i.e., the performance of the product shows small fluctuations. The manufacturing production errors usually follow some deterministic distribution.

12.2.1.1 Definition

The mathematical formal expression of the robust optimization with uncertainties in decision variables can be written as:

$$\begin{aligned} \min \ & f(x + \delta) \\ \text{s. t. } & g(x + \delta) \le 0 \end{aligned} \tag{12.2}$$

where $f(x + \delta)$ is the vector of objective functions and δ is a vector of noise/disturbance described by a deterministic distribution with a priori knowledge about the uncertainties, that is, $\delta \sim \mathcal{D}(0, \Sigma_x)$.

After the noise is introduced into decision variables, each value of the function follows a deterministic distribution by mapping the noise through $f(*)$, and the distribution can be written as $\mathcal{D}(\mu_f, \Sigma_f)$. Similarly, the distribution of constraint values can be written as $\mathcal{D}(\mu_g, \Sigma_g)$.

From the definition, both the objective functions and the constraint functions are affected by the noise. Based on different scenarios, robustness can be divided into performance robustness and feasibility robustness. Normally, the mean value (μ) and variance (Σ) of the functions or the constraints are used to reflect the performance robustness or feasibility robustness. If the robustness is defined as performance robustness, then $\mu = \mu_f$ and $\Sigma = \Sigma_f$. If the robustness is defined as feasible robustness, then $\mu = \mu_g$ and $\Sigma = \Sigma_g$.

12.2.1.2 Measures

From the definition, the aggregation approach can be used to calculate the robustness. One of the integral measures of robustness can be defined by

$$\mu_{f_i} = \int_{-\infty}^{+\infty} f_i(\boldsymbol{x} + \boldsymbol{\delta})p(\boldsymbol{\delta})d\boldsymbol{\delta} \qquad i = 1, 2, \ldots, M \tag{12.3}$$

where $p(*)$ is the distribution density function of $\boldsymbol{\delta}$. Accordingly, the variance can be used to find plateau-like regions in fitness landscapes by

$$\sigma_{f_i} = \int_{-\infty}^{+\infty} (f_i(\boldsymbol{x} + \boldsymbol{\delta}) - f_i(\boldsymbol{x}))^2 p(\boldsymbol{\delta})d\boldsymbol{\delta} \qquad i = 1, 2, \ldots, M \tag{12.4}$$

Here, we assume that the noise from reality follows a certain distribution, such as the normal distribution, Cauchy distribution, and so on. Take the normal distribution as an example ($\boldsymbol{\delta} \sim \mathcal{N}(\boldsymbol{0}, \Sigma_x)$), and suppose that the noise in each dimension is independent of each other. As shown in Fig. 12.2, almost 99% points are distributed over the interval $[-2.58\sigma, 2.58\sigma]$ in one dimension ($\delta \sim \mathcal{N}(0, \sigma)$), which means that the points in this interval can be used to estimate the mean value and variance of the distribution.

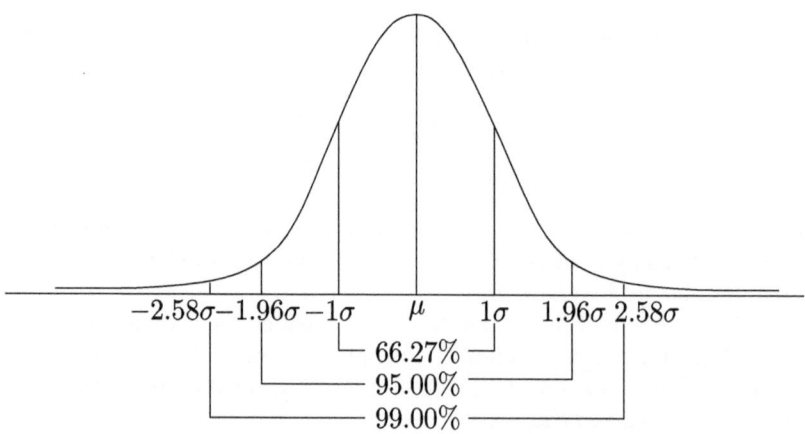

Fig. 12.2 Normal distribution

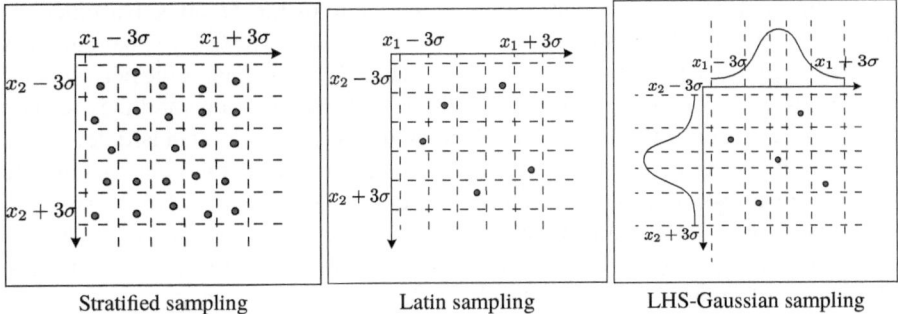

Fig. 12.3 Statistical space sampling methods

Then how to efficiently sample within a bounded area is a key problem in robustness evaluation. As shown in Fig. 12.3, one of the classical methods is the stratified sampling method, which divides the areas into several grids with equal sizes and then randomly samples in each grid. To further reduce the number of samplings, Latin hypercube sampling method (viz., LHS) only selects one grid in each row and each column to sample. Suppose that there are $N * N$ divided grids in the area, LHS only creates N sample points while stratified sampling needs $N * N$ sample points. If the noise distribution is known, the size of grids can be adjusted according to the distribution, such as LHS-Normal sampling, where small grids are designed in the dense data area, while large grids are designed in the sparse data area.

The definition shows that there are several metrics of robustness, and different metrics represent different requirements for robustness. For example, the expected objective function μ_f indicates the performance of a solution, while the standard deviation σ_g of the constraint function indicates the feasibility robustness. A general robust formulation is introduced here that can be applied to different robust optimization problems with different robustness requirements [1].

$$\min_{(\boldsymbol{x},\boldsymbol{\delta})} \mu_{f_i(\boldsymbol{x}+\boldsymbol{\delta})}, i = 1, 2, \ldots, M \tag{12.5}$$

$$\min_{(\boldsymbol{x},\boldsymbol{\delta})} f_{M+1}(\boldsymbol{x} + \boldsymbol{\delta}) = \max(\sigma_g, R_c) \tag{12.6}$$

$$\max_{(\boldsymbol{x},\boldsymbol{\delta})} f_{M+2}(\boldsymbol{x} + \boldsymbol{\delta}) = \min(\sigma_f, R_f) \tag{12.7}$$

$$\text{s. t. } \sigma_g \equiv \max_j(\mu_{g_j(\boldsymbol{x}+\boldsymbol{\delta})}/\sigma_{g_j(\boldsymbol{x}+\boldsymbol{\delta})}) \leq 0 \tag{12.8}$$

where

$$\sigma_f \equiv \min_i(\sigma_{\bar{f}_i(\boldsymbol{x}+\boldsymbol{\delta})}/\sigma_{f_i(\boldsymbol{x}+\boldsymbol{\delta})}) \tag{12.9}$$

where the first M objectives are the expected value of the M objective functions and there are two added objectives; the first added objective (Eq. (12.6)) indicates the feasibility robustness, and the second one (Eq. (12.7)) indicates the performance robustness; σ_g refers to the maximum ratio of the expected value and the standard deviation of all constraint functions, which indicates how many standard deviations can be fit between the constraint boundary (0) and the given solution; σ_f refers to the ratio of a user-defined acceptable deviation $\sigma_{\bar{f}}$ and the standard deviation of objective f; R_c is the feasibility robustness level provided by the user, i.e., solutions with values of σ_g smaller than R_c satisfy the feasibility robustness level; similarly, R_f is the performance robustness level.

Then the robust optimization problem is formulated as a multi-objective constraint optimization problem with $M + 2$ objectives and one constraint. Thus, multi-objective optimization algorithms and constrained optimization algorithms can be applied to this problem.

12.2.2   Uncertainties in Objective Function

Noise is introduced into objective function due to the inaccuracy of the sensor measurement errors, which can be written as [9]

$$\min f(x) + \delta \tag{12.10}$$

where $\delta \sim D(0, \Sigma_f)$.

Similarly to robust optimization with uncertainties of decision variables, it can also use the expectation of the mean value ($E(f|x)$) and variance (Σ_f) as its robustness value. Similar to the robust optimization problem with uncertainties in decision variables, this robust optimization problem with uncertainties can also be formulated into multi-objective problems. It is simpler because the robustness of a solution can be easily evaluated with sufficient evaluations of the same solution.

Research on this topic is paid more attention to the evaluation of the robustness. With noise in the objective function, the selection of better solutions in EC becomes more difficult, because the evaluation of the solutions is unreliable. To obtain a more accurate evaluation of a solution, several strategies were proposed [9]:

(1) Fitness sampling of individual trial solution. Fitness sampling, also known as sampling, tries to eliminate the impact of noise by evaluating a solution many times, and many strategies have been proposed to improve the sampling efficiency by adjusting the sample size to each solution.
(2) Fitness estimation of noisy samples. Fitness estimation attempts to approximate the evaluation of a solution with a surrogate model built on noisy fitness samples. The surrogate model will be introduced in detail in Chap. 14.
(3) Dynamic population size over generations. This method increases the population size to reduce the effects of noise. Generally, the fitness landscape is

continuous, and, thus, the solutions that are close to each other have similar objective values. With a large population size, more solutions will be generated in the same area, and thus the promising area near the global optimum can be found by EAs more easily.

(4) Adaptation of the evolutionary search strategy. The evolutionary search strategies are modified to overcome two challenges in the robust optimization with uncertainties in variables decision: ① to determine the promising area near the global optimal and ② to avoid premature convergence caused by the inaccurate evaluation of some certain solutions.

(5) Modification in the selection strategy. In traditional EAs, the selection strategy tends to select the solutions with higher fitness values, while in robust optimization with uncertainties in the objective function, the evaluation of a solution may be inaccurate due to the contamination of noise. The proposed selection strategies aim to select solutions with a higher degree of uncertainty to resample and select the truly good solutions to promote evaluation.

Here is a simple example of a sampling method for a single objective problem. A simple idea for sampling size assignment is to adjust the sample size according to the degree of uncertainty. If the solution is located in an area with high uncertainty, a large sample size is needed to evaluate the solution more accurately. The standard error is one of the most effective indicators of the contamination level of noise, which is used in standard error dynamic resampling method, namely, SEDR [8].

Algorithm 12.1: SEDR

Set $n(x) = n^{min}$;
Sample $f(x)$ for $n(x)$ times;
while *true* **do**

> Evaluate the estimated mean fitness $\hat{f}(x) = \frac{1}{n(x)} \sum_{j=1}^{n(x)} f(x)_j$;
> Calculate the estimated sample standard deviation
> $\hat{\sigma} = \sqrt{\frac{1}{n(x)-1} \sum_{j=1}^{n(x)} (f(x)_j - \hat{f}(x))^2}$;
> Determine the standard error of the mean fitness estimate $se(\hat{f}(x)) = \hat{\sigma}/\sqrt{n(x)}$;
> **if** $se(\hat{f}(x)) \geq se^{th}$ **then**
>> $n(x) = n(x) + 1$;
>> Resample the fitness;
>
> **else**
>> break;

In Algorithm 12.1, $n(x)$ is the sample size for one solution. Initially, the solution is given a small sample size n^{min}, which is a user-defined parameter. Then the solution is re-evaluated for $n(x)$ times and then its standard deviation $\hat{\sigma}$ and the standard error of the mean fitness estimate $se(\hat{f}(x))$ is determined. Intuitively, $se(\hat{f}(x))$ becomes smaller with more fitness samples. The procedure is repeated until $se(\hat{f}(x))$ is smaller than a threshold se^{th}.

12.2.3   Uncertainties in Environments

In industrial production, the optimization problem is also influenced by many other factors from the environment, which are difficult to control, such as the temperature and humidity of the environment or the opponent's strategies in a game problem, which are called uncertainty factors. Note that a value of the uncertainty factors is called a scene in a robust problem.

The robust optimization with uncertainties in the environment can be written as:

$$\begin{aligned} \min\ & f(x, s) \\ \text{s. t.}\ & x \in \mathbb{X} \\ & s \in \mathbb{S} \end{aligned} \tag{12.11}$$

where $\mathbb{S}$ is the set of all uncertainty factors.

12.2.3.1   Min-Max Optimization Problem

The goal of the problem is to optimize the worst-case performance of a solution. For example, to win a game problem, one needs to consider all possible strategies from his opponent and find a relatively good strategy that can react well to all these possible strategies.

In this problem, the worst-case fitness value is generally used as the optimization goal for the problem

$$\min_{x \in \mathbb{X}} \max_{s \in \mathbb{S}}\ f(x, s) \tag{12.12}$$

This type of problem is also known as a min-max optimization problem according to its expression.

12.2.3.2   Coevolution for Min-Max Optimization Problem

According to the definition of the min-max optimization problem, a solution x to be evaluated needs to be sampled in the entire uncertain set $\mathbb{S}$. However, the sampling would be very time-consuming with a large $\mathbb{S}$. In fact, there may be some typical scenarios in the uncertainty set in which most solutions will get the worst performance. For example, in practical applications, the performance of an instrument generally declines under high temperatures. Based on this assumption, a simple idea is to divide the problem into two subproblems and optimize the solution

x in some typical asymmetric scenarios s, so that some typical scenarios can be collected to accurately evaluate a solution.

$$\min_{x \in \mathbb{X}} h_1(x) = \min_{x \in \mathbb{X}} \max_{s \in \mathbb{P}_s} f(x, s) \tag{12.13}$$

$$\max_{s \in \mathbb{S}} h_2(s) = \max_{s \in \mathbb{S}} \min_{x \in \mathbb{P}_x} f(x, s) \tag{12.14}$$

where $\mathbb{P}_s$ and $\mathbb{P}_x$ are the set of the representative decision variables or representative scenarios, respectively.

Thus, the idea of coevolution introduced in Sect. 13.2 can be applied to this problem. But there are some differences between the origin CC and the CC for the min-max optimization problem:

(1) There is no regrouping in the coevolution for min-max optimization problem, because the variables are naturally divided into two groups, one group for the decision variables and the other group for the scenarios;
(2) The output solution is the best solution from the subpopulation that represents the decision variables, while the coevolution needs to combine the best solutions from all the subpopulations to get a complete solution;
(3) With more samples in $\mathbb{S}$ or $\mathbb{X}$, the evaluation of the decision variables x or the uncertainty factors s is more accurate. Therefore, it normally takes all individuals from the other subpopulation to evaluate the individual in the current subpopulation.

12.2.3.3   Limitation and Improvements

Although there is much research about coevolution for min-max optimization problems, there is a limitation in the framework of coevolution for the min-max problems that cannot be avoided, that is the asymmetry problem as shown in Fig. 12.4. Different solutions will obtain their worst performance at different

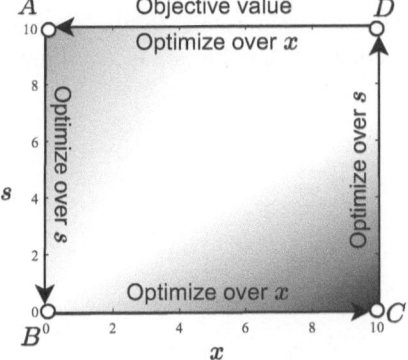

Fig. 12.4 Asymmetrical situations

scenarios, and thus the two subpopulations are not independent of each other, causing the algorithm to fall into an infinite loop.

For example, in the case with one objective, as shown in Fig. 12.4, suppose that the algorithm has converged to the optimal position $B = (x, s) = (0, 0)$, which means that all the individuals from the two subpopulations are concentrated near the position B. Then the algorithm continues to optimize the subpopulation in the decision space $\mathbb{X}$. At the same time, the best individuals in the subpopulation $\mathbb{S}$ are near zero, that is, $\mathbb{P}_s = \{0\}$.

$$\min_{x \in \mathbb{X}} h_1(x) = \min_{x \in \mathbb{X}} \max_{s \in \mathbb{P}_s = \{0\}} f(x, s) = \min_{x \in \mathbb{X}} f(x, 0) = f(10, 0)$$

then the algorithm pushes the subpopulation in $\mathbb{X}$ to converge to $C = (10, 0)$ and thus $\mathbb{P}_x = \{10\}$. The algorithm then switches to evolve the subpopulation in space $\mathbb{S}$.

$$\max_{s \in \mathbb{S}} h_2(s) = \max_{s \in \mathbb{S}} \max_{x \in \mathbb{P}_x = \{10\}} f(x, s) = \max_{s \in \mathbb{S}} f(10, s) = f(10, 10)$$

Similarly, the algorithm pushes the $\mathbb{S}$ subpopulation to converge to $D = (10, 0)$ and thus $\mathbb{P}_x = \{10\}$. During evolution, the whole population will be pushed along B to C to D to A and then again.

The key problem with this situation is that some typical scenarios are lost in the selection process of EAs. To save these typical scenarios, some fitness assignment methods were proposed to handle this situation, which evaluates the scenarios in a different way [3]. However, another limitation of the coevolution for the min-max problems is the high computational cost. The algorithm not only needs to completely sample in the $\mathbb{P}_s$ to evaluate a solution x but also needs to evolve the scenarios set at the same time. To avoid the asymmetric situation and improve the efficiency, Rehman et al. [11] attempted to use the surrogate models which will be introduced in Chap. 14 to evaluate the solutions.

12.2.4 Uncertainties Over Time

The dynamic refers to the time-dependent change of a problem. Its general mathematical definition is as follows

$$f(x, \phi(t)) \tag{12.15}$$

The survival time of a solution as an objective for the robust optimization over time with a single objective is defined as follows [5]:

$$R_{sur}(x, t) = \begin{cases} 0 & \text{if } f(x, \phi(t)) < \epsilon \\ \max(l|t \le i \le t + l : f(x, \phi(i)) \ge \epsilon) & \text{otherwise} \end{cases}$$

(12.16)

where t is the current time stamp; ϵ is a predefined acceptable fitness threshold.

The survival time stands for the maximum time for the solution with its objective value greater than ϵ in successive environments. Then the bi-objective optimization algorithm can be used to solve the problem. However, the multi-objective optimization algorithm will generate a Pareto set of solutions, which causes difficulty in how to choose one solution.

In practical problems, there is a certain cost to change the solution. Paper [12] approximates the distance between solutions as the cost to switch between the two solutions and takes it as a new objective for the robust optimization over time, which is defined as follows

$$R_{SC}(x, t) = \omega * \frac{\|x - \check{x}(t)\|}{\sqrt{D}}$$

(12.17)

where $\check{x}(t)$ is the solution executed at time t and D is the dimension of the problem.

There are several indicators for robustness in a dynamic environment, and each represents different types of robustness. The robust optimization is formulated as a multi-objectives problem with different types of robustness definition as objectives. For example, Huang [6] took the survival time and switching cost as two objectives in the robust mathematical model

$$\min(R_{sur}(x, t), R_{SC}(x, t))$$

(12.18)

and then a multi-objective algorithm was used as the optimizer.

Another key question to solve the problem is how to estimate the robustness over time. From the definition of the robustness of survival time, the fitness value of the solutions in the future time span $[t, t + l]$ should be obtained. To get the fitness value of a solution in the future, a predictor based on the historical data is needed. However, if a solution is newly constructed, it has little or no historical data, which is not sufficient to build a prediction model. Then a local approximator is built on the whole fitness landscape based on all historical data at time t to estimate the fitness value of any solution at time t.

In a general framework [7] for one objective problem, the radial basis function model (RBF) with a linear kernel is selected as the approximation model

$$\hat{f}(x, \phi(t)) = \sum_{i=1}^{n_c} \omega_i \|x - c_i\|_2$$

(12.19)

where $\hat{f}(x, \phi(t))$ is the estimated objective value of x at time t; n_c is the number of the selected training data; $\omega = [\omega_1, \omega_2, \ldots, \omega_{n_c}]^\mathsf{T}$ is the weight vector, which can be calculated using the least square method as follows

$$\omega^* = (K^T K)^{-1} K^T z_t \tag{12.20}$$

where $z_t = [f(c_1, \phi(t)), f(c_2, \phi(t)), \ldots, f(c_{n_c}, \phi(t))]^\mathsf{T}$ and the ijth element of matrix K is calculated as $\|c_i - c_j\|_2$.

Then the autoregressive model is selected as the predictor. With the RBF, the time series of fitness values of a solution x can be obtained

$$p(x, t) = [f(x, \phi(t - \psi)), f(x, \phi(t - \psi + 1)), \ldots, f(x, \phi(t))]^\mathsf{T} \tag{12.21}$$

The predictor aims to use the time series fitness value to predict the future fitness value of the solution x at time t. The predictor is built as follows

$$q(x, t) = \hat{f}(x, \phi(t + 1)) = \epsilon_{t+1} + \sum_{i=0}^{\psi} \eta_i f(x, \phi(t - i)) \tag{12.22}$$

where ϵ_{t+1} is the noise ; $\eta_0, \eta_1, \ldots, \eta_\psi$ are the parameters of the model, which indicates the coefficient of the variables.

In the predictor, the input-output pair of solution x_k at time t forms as

$$(p(x_k, t), q(x_k, t)) \tag{12.23}$$

Then the parameters of the model η can be calculated using the least square method

$$\eta^* = (\chi^\mathsf{T} \chi)^{-1} \chi^\mathsf{T} \gamma \tag{12.24}$$

where $\chi = [p(x_1, t), p(x_2, t), \ldots, p(x_{n_s}, t)]^\mathsf{T}$;$\gamma = [q(x_1, t), q(x_2, t), \ldots, q(x_{n_s}, t)]^\mathsf{T}$; n_s is the number of the selected training data.

In fact, the predictor in the framework is not only used to predict the future fitness value but also to act as a strategy to handle changes in the environment.

12.3 Discussions

Although much research has been done on robust optimization, there is still a lot of work to do.

The robust optimization aims to satisfy the practical demand for robustness in reality. However, there exists a gap between the mathematical definition of robustness and the true demand for robustness in actual production and daily life, especially for robust optimization with dynamic changes. The traditional definition of robustness in a dynamic environment does not consider the robustness of the space. In practice, minor changes to a solution should be considered acceptable. These changes can help enhance the performance of the current solution or allow for necessary adjustments in response to emergencies.

Robust optimization methods play an important role in addressing the challenges posed by the uncertainties that exist widely in theoretical research and practical applications. For example, robust optimization was used to enhance the local stability of neural nets in [10].

References

1. Asafuddoula, M., Singh, H.K., Ray, T.: Six-sigma robust design optimization using a many-objective decomposition-based evolutionary algorithm. IEEE Trans. Evol. Comput. **19**(4), 490–507 (2015)
2. Beyer, H.G., Sendhoff, B.: Robust optimization – a comprehensive survey. Comput. Methods Appl. Mech. Eng. **196**(33), 3190–3218 (2007)
3. Branke, J., Rosenbusch, J.: New approaches to coevolutionary worst-case optimization. In: Rudolph, G., Jansen, T., Beume, N., Lucas, S., Poloni, C. (eds.) Lecture Notes in Computer Science, pp. 144–153. Springer, Berlin (2008)
4. Chen, W., Allen, J., Tsui, K.L., Mistree, F.: A procedure for robust design: minimizing variations caused by noise factors and control factors. ASME J. Mech. Des. **118**, 478–485 (1996)
5. Fu, H., Sendhoff, B., Tang, K., Yao, X.: Robust optimization over time: problem difficulties and benchmark problems. IEEE Trans. Evol. Comput. **19**(5), 731–745 (2015)
6. Huang, Y., Ding, Y., Hao, K., Jin, Y.: A multi-objective approach to robust optimization over time considering switching cost. Inf. Sci. **394**, 183–197 (2017)
7. Jin, Y., Tang, K., Yu, X., Sendhoff, B.: A framework for finding robust optimal solutions over time. Memetic Comput. **5**, 3–18 (2012)
8. Pietro, A.D., While, L., Barone, L.: Applying evolutionary algorithms to problems with noisy, time-consuming fitness functions. In: IEEE Congress on Evolutionary Computation, pp. 1254–1261. IEEE, Portland (2004)
9. Rakshit, P., Konar, A., Das, S.: Noisy evolutionary optimization algorithms – a comprehensive survey. Swarm Evol. Comput. **33**, 18–45 (2017)
10. Shaham, U., Yamada, Y., Negahban, S.: Understanding adversarial training: increasing local stability of supervised models through robust optimization. Neurocomputing **307**, 195–204 (2018)
11. Ur Rehman, S., Langelaar, M., van Keulen, F.: Efficient kriging-based robust optimization of unconstrained problems. J. Comput. Sci. **5**(6), 872–881 (2014)
12. Yazdani, D., Branke, J., Omidvar, M.N., Nguyen, T.T., Yao, X.: Changing or keeping solutions in dynamic optimization problems with switching costs. In: 2018 Genetic and Evolutionary Computation Conference, pp. 1095–1102. Association for Computing Machinery, Kyoto (2018)

Chapter 13
Large-Scale Global Optimization

Abstract With the development of computing capability and optimization technologies, research on optimization has shifted from simple optimization problems to complex optimization problems. One of the challenges posed by this complexity is the dimensionality disaster, which is caused by increases in the number of variables. Large-scale global optimization (LSGO) problems, which usually contain more than 1000 variables, exist in many applications, e.g., large-scale power system design, vehicle routing problems, genetic identification, inverse chemical kinetics problem, and so on. This chapter introduces the definition of large-scale optimization problems, some classical algorithms, and some discussion of LSGO.

13.1 Large-Scale Global Optimization Problems

To effectively solve the large-scale global optimization problem, we normally do not search in the original search space. The nature of the problem, whether it is separable, partially separable, or non-separable, depends on the interrelationships between variables (see Sect. 2.2.5).

$f(x)$ is a **fully separable** function if (if and only if)

$$\min / \max f(x) = (f(x_1), f(x_2), \cdots, f(x_n)) \qquad (13.1)$$

where $x = (x_1, x_2, \cdots, x_n)$ is a D-dimensional decision vector.

$f(x)$ is a **partially separable** function with m independent subcomponents iff

$$\min / \max f(x) = (f(x_1, \cdots), \cdots, f(\cdots, x_m)) \qquad (13.2)$$

where $x_1, \cdots, x_m$ are disjoint sub-vectors of x and $2 \leq m < n$. Note, $f(x)$ is a **fully non-separable** function, if every pair of its decision variables interacts with each other.

$f(x)$ is a **partially additively separable** function if

$$f(x) = \sum_{i=1}^{m} f_i(x_i) \tag{13.3}$$

where x_i are mutually exclusive decision vectors of f_i ; m is the number of separable subcomponents.

Non-separable and overlapping functions are the most difficult to solve.

(1) As the number of decision variables increases, the search space generally increases exponentially, and the algorithm might get trapped in local optima more easily.
(2) The objective function generally exhibits the characteristics of nonlinear, non-convex, multimodal, and non-differentiable.
(3) Algorithm performance relies on search space decomposition strategies in problems where variables are partially separable or completely non-separable.

It is very difficult for existing mathematical methods to efficiently search in such a large search space due to the unaffordable computational complexity. With these difficulties, it is also challenging for EAs to explore the entire search space effectively.

In general, meta-heuristics that solve large-scale optimization problems have two branches: space decomposition methods (coevolution) and non-decomposition-based methods, which are introduced as follows.

13.2   Coevolution Methods

The coevolutionary method decomposes the large-scale global optimization problem into several low-dimensional subproblems and then uses the divide-and-conquer method to solve the large-scale optimization problem. Due to the interaction between variables in indivisible problems that have a great influence on algorithm performance, most decomposition methods try to identify the interacting variables and assign them to the same sub-question.

13.2.1   Cooperative Coevolution

In order to solve high-dimensional problems, Potter and De Jong proposed the cooperative coevolutionary (CC) approach [13] in 1994. The CC algorithm was designed to improve the performance of GAs. They provided the possible decomposition methods and the cooperative optimization method of subproblems. Two simple collaboration methods among subcomponents were suggested, i.e., the best and random collaboration methods. The framework of the classic CC algorithm has three steps:

(1) Problem decomposition. The n-dimensional decision variables of the objective function are decomposed into m subdimensional components.
(2) Subcomponent optimization. A traditional optimization algorithm is used to optimize the m subdimensional components in turn.
(3) Cooperative combination. Merging of an n-dimensional solution vector from the best individuals of the current component and the best individuals of the other components.

Based on different grouping strategies, the coevolution algorithms used to solve LSGO problems are mainly divided into two categories: static grouping methods and dynamic grouping methods.

13.2.2 Static Grouping

The previously mentioned CC algorithms [13] were tested only for the maximum dimension of 30. The results show that the proposed algorithms are more effective than classical GAs in separable problems, but less effective in non-separable problems. To address problems on a larger scale, Liu et al. [8] combined the CC framework with FEP (fast evolutionary programming) to solve the problem of 100–1000 dimensional continuous optimization, confirming the inadequacy of Potter and De Jong's decomposition method in dealing with non-decomposition problems.

Thereafter, scholars tried to combine the idea of CC with algorithms based on swarm intelligence, such as CPSO-SK and CPSO-HK [1]. However, the coevolutionary algorithm based on static grouping is only effective in low-dimensional problems due to imprecise grouping, so scholars have tried to develop new decomposition strategies with dynamic grouping ideas to deal with large-scale optimization problems more efficiently.

13.2.3 Dynamic Grouping

Some scholars try to propose new methods to detect the interacting relationship between variables and assign the interacting variables to the same subcomponent. In static grouping, the number of subcomponents (k) is fixed, while in dynamic grouping, the structure of subcomponents can be dynamically adjusted.

Dynamic grouping methods can be divided into two categories: random dynamic grouping and learning-based dynamic grouping.

13.2.3.1 Random Dynamic Grouping

An excellent DE-based coevolution algorithm (DECC-G) using a random grouping strategy was proposed by Yang et al. [22]. This algorithm attempts to solve the non-decomposing LSGO problem of 500–1000 dimensions. Decompose a n dimensional target vector into multiple low-dimensional subcomponents, each of which is evolved by a self-adaptive DE with a neighborhood search algorithm. Then, a random weight is assigned to each subcomponent. They proposed an adaptive weighting framework to further improve solutions, which assigns a weight to each of the subcomponents after each cycle. An optimization algorithm is used to optimize these weights, and the dimension of the optimization problem is much lower than before. Algorithm 13.1 presents the pseudocode of the framework of the DECC-G algorithm.

Algorithm 13.1: DECC-G algorithm

---

while *termination criterion is not fulfilled* **do**
    Set $i = 0$;
    The n-dimensional object vector is randomly divided into m s-dimensional subcomponents;
    while $i < m$ **do**
        $i + +$;
        Evolve the ith subcomponent with a certain EA;
    Apply a weight to each of the subcomponents;
    Optimize the weight vectors via a certain EA ;

The random decomposition can be extended to non-separable benchmark functions, but the performance of the algorithm will deteriorate with the increase of the number of strongly interacting variables [11].

To improve the performance of DECC-G, a new multilevel CC algorithm (MLCC) was proposed [23]. To reduce the impact of the group size on the evolution of the objective function, MLCC uses a decomposer pool. Each decomposer specifies a group size. The algorithm selects the current decomposer to group objective vectors based on the performance of the decomposers in the evolution process and updates the decomposer. This method is more applicable to most real-world problems [10].

13.2.3.2 Learning-Based Dynamic Grouping

In general, decomposing problems with interactions between variables requires prior knowledge of the problem. In such methods, the identification of interactions between variables is learned before or during the optimization process. The purpose of this type of method is to increase the chance that interacting variables are assigned to the same subcomponent.

Ray and Yao proposed a coevolutionary algorithm based on a correlation matrix in [14]. In the first several cycles of running, a subcomponent contains all the variables, so the evolution process is similar to the standard EA. In the following cycle, the correlation matrix of the first 50% solution in the population is calculated, and the decision variables are divided into several subcomponents according to the correlation between decision variables, that is, the variables whose correlation coefficient is greater than a certain threshold are placed in the same subcomponent.

Inspired by [14], CCEA-AVP with an adaptive grouping method was proposed in [16]. However, the methods based on the correlation coefficient require a large amount of calculations, and it cannot identify the nonlinear dependence between variables. Based on the above analysis, a coevolutionary algorithm, contribution-based cooperative coevolution (CBCC) [12], was proposed to allocate computing resources according to the contribution of each subcomponent. The results show that if there is an imbalance among the separable and non-separable parts of the fitness value in the LSGO problem, this method can significantly reduce the computational resource.

To decompose an LSGO problem, you need to know if there is any interaction between the variables. A simple method for identifying interactions between variables is described below [21]. Suppose *"best"* is the best solution currently obtained, *"new"* represents the best individual obtained by the CC optimizer for dimension i, and *"rand"* represents an individual randomly selected in the population. According to the following principle, two new individuals will be generated from the above three vectors

$$
x_j = \begin{cases} new_i & \text{if } j = i \\ best_j & \text{otherwise} \end{cases} \qquad x'_j = \begin{cases} new_i & \text{if } j = i \\ rand_k & \text{if } j = k \\ best_j & \text{otherwise} \end{cases} \qquad (13.4)
$$

If $f(x')$ is better than $f(x)$, the probability of interaction between dimensions i and k is relatively high. Based on the above ideas, Chen et al. [3] proposed a method called CCVIL (i.e., CC method with variable interaction learning) that can adaptively adjust the size of the group. This approach consists of two phases: learning and optimization. In the learning phase, the main purpose is to detect the interaction between variables, and the method is the same as [21].

The theorem used to identify the interaction between two variables is defined as follows [10].

Theorem 13.1 *Suppose $f(x)$ is an additively separable function, $\forall a, b_1 \neq b_2, \delta \in \mathbb{R}, \delta \neq 0$, if the following condition holds, then x_p and x_q are nonseparable*

$$
\Delta_{\delta,x_p}[f](x)|_{x_p=a,x_q=b_1} \neq \Delta_{\delta,x_p}[f](x)|_{x_p=a,x_q=b_2} \qquad (13.5)
$$

where $\Delta_{\delta,x_p}[f](\boldsymbol{x}) = f\left(\ldots, x_p + \delta, \ldots\right) - f\left(\ldots, x_p, \ldots\right)$ *is the forward differ-ence of* f *with respect to variable* x_p *with the interval* δ.

To find the interaction of decision variable i, the following loop is repeated: each decision variable i is checked for mutual use with all other variables using Theorem 13.1. If an interaction is detected, the variable is assigned to subcomponent j. If no interaction is identified between all other variables and variable i, then variable i is identified as separable.

The third dependency identification (DI) technique used to decompose the LSGO problem was proposed in [15]. The definition of the partially separable problem is as follows

$$f(\boldsymbol{x}) = \sum_{k=1}^{m} f_k\left(x_v\right) \quad v = [1, V] \tag{13.6}$$

where the problem $f(\boldsymbol{x})$ is decomposed into m subcomponents which each subcomponent has v dependent variables.

The DI technique decomposes the decision variables into several subgroups to minimize the least square difference sq_{diff} between $f(\boldsymbol{x})$ and $f(\boldsymbol{x}) = \sum_{k=1}^{m} f_k\left(x_v\right)$, $v = [1, V]$ which is calculated by the following equation

$$sq_{diff} = \left[f(\boldsymbol{x}) - \sum_{k=1}^{m} f_k\left(x_v\right) \right]^2 \quad v = [1, V] \tag{13.7}$$

13.3   Non-decomposition-Based Methods

The non-decomposed method mainly attempts to improve the performance of the standard meta-heuristic algorithm to solve the LSGO problem. Such algorithms focus on defining new mutations, selection and crossover operators, hybridization, opposition-based learning, designing and employing local search, sampling operators, or variable population size methods.

13.3.1   PSO-Based Algorithms

Unlike standard PSO (see Sect. 5.4), a PSO with velocity modulation and restarting strategy [6] was proposed. The velocity modulation controls the directional motion of particles in a finite range. Restart strategies are used to prevent premature. If the overall change in the standard deviation of particle fitness in the whole population is very small, the restart strategy is used.

An incremental particle swarm optimizer with local search (IPSOLS) was proposed to solve the LSGO problem [9]. They used a tuning-in-the-loop approach to redesign the IPSOLS. The redesign process has six stages: selecting the local search method, alteration of calling and controlling the local search method, using vector PSO rules, penalizing bound constraints violation, and fighting stagnation with restarting. The information obtained in the previous stages is used to guide the development process in the latter stages.

In [4], a competitive PSO algorithm was proposed to solve the LSGO problem (up to 5000 variables). In each generation of the algorithm, the particles in the current particle swarm engage in random pairwise competitions. The winning particles are selected to survive in the next generation, while the losing particles learn from the position and velocity of the winners.

13.3.2   EDA-Based Algorithms

EDA is a population-based optimization algorithm based on statistical principles (see Sect. 5.7). In [18], a univariate EDA (LSEDA-GL) is proposed to solve LSGO problems. The Gaussian sampling, the Levy probability distribution, and a restart strategy are adopted to prevent premature. Then, they analyzed the performance of univariate EDAs mixed with different kernel probability densities based on fitness landscape analysis and proposed an adaptive mixed distribution-based univariate EDA and mixed kernel [19].

In order to reduce the computational complexity of EDAs, an EDA framework with model complexity control was proposed [5] for solving continuous LSGO problems. The authors first identified the weakly correlated variables of the problem and then converted the problem model to a univariate Gaussian model. To obtain better overall performance, they divided the n-dimensional search space into multiple subspaces, and then a multivariate model is constructed in each subspace.

13.3.3   DE-Based Algorithms

Since the DE algorithm has the characteristics of simple structure and strong robustness (see Sect. 5.6), many researchers have applied DE to LSGO problems.

A new mutation operation DE/current-to-pbest [25] (see Eq. (5.31)) was applied in [24] for optimizing each subcomponent with a random dynamic grouping method [22] and an adaptive weighting process.

A DE based on detection of landscape modality was proposed [17]. Some points on a line connecting the centroid of search points and a search point are sampled. When the objective values of the sampled points are changed decreasingly and then increasingly, it is thought that one valley exists. If there exists only one valley, according to the modality detection technique, a scale factor selection method based on landscape modality detection was proposed to deal with LSGO problems.

An adaptive DE method, which uses a population size reduction method and the sign alteration technique of the scale parameters, was proposed [2]. In the search process, the sign alteration of scale parameter F is probabilistically exchanged based on the fitness value of the randomly selected vector, which is used in the mutation operation of the DE algorithm.

A shuffle parallel DE based on multi-population was proposed [20], in which two random strategies were used. The algorithm divides the population into several subpopulations, with each corresponding to a scale factor. The first strategy, known as the shuffle strategy, repartitions the population into several subpopulations with a certain probability. The second strategy replaces the scale factors of all subpopulations with random samples between 0.1 and 1.

13.4   Learning-Based Algorithms

Combination optimization problems related to practical engineering are often large-scale problems, such as the vehicle routing problem (VRP) and bin packing problem. Algorithms for solving this kind of problem are divided into two types: exact algorithms and heuristic algorithms. The exact algorithm provides the exact optimal solution of the problem but requires a large amount of computation and is only suitable for solving small-scale combinatorial optimization problems. Although the heuristic algorithm is fast, it can only acquire the approximate optimal solution. Engineers require a lot of experience or expertise to design special search strategies for different types of problems in order to obtain good search results and reduce the computational resources of the algorithms.

With the rise of machine learning, learning methods have been used to solve these difficult optimization problems. With the help of learning models, the specific prior knowledge of the problem can be learned, and many effective heuristic strategies can be auto-designed. The following is an illustration of the project of Noah's Ark Lab of Huawei on the integration of vehicle path planning and 3D packing problem [7].

A complete logistics process generally involves loading and transportation. The vehicle routing problem and bin packing problem are two major problems in actual logistics engineering. In order to solve the above difficulties, Noah's Ark Lab divides the logistics process into three layers: vehicle route planning, cargo splitting, and container loading. Each problem has a large solution space due to the permutation and combination, and they are difficult to solve.

Noah's Ark Lab uses the expert knowledge generated by historical data, probability model, and prediction model to guide the EDA to search the solution space of the problem more effectively.

The vehicle routing problem is the outermost layer of the logistics problem. The VRP solution space is efficiently represented by a generator distribution to improve the search efficiency. Human expert knowledge-based historical data are fused into this process as prior knowledge to improve the quality of initial candidate solutions.

The problem in the middle layer is the cargo splitting, the purpose of which is to allocate goods for the vehicles that have planned the route. In this step, the algorithm clusters boxes according to their length, width, height, and weight and then, for each collection point, establishes the probability of each category of boxes being assigned to each route.

After the first two layers of processes are completed, the innermost cargo loading problem needs to be solved. This link is actually a bin packing problem. The solution to this problem is to combine the existing heuristic packing algorithm with supervised learning and use the offline regression model to enhance the exploration of the solution space.

13.5   Discussions

The large-scale optimization is a significant part of big data optimization. The difficulty of the problems increases as the growth of the dimensions of the problem. How to reduce the difficulty of solving large-scale optimization problems is an important research direction. The efficiency of the search will drop significantly as the number of dimensions increases. To address this challenging issue, three promising approaches are problem decomposition, dimensionality reduction, and learning-based search.

The dimensionality reduction aims to identify the most significant variables for the problem and hence transforms the problem from high to low dimensions while preserving the major features. The search will be carried out in the transformed low-dimensional space, and the solution will be transformed back to the large-scale space for the objective evaluation.

The problem decomposition aims to put the most relevant variables into one group and hence divides the problem into a set of small-scale problems, and the search is carried out on these small-scale problems. The solutions to these small problems will be assembled together for the objective evaluation.

The learning-based method aims to extract useful information about the problem and uses it as prior knowledge of the problem to speed up the search. This is particularly useful for real-world applications.

References

1. Van den Bergh, F., Engelbrecht, A.P.: A cooperative approach to particle swarm optimization. IEEE Trans. Evol. Comput. **8**(3), 225–239 (2004)
2. Brest, J., Zamuda, A., Boskovic, B., Maucec, M.S., Zumer, V.: High-dimensional real-parameter optimization using self-adaptive differential evolution algorithm with population size reduction. In: IEEE Congress on Evolutionary Computation, pp. 2032–2039. IEEE, Hong Kong (2008)

3. Chen, W., Weise, T., Yang, Z., Tang, K.: Large-scale global optimization using cooperative coevolution with variable interaction learning. In: Schaefer, R., Carlos, C., Kolodziej, J., Rudolph, G. (eds.) Parallel Problem Solving from Nature, pp. 300–309. Springer, Berlin (2010)
4. Cheng, R., Jin, Y.: A competitive swarm optimizer for large scale optimization. IEEE Trans. Cybern. **45**(2), 191–204 (2014)
5. Dong, W., Chen, T., Tiňo, P., Yao, X.: Scaling up estimation of distribution algorithms for continuous optimization. IEEE Trans. Evol. Comput. **17**(6), 797–822 (2013)
6. Garcia-Nieto, J., Alba, E.: Restart particle swarm optimization with velocity modulation: a scalability test. Soft Comput. **15**(11), 2221–2232 (2011)
7. Li, X., Yuan, M., Chen, D., Yao, J., Zeng, J.: A data-driven three-layer algorithm for split delivery vehicle routing problem with 3d container loading constraint. In: Proceedings of the 24th ACM SIGKDD International Conference on Knowledge Discovery & Data Mining, pp. 528–536. Association for Computing Machinery, London (2018)
8. Liu, Y., Yao, X., Zhao, Q., Higuchi, T.: Scaling up fast evolutionary programming with cooperative coevolution. In: Proceedings of the 2001 Congress on Evolutionary Computation, pp. 1101–1108. IEEE, Seoul (2001)
9. Montes de Oca, M.A., Aydn, D., Stützle, T.: An incremental particle swarm for large-scale continuous optimization problems: an example of tuning-in-the-loop (re) design of optimization algorithms. Soft Comput. **15**(11), 2233–2255 (2011)
10. Omidvar, M.N., Li, X., Mei, Y., Yao, X.: Cooperative co-evolution with differential grouping for large scale optimization. IEEE Trans. Evol. Comput. **18**(3), 378–393 (2013)
11. Omidvar, M.N., Li, X., Yang, Z., Yao, X.: Cooperative co-evolution for large scale optimization through more frequent random grouping. In: IEEE Congress on Evolutionary Computation, pp. 1–8. IEEE, Barcelona (2010)
12. Omidvar, M.N., Li, X., Yao, X.: Smart use of computational resources based on contribution for cooperative co-evolutionary algorithms. In: Proceedings of the 13th Annual Conference on Genetic and Evolutionary Computation, pp. 1115–1122. Association for Computing Machinery, Dublin (2011)
13. Potter, M.A., De Jong, K.A.: A cooperative coevolutionary approach to function optimization. In: Davidor, Y., Schwefel, H.-P., Männer, R. (eds.) Parallel Problem Solving from Nature – PPSN III, pp. 249–257. Springer, Berlin (1994)
14. Ray, T., Yao, X.: A cooperative coevolutionary algorithm with correlation based adaptive variable partitioning. In: IEEE Congress on Evolutionary Computation, pp. 983–989. IEEE, Trondheim (2009)
15. Sayed, E., Essam, D., Sarker, R.: Dependency identification technique for large scale optimization problems. In: IEEE Congress on Evolutionary Computation, pp. 1–8. IEEE, Brisbane (2012)
16. Singh, H.K., Ray, T.: Divide and conquer in coevolution: a difficult balancing act. In: Sarker, R.A., Ray, T. (eds.) Agent-Based Evolutionary Search, pp. 117–138. Springer, Berlin (2010)
17. Takahama, T., Sakai, S.: Large scale optimization by differential evolution with landscape modality detection and a diversity archive. In: IEEE Congress on Evolutionary Computation, pp. 1–8. IEEE, Brisbane (2012)
18. Wang, Y., Li, B.: A restart univariate estimation of distribution algorithm: sampling under mixed Gaussian and Lévy probability distribution. In: IEEE Congress on Evolutionary Computation, pp. 3917–3924. IEEE, Hong Kong (2008)
19. Wang, Y., Li, B.: A self-adaptive mixed distribution based uni-variate estimation of distribution algorithm for large scale global optimization. In: Chiong, R. (ed.) Nature-Inspired Algorithms for Optimisation, pp. 171–198. Springer, Berlin (2009)
20. Weber, M., Neri, F., Tirronen, V.: Shuffle or update parallel differential evolution for large-scale optimization. Soft Comput. **15**(11), 2089–2107 (2011)
21. Weicker, K., Weicker, N.: On the improvement of coevolutionary optimizers by learning variable interdependencies. In: IEEE Congress on Evolutionary Computation, pp. 1627–1632. IEEE, Washington (1999)

22. Yang, Z., Tang, K., Yao, X.: Large scale evolutionary optimization using cooperative coevolution. Inf. Sci. **178**(15), 2985–2999 (2008)
23. Yang, Z., Tang, K., Yao, X.: Multilevel cooperative coevolution for large scale optimization. In: IEEE Congress on Evolutionary Computation, pp. 1663–1670. IEEE, Hong Kong (2008)
24. Yang, Z., Zhang, J., Tang, K., Yao, X., Sanderson, A.C.: An adaptive coevolutionary differential evolution algorithm for large-scale optimization. In: IEEE Congress on Evolutionary Computation, pp. 102–109. IEEE, Trondheim (2009)
25. Zhang, J., Sanderson, A.C.: Jade: adaptive differential evolution with optional external archive. IEEE Trans. Evol. Comput. **13**(5), 945–958 (2009)

Chapter 14
Expensive Optimization

Abstract In this chapter, we will first introduce the concept of expensive optimization. Some surrogate models and model management strategies in surrogate-assisted EAs are then described in detail. Finally, some challenges in the field of expensive optimization are discussed.

14.1 Introduction

In general, the objective or constraint functions of a solution can be evaluated using an explicit objective or constraint function, a computational simulation, or a physical experiment. We focus on one special type of problem, where the evaluation of their solutions is time-consuming by means of the evaluation method.

14.1.1 Expensive Optimization Problems

In practical terms, the evaluation of objective or constraint functions may be nontrivial, especially for engineering applications. These evaluations typically involve a highly time-consuming simulation, a prohibitive physical experiment, or the unavailable objective or constraint functions. These types of optimization problems with such kinds of characteristics are often referred to as expensive optimization problems (EOP).

One example is the aerodynamic design optimization problem. The performance of a given structure usually needs to be evaluated using computational fluid dynamics simulations. However, the evaluation process is computationally expensive. For example, for a three-dimensional simulation, it will take over 10 hours on a high-performance computer for one calculation.

For problems without explicit formulas for objective or constraint functions, intensive human interventions are needed for the evaluation. In many situations, such as in art design and music composition, as well as in some fields of industrial design, the evaluation of objective or constraint functions depends merely on the experience of human users.

14.1.2   Surrogate-Assisted Evolutionary Algorithms

To reduce the computational cost of evaluating functions in EAs, surrogate models (also known as meta-models) are used in EAs. This type of EAs is termed surrogate-assisted evolutionary algorithms (SAEA) [10]. In SAEAs, surrogate models built on a small amount of data are applied to replace the original objective or constraint functions to reduce the cost of the original evaluations. In this way, the total number of original function evaluations can be reduced.

SAEAs are mainly motivated by reducing computational costs for solving expensive optimization problems, where complex computational simulations and experiments are involved. The research on SAEAs was first reported in the study using approximate fitness evaluations [5]. After that, an increasing number of publications on this topic have been proposed, and special sessions on expensive optimization have also been held at the annual conferences of evolutionary computation.

14.2   Surrogate Models

When designing a SAEA, one crucial issue is how to select an appropriate surrogate model. There are several methods used for building surrogate models, including response surface methods (RSM), Gaussian processes (GP)(or Kriging), artificial neural networks (ANN), radial basis function (RBF) networks, support vector machines (SVM), and ensemble methods based on these models.

14.2.1   Response Surface Methods

The RSMs, also known as polynomial regressions (PR), employ statistical techniques of regression and analysis of variance in order to obtain the minimum variance. RSMs are usually used as surrogate models for regression and prediction analysis. There are two main advantages to RSMs. One is that RSMs derive from powerful statistical techniques. The other is that the minimum variances of the prediction can be obtained using the design of experiments with a small number of training data.

In general, a RSM is a linear regression function. The second-order RSMs (also known as quadratic polynomials) are widely adopted in regression, because they are simple. Other types of function are also possible. For the quadratic polynomial, the formulation is described as follows

$$\hat{f}(\boldsymbol{x}^*) = \beta_0 + \Sigma_{i=1}^{D}(\beta_i \cdot x_i^*) + \Sigma_{i,j=1,i<=j}^{D}(\beta_{i,j} \cdot x_i^* \cdot x_j^*) \tag{14.1}$$

where D is the dimension of decision variables and β_0, β_i, and $\beta_{i,j}$ are the unknown coefficients to be estimated.

The least-squares method and the gradient method (see Sect. 3.1) are usually used to estimate these unknown coefficients. Note that the size of the training data must be larger than the number of coefficients.

Although RSMs have many advantages in approximations, they also suffer numerically unstable issues. When the order of polynomial functions is high, the computational cost will increase enormously as the order increases. Otherwise, if the order is too low, it will not accurately approximate the original objective or constraint functions. Therefore, some methods [23] are proposed to modify the order of polynomial functions for the trade-off between approximate accuracy and computational cost.

14.2.2 Gaussian Processes

When building a cheap surrogate model for approximating an expensive function $y = f(\boldsymbol{x}), \boldsymbol{x} = [x_1, \cdots, x_D]^{\mathrm{T}} \in \mathbb{R}^D$, the GPs usually make the following assumptions [21].

(1) $F(\boldsymbol{x}) \sim \mathcal{N}(\mu, \sigma^2)$ is a Gaussian random variable, and μ and σ^2 are two constants independent from $\boldsymbol{x}$. For any $\boldsymbol{x}$, $F(\boldsymbol{x})$ is a sample of $\mu + \epsilon(\boldsymbol{x})$, where $\epsilon(\boldsymbol{x}) \sim \mathcal{N}(0, \sigma^2)$.
(2) For any $\boldsymbol{x}, \boldsymbol{x}' \in \mathbb{R}^D$, the correlation function $c(\boldsymbol{x}, \boldsymbol{x}')$ between $\epsilon(\boldsymbol{x})$ and $\epsilon(\boldsymbol{x}')$, depends on $\|\boldsymbol{x} - \boldsymbol{x}'\|$. The correlation function can be arbitrary types about distance. For example, the correlation function can be

$$c(\boldsymbol{x}, \boldsymbol{x}' \mid \boldsymbol{\theta}) = \exp\left(-\sum_{i=1}^{D} \theta_i |x_i - x_i'|^p\right) \tag{14.2}$$

where D is the dimension of problems; $\boldsymbol{\theta}$ can be interpreted as the measure of the importance or activity of variable $\boldsymbol{x}$.

Generally, when two variables are close, the distance between the two variables is small. Thus, the correlation between the two variables is considered to be high and vice versa. Figure 14.1 shows two correlation functions with $\theta = 1$ and $\theta = 4$. Comparing with $\theta = 1$, the $\theta = 4$ relates to a more active variable, since the correlation drops more rapidly as the increase in distance of $|x - x'|$. The exponent p is interpreted as measuring the smoothness of the function, e.g., functions with $p = 2$ are more smooth than that of $p = 1$ [18].

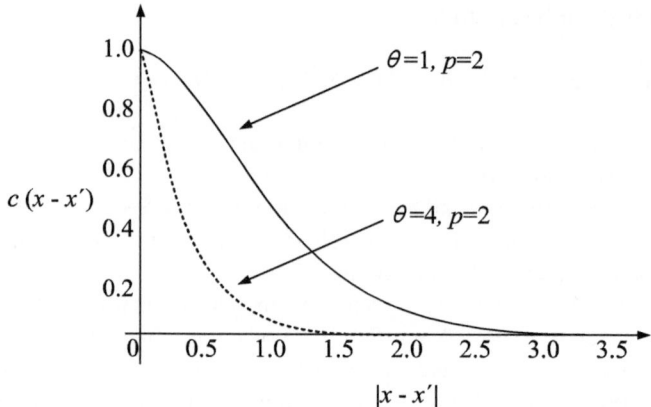

Fig. 14.1 An example for interpreting correlation functions

With the above assumption, the GPs formulation can be written as

$$F(x) = \mu + \epsilon(x) \tag{14.3}$$

Hyper Parameter Estimation
With the above assumption, the next question to be solved is to determine the hyper parameters μ, σ^2, and $\boldsymbol{\theta} = [\theta_1, \cdots, \theta_D]^T$. These hyper parameters can be estimated by maximizing the log likelihood function based on $f(\boldsymbol{x}_i) = y_i$ at $\boldsymbol{x}_i (i = 1, \cdots, N)$ [14]

$$-\frac{1}{2}[N \lg(2\pi\sigma^2) + \lg(\det(\boldsymbol{C})) + (\boldsymbol{y} - \mu\mathbf{1})^T \boldsymbol{C}^{-1}(\boldsymbol{y} - \mu\mathbf{1})/\sigma^2] \tag{14.4}$$

where $\boldsymbol{C}$ is a $N \times N$ correlation matrix whose (i, j) element is $c(\boldsymbol{x}_i, \boldsymbol{x}_j)$, $\boldsymbol{y} = [y_1, \cdots, y_N]^T$ is a N-dimensional column vector, and $\mathbf{1}$ is a N-dimensional column vector of ones.

To maximize Eq. (14.4), the values of μ and σ^2 must be

$$\hat{\mu} = \frac{\mathbf{1}^T \boldsymbol{C}^{-1} \boldsymbol{y}}{\mathbf{1}^T \boldsymbol{C}^{-1} \mathbf{1}} \tag{14.5}$$

$$\hat{\sigma}^2 = \frac{(\boldsymbol{y} - \mathbf{1}\hat{u})^T \boldsymbol{C}^{-1}(\boldsymbol{y} - \mathbf{1}\hat{u})}{N} \tag{14.6}$$

Substitute Eqs. (14.5) and (14.6) into Eq. (14.4), then we can estimate the unknown parameters $\boldsymbol{\theta}$ by maximizing the likelihood function Eq. (14.4).

Best Linear Unbiased Prediction and Predictive Distribution

When hyper parameters $\boldsymbol{\theta}$, μ, and σ^2 are determined, then the prediction $y = f(\boldsymbol{x}^*)$ at any a untested point $\boldsymbol{x}^*$ based on the function values y_i at $\boldsymbol{x}_i$ for $i = 1, \cdots, N$, can be obtained by the best linear unbiased predictor [14, 21]

$$\hat{f}(\boldsymbol{x}^*) = \hat{u} + \boldsymbol{r}^{\mathrm{T}} \boldsymbol{C}^{-1} (\boldsymbol{y} - \boldsymbol{1}\hat{u}) \tag{14.7}$$

and its mean squared error is

$$\hat{s}(\boldsymbol{x}^*)^2 = \hat{\sigma}^2 \left[1 - \boldsymbol{r}^{\mathrm{T}} \boldsymbol{C}^{-1} \boldsymbol{r} + \frac{(1 - \boldsymbol{1}^{\mathrm{T}} \boldsymbol{C}^{-1} \boldsymbol{r})^2}{\boldsymbol{1}^{\mathrm{T}} \boldsymbol{C}^{-1} \boldsymbol{1}} \right] \tag{14.8}$$

where $\boldsymbol{r} = [c(\boldsymbol{x}^*, \boldsymbol{x}_1), \cdots, c(\boldsymbol{x}^*, \boldsymbol{x}_N)]^{\mathrm{T}}$, $N(\hat{f}(\boldsymbol{x}^*), \hat{s}(\boldsymbol{x}^*)^2)$ can be regarded as a predictive distribution for $F(\boldsymbol{x}^*)$ given the function values y_i at $\boldsymbol{x}_i$ for $i = 1, \cdots, N$.

GPs are widely used in various fields in practice. It provides not only the prediction for any untested points but also the uncertainty degree of the prediction. Unfortunately, the computational cost of GPs is very high, with $O(N^3)$, N is the number of training data.

14.2.3 Artificial Neural Networks

Artificial neural networks (ANN) are inspired by biological nervous systems, such as the brain, to process information. They are widely used to learn the unknown domain knowledge, for example, pattern recognition, data classification, and regression.

In ANNs, there are three types of neurons, including input nodes, hidden nodes, and output nodes. The input nodes, as the input layer, are mainly to take in information. The information is then presented as activation values in hidden nodes. Finally, the output nodes are used to collect the results obtained by the network. The primary element in ANNs is how to construct a network, involving selecting inputs and outputs, the type of activation functions in hidden nodes, and the number of activation functions in each layer and the number of hidden layers.

In ANNs, the multilayer perception (MLP) as a multilayered feed-forward neural network has been widely used for regression and classification based on the discovery of practical experience that the MLP can provide a good approximation for optimization problems, as shown in Fig. 14.2. The MLP formulation is

$$\hat{f}(\boldsymbol{x}^*) = \phi(\Sigma_{i=1}^{N} \omega_i \cdot x_i^* + b) \tag{14.9}$$

where x_i^* are the inputs of the neuron; ω_i is the weight corresponding to the i-th input; and ϕ is the activation function with nonlinear function.

Fig. 14.2 A graphical
representation of an ANN

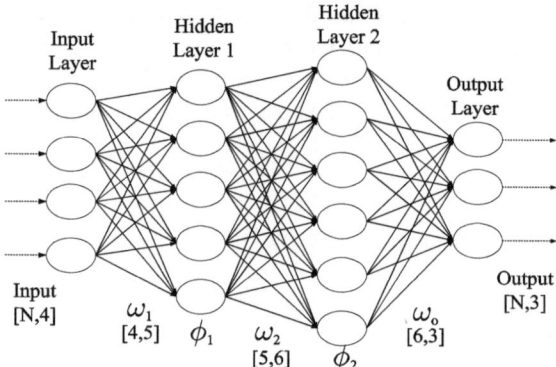

As we all know, training an ANN needs a large number of samples. Then the
high computational cost will become a bottleneck of using ANNs. However, good
performance can be obtained by sacrificing more computational cost. Therefore, the
computational cost and performance of the model need to be balanced by users.

14.2.4   Radial Basis Function Networks

The radial basis function (RBF) networks are three-layer feed-forward neural
networks, where the RBF is a function whose value depends only on the distance
from each point x to the origin, for instance, $\phi(x) = \psi(\|x\|)$, or can be generalized
to the distance from some other center point c, that is $\phi(x) = \psi(\|x - c\|)$. Any
function ϕ that satisfies the property $\phi(x) = \psi(\|x\|)$ is called RBF. The norm is
usually the Euclidean distance, although other distance functions are also possible.

The RBFNs have three layers, namely, the input layer, the hidden layer, and the
output layer. It uses an RBF as its activation function. In RBFNs, the input layer is
directly connected to the hidden layer, and the output can be shown in Eq. (12.19).
Note that the time t will not be considered in this section. The weights in the RBFNs
can be determined using the least-squares method, which is shown in Eq. (12.20).

Many types of RBFs can be chosen as shown in Table 14.1. RBFNs are simple,
flexible, and attractive surrogate models. Also, the computational cost is low due to
their simple structure. However, the precision of the RBFNs is still to be improved.

14.2.5   Support Vector Machines

The support vector machine (SVM) is proposed based on the inspiration from
statistical learning theory. The SVM can be used for classification and regression
by constructing a hyperplane or a set of hyperplanes in a high-dimensional space.

Table 14.1 Types of the radial basis function

Types	Radial basis function
Linear	$\psi(r) = r$
Cubic	$\psi(r) = r^3$
Thin plate spline	$\psi(r) = r^2 \lg(r)$
Gaussian	$\psi(r) = \exp(-r^2)$
Multiquadric	$\psi(r) = (r^2 + c^2)^{\frac{1}{2}}$
Inverse multiquadric	$\psi(r) = (r^2 + c^2)^{-\frac{1}{2}}$

The SVM at an untested x^* is expressed as

$$\hat{f}(x^*) = \omega^T \phi(x^*) + b \tag{14.10}$$

where $\phi(x^*)$ is feature function, and ω and b are coefficients to be estimated.

The unknown parameter ω and b can be obtained by optimizing a constrained optimization function based on given function values y_i at x_i for $i = 1, \cdots, N$.

$$\min \quad \frac{1}{2}\|\omega\|^2 + L \sum_{i=1}^{N}(\xi_i + \xi_i^*)$$

$$\text{s.t.} \begin{cases} y_i - \omega^T\phi(x_i) - b \le \varepsilon + \xi_i \\ \omega^T\phi(x_i) + b - y_i \le \varepsilon + \xi_i^* \\ \xi_i^*, \xi_i \ge 0 \end{cases} \tag{14.11}$$

where L and ε are prespecified values by users and ξ_i and ξ_i^* are slack variables representing upper and lower constraints.

The SVM has good generalization performance. However, when a large number of training samples are involved, it will take a lot of computational overhead.

14.2.6 Model Ensembles

In machine learning, there is a suggestion that an ensemble model can provide more accurate predictions than any of its members alone [2]. In addition, some observations demonstrate that a heterogeneous ensemble model performs more robust and reliable on various problems by comparing with using one single surrogate model for GPs. The model ensembles have been widely used in expensive optimization [12, 16, 27]. An ensemble model consisting of three types of surrogate model [26], i.e., a PR model, a RBF model, and a GP model, shows more robust performance than the three models in solving optimization problems. Likewise, the heterogeneous ensembles were applied to replace GPs [6], which show that the heterogeneous ensembles are comparable to the GPs in solving expensive multi-objective problems.

Generally, model ensembles show better performance than one single model. On the one hand, they efficiently use the merit of other surrogate models. On the other hand, they can reduce computational complexity. For instance, model ensembles are good alternatives to GPs due to their relatively small computational complexity.

14.3 Model Management

Model management is a critical component in SAEAs. It is used to select samples in the database that are used to train surrogate models. In addition, it is also used to select candidate samples to be evaluated by the original objective or constraints to balance the search between exploration and exploitation (see Chap. 7). Model management is an important component in both offline SAEAs and online SAEAs, which will be introduced in this section.

14.3.1 Model Management in Offline SAEAs

The diagram of the offline SAEAs is shown in Fig. 14.3. In the offline optimization process, no new data can be generated [27]. Thus, the model management heavily relies on the quality and amount of the pre-given data, and the surrogate models only depend on the samples from model management.

In offline SAEAs, the quality and amount of the data in the database are essential because the accuracy of the surrogate models highly depends on the data. However, these data are often non-ideal in practice. Many challenges can be encountered, e.g., the data are incomplete, imbalanced, or noisy. Besides, the amount of data may be huge or small, which will result in the high computational costs for processing these data and building surrogate models and the low accuracy of the surrogate models, respectively.

There are also some methods to address the above challenges in offline SAEAs, including data pre-processing, data mining, and synthetic data generation [13], which are shown in Fig. 14.4.

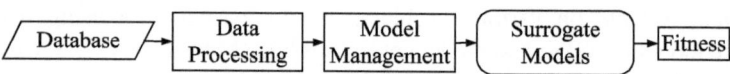

Fig. 14.3 The diagram of offline SAEAs

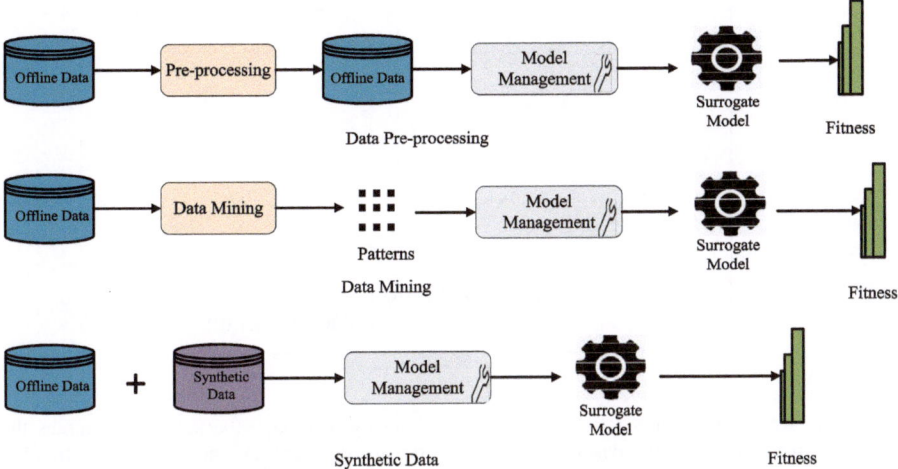

Fig. 14.4 An explanation of data processing in offline optimization

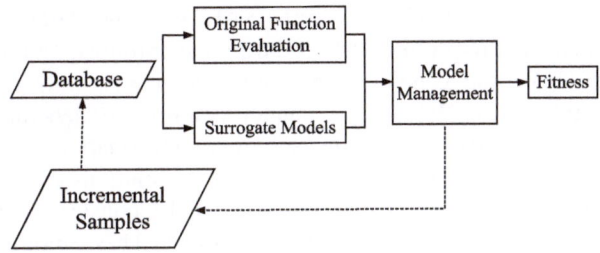

Fig. 14.5 The diagram of online SAEAs

14.3.2   Model Management in Online SAEAs

Model management combines surrogate models and the original fitness functions in online SAEAs to improve the performance of algorithms and the accuracy of surrogate models. Compared with offline SAEAs, online SAEAs allow the generation of candidate solutions to update the surrogate models during optimization, as shown in Fig. 14.5. Thus, it is more flexible than offline SAEAs.

Note that no surrogate model is globally accurate, especially in the high-dimensional search space. For example, a false optimum is the optimum of the surrogate models, but it is not the optimum of the original fitness functions, as shown in Fig. 14.6. Generally, selecting some candidate solutions by the model management is to constantly update the surrogate model. In this way, the accuracy of the surrogate model can be improved effectively.

It should be noted that offline SAEAs can be deemed as a special case of online SAEAs. Based on this case, methodologies developed for offline SAEAs discussed

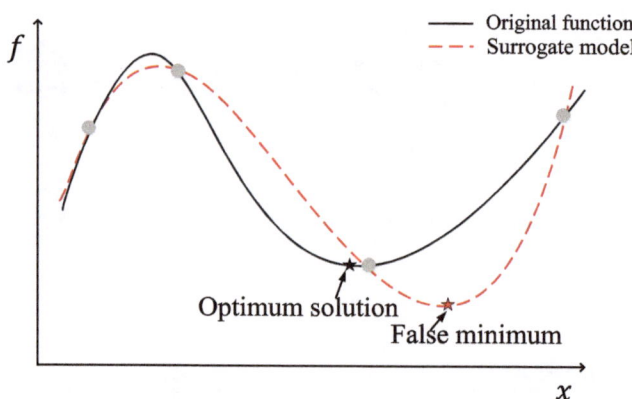

Fig. 14.6 An example of a false minimum in the surrogate model, the solid line denotes the original fitness function, the dashed line denotes the surrogate model, and the dots are training samples

in Sect. 14.3.1 can also be applied to online SAEAs. In the following, we discuss the model management for online SAEAs during the optimization process.

In online SAEAs, the generation of solutions may or may not be actively controlled by EAs [13]. When EAs cannot control the generation, the main challenge is to efficiently capture the solutions' information. In the case that solution generation is controlled by EAs, candidate solutions can be obtained by model management, and the surrogate model can be constantly updated. Many model management strategies have been developed. These model management strategies can be categorized into two types: generation-based and individual-based. Generation-based strategies adjust the sampling frequency generation by generation, while individual-based strategies select a small number of candidate solutions at each generation.

For generation-based model management strategies, candidate solutions are sampled in η generations of the optimization process, then these solutions are added to the database, and the surrogate models are updated based on the database, as shown in Fig. 14.7. The parameter η is involved in generation-based model management strategies. There are several methods for setting the parameter. For example, the parameter η was predefined in [3, 20] or adaptively tuned according to the quality of the surrogate models [10].

Compared with generation-based strategies, individual-based strategies are more flexible, as shown in Fig. 14.8. Generally speaking, training samples are selected based on their predicted fitness [10, 11] and their uncertainty according to the current surrogate model [4, 9].

The samples with promising predicted fitness can accelerate the convergence of algorithms and enhance the accuracy of surrogate models. For the samples with a large degree of uncertainty, there are two reasons for considering them. First, the fitness landscape around these samples has not been well explored, and therefore evaluating these samples can improve the exploration ability of algorithms [1].

Evolutionary progress

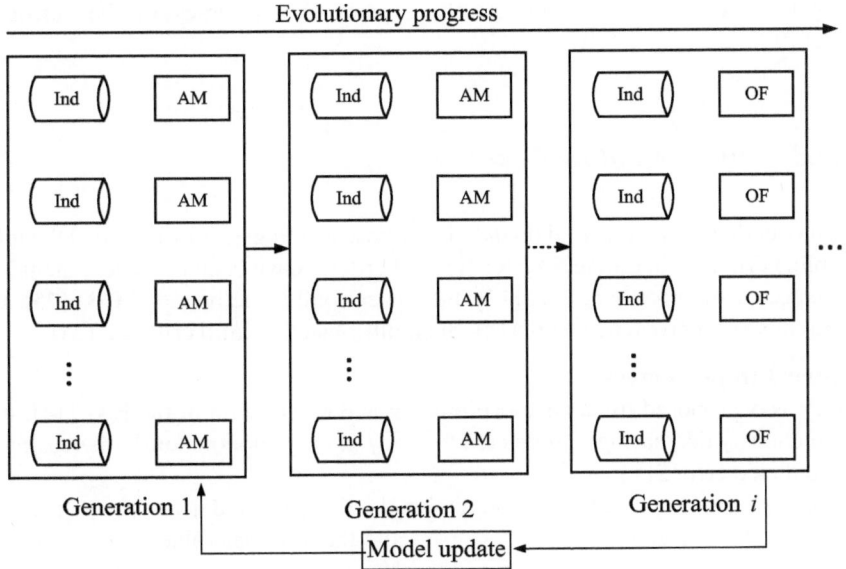

Fig. 14.7 Generation-based strategies, where AM denotes fitness evaluation by the surrogate model and OF by the original function

Evolutionary progress

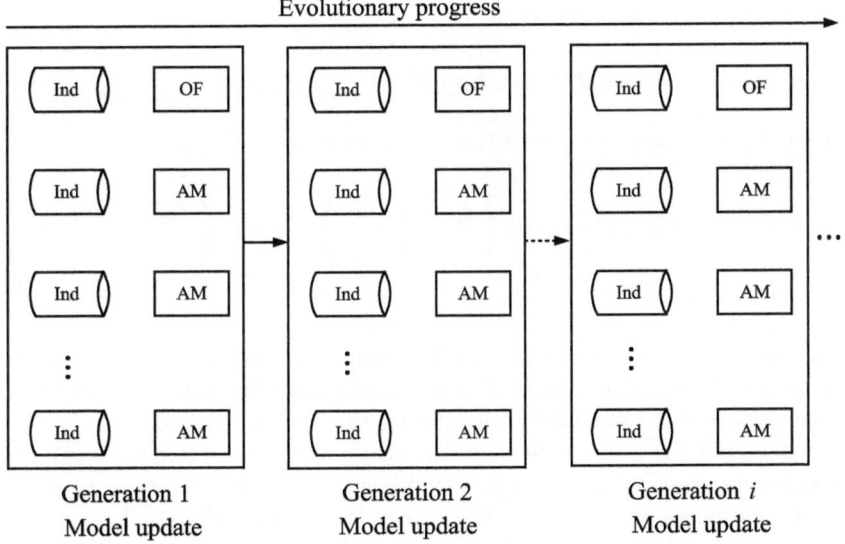

Fig. 14.8 Individual-based strategies, where AM denotes fitness evaluation by the surrogate model and OF by the original function

Second, these samples can effectively improve the accuracy of the surrogate model [9].

14.3.3 Infill Sampling Criterion

The model management based on individual-based strategies is often called the infill sampling criterion in online SAEAs. Several criteria have been proposed, including the expected improvement (EI) [14], the lower confidence bound (LCB) [25], the probability of improvement (PoI) [8], the multi-objective infill criterion [24].

Expected Improvement
The EI was proposed to balance exploitation and exploration of the EAs [14]. It is monotonously decreasing in predicted fitness $\hat{f}(x)$ and monotonously increasing in the predicted error $\hat{s}(x)$.

Suppose $\mathcal{N}(\hat{f}(x), \hat{s}(x)^2)$ is a predictive distribution model for $f(x)$ given the function values y_i at x_i for $i = 1, \cdots, N$, and the minimal value of $f(x)$ over all the evaluated points in y is f_{min}, then the improvement of $f(x)$ at a untested point x is

$$I(x) = \max\{f_{min} - \hat{f}(x), 0\} \tag{14.12}$$

Thus, the expected improvement is

$$E[I(x)] = E[\max\{f_{min} - \hat{f}(x), 0\}] \tag{14.13}$$

Then expected improvement [14] can be written as

$$E[I(x)] = (f_{min} - \hat{f}(x))\Phi\left(\frac{f_{min} - \hat{f}(x)}{\hat{s}(x)}\right) + \hat{s}(x)\phi\left(\frac{f_{min} - \hat{f}(x)}{\hat{s}(x)}\right) \tag{14.14}$$

where the prediction mean $\hat{f}(x)$ and the prediction variance $\hat{s}(x)$ are calculated according to the surrogate model; Φ is the standard normal cumulative distribution function; and ϕ is the standard normal probability density function.

Equation (14.14) is monotonically decreasing with respect to $\hat{f}(x)$ and increasing with respect to $\hat{s}(x)$; therefore, maximizing Eq. (14.14) balances exploitation and exploration to some extent. The principle of EI is illustrated as shown in Fig. 14.9. As can be seen from the left figure, there are two peaks, one at $x = 3$ and the other at $x = 13$. The peak at $x = 3$ is higher than at $x = 13$, so sample $x = 3$ is selected to enhance exploitation capacity. But on the next iteration, as shown in the right figure, the expected improvement is maximized at $x = 13$, and sample $x = 13$ is selected to enhance capacity of exploration.

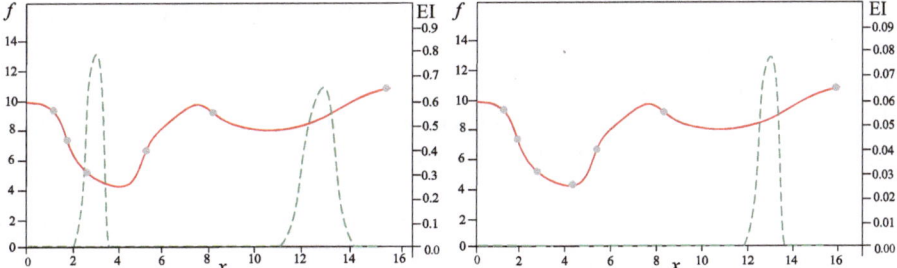

Fig. 14.9 An example for the illustration of the EI principle, where the left scale is for the original objective function and the right is for the expected improvement function. The solid line with samples represents the objective function curve, and the dashed line is the expected improvement function curve

The above EI formulation in Eq. (14.14) is for unconstrained optimization problems. For constrained optimization problems, the penalized form of EI is normally used [22], known as constrained expected improvement (CEI). Assuming the objective function $F(x)$ and constrained functions $G_j(x)(j = 1, \ldots, m)$ are Gaussian processes. When assuming objective and constraint functions are mutually independent at tested function values $f(x_i), g_j(x_i), i = 1, \cdots, N, j = 1, \cdots, m$. The improvement of the objective under satisfying constraints is defined as

$$
I_c(x) = \begin{cases} f_{\min} - \hat{f}(x) & \text{if} \quad \hat{f}(x) \le f_{\min} \quad \text{and} \quad l_j \le g_j(x) \le u_j \\ 0 & \text{if} \quad \text{otherwise} \end{cases} \tag{14.15}
$$

where $f_{\min}$ is the best objective function value over the feasible points in the tested N points; $\hat{f}(x)$ is prediction value at x; $g_j(x)$ is constraint value; and l_j and u_j are the lower and upper bound of constraint function, respectively.

Then CEI can be written as

$$
E\{I_c(x)\}
$$

$$
= E[I(x)] \times \prod_{j=1}^{m} P(l_j \le g_j(x) \le u_j)
$$

$$
= (f_{\min} - \hat{f}(x))\Phi\left(\frac{f_{\min} - \hat{f}(x)}{\hat{s}_f(x)}\right) + \hat{s}_f(x)\phi\left(\frac{f_{\min} - \hat{f}(x)}{\hat{s}_f(x)}\right) \tag{14.16}
$$

$$
\times \prod_{j=1}^{m}\left(\Phi\left(\frac{u_j - \hat{g}_j(x)}{\hat{s}_{g_j}(x)}\right) - \Phi\left(\frac{l_j - \hat{g}_j(x)}{\hat{s}_{g_j}(x)}\right)\right)
$$

where $\phi(\cdot)$ is the standard normal probability density function; $\Phi(\cdot)$ is the standard normal cumulative distribution function; and $\hat{f}(x), \hat{g}_j(x), \hat{s}_f(x), \hat{s}_{g_j}(x)$ are means and variances of the objective and constraints, respectively, which can be obtained by surrogate models.

Lower Confidence Bound

The LCB is inclined to guide the search of the algorithm toward less explored regions in the search space. It can be written

$$f(x) = \hat{f}(x) - \omega \hat{s}(x) \tag{14.17}$$

where $\hat{f}(x)$ is prediction mean; the $\hat{s}(x)$ is prediction variance; and the parameter ω is a constant.

The LCB is widely suggested in the literature [4], especially in solving multimodal optimization problems. In LCB, the parameter ω is a burden for the user. The performance of the criterion is highly dependent on ω, especially for high-dimensional optimization problems. There is a reasonable choice for parameter $\omega = 2$, which leads to a high confidence probability (around 97%).

Probability of Improvement

In order to get rid of the parameter ω, the PoI comes into play without parameter selection. Areas with a high PoI have a high probability of being sampled because of the samples in these areas with objective values smaller than f_{min}, which is illustrated in Fig. 14.10. Instead, areas with a low PoI have a low probability.

The PoI can be expressed

$$\text{PoI}(x) = \Phi \left(\frac{f_{min} - \hat{f}(x)}{\hat{s}(x)} \right) \tag{14.18}$$

where f_{min} is the current best fitness value and $\Phi(\cdot)$ is the standard normal cumulative distribution function.

Fig. 14.10 An example for the illustration of the POI principle, where the gray filled area represents the probability that a surrogate model output value $f(x^*)$ is sampled at point x^*, which is smaller than f_{min}

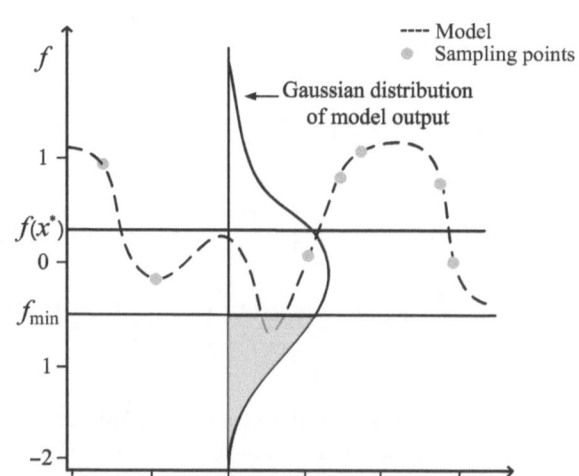

14.4 Discussions

This section will summarize the research field and discuss some challenges that exist in the research field, especially online SAEAs. In online SAEAs, the surrogate model and the model management are critical to improve the performance. Besides, there are some other challenges in solving real-world optimization problems, such as the potential applications, the choice of EAs, termination criterion and performance metric, etc. The main challenges in online SAEAs are discussed as follows.

- **Selection of surrogate models** The appropriate surrogate model is very important for different optimization problems. For example, the RBFN is often considered to solve coastal aquifer management problems because of their popularity for groundwater applications [15]. The GPs are the most widely used surrogate model, because they can provide a good prediction from a small amount of data and the degree of uncertainty for the prediction. However, different types of GPs are needed to consider for problems with different characteristics, e.g., the stationary GPs can be used to solve simple problems with stationarity. Instead, the nonstationary GPs can solve complex optimization problems.
- **Using of surrogate models** The surrogate models can be used for different types of applications. It is worth studying how to take full advantage of them. The surrogate models are often used to approximate the original objective or constraint functions. They can also be used to solve robust optimization problems with uncertainties (see Chap. 12). Besides, they can also be used to deal with different tasks, e.g., the PF approximation [7], ranks prediction [17], hypervolume prediction [19], etc.
- **Model management** The model management is very important with the aim of effectively selecting training data. The strategies are usually defined as an aggregate function. Although the balance between the two types of samples for exploration and exploitation has been considered in these criteria, there are some limitations.

  For the EI criterion, it is highly multimodal, and there are large areas where the expected improvement is essentially zero and even appears relatively flat areas. Both of these limitations increase the difficulty to search for the optimal EI value, and the optimum EI value obtained may be potentially unreliable [25]. For the LCB criterion, parameter ω is used to balance the two types of samples. The main limitation of the criterion is the selection of parameter ω that can be seen as a burden on users, attributed to the performance of the criterion is highly dependent on ω especially for high-dimensional optimization problems [25]. For the PoI criterion, a significant limitation is that it equally promotes solutions, which is likely to give very small improvement with high probability and those expected to lead to considerable improvement with a small probability.

  Although some strategies are proposed for separating the two types of samples, controlling the selection of the two types of samples is still unavoidable.
- **Using of EAs** EAs are usually used to find optimal parameters values in surrogate models and the optimal solutions for optimization problems. Different EAs

have different advantages and disadvantages. Thus, the selection of EAs should be considered according to the properties of optimization problems, such as domination-based, indicator-based, and decomposition-based algorithms can be used to solve multi-objective optimization problems (see Chap. 9).

- **Termination criterion and performance metric** The evaluation for original objective or constraint functions is costly. Thus, the termination criterion and performance metric of the algorithms are also critical. When there is no improvement, the algorithms still run, which will waste computational resources. Several performance metrics, such as the IGD and hypervolume indicators, can be used. However, these metrics are influenced by several parameters, such as the size of the reference set in the calculation of the IGD.

References

1. Branke, J., Schmidt, C.: Faster convergence by means of fitness estimation. Soft Comput. **9**(1), 13–20 (2005)
2. Brown, G., Wyatt, J.L., Tiňo, P.: Managing diversity in regression ensembles. J. Mach. Learn. Res. **6**(9), 1621–1650 (2005)
3. Bull, L.: On model-based evolutionary computation. Soft Comput. **3**(2), 76–82 (1999)
4. Emmerich, M., Giotis, A., Özdemir, M., Bäck, T., Giannakoglou, K.: Metamodel-assisted evolution strategies. In: International Conference on Parallel Problem Solving from Nature, pp. 361–370. Springer, Granada (2002)
5. Grefenstette, J., Fitzpatrick, J.: Genetic search with approximate fitness evaluations. In: International Conference on Genetic Algorithms and Their Applications, pp. 112–120. L. Erlbaum Associates, Hillsdale (1985)
6. Guo, D., Jin, Y., Ding, J., Chai, T.: Heterogeneous ensemble-based infill criterion for evolutionary multiobjective optimization of expensive problems. IEEE Trans. Cybern. **49**(3), 1012–1025 (2018)
7. Hartikainen, M., Miettinen, K., Wiecek, M.M.: Paint: Pareto front interpolation for nonlinear multiobjective optimization. Comput. Optim. Appl. **52**(3), 845–867 (2012)
8. Holger, U., Felix, S., Andreas, Z.: Evolution strategies assisted by Gaussian processes with improved preselection criterion. In: IEEE Congress on Evolutionary Computation, pp. 692–699. IEEE, Canberra (2003)
9. Jin, Y.: Surrogate-assisted evolutionary computation: recent advances and future challenges. Swarm Evol. Comput. **1**(2), 61–70 (2011)
10. Jin, Y., Olhofer, M., Sendhoff, B.: On evolutionary optimization with approximate fitness functions. In: Proceedings of the 2nd Annual Conference on Genetic and Evolutionary Computation, pp. 786–793. Morgan Kaufmann, San Francisco (2000)
11. Jin, Y., Olhofer, M., Sendhoff, B.: A framework for evolutionary optimization with approximate fitness functions. IEEE Trans. Evol. Comput. **6**(5), 481–494 (2002)
12. Jin, Y., Sendhoff, B.: Reducing fitness evaluations using clustering techniques and neural network ensembles. In: Genetic and Evolutionary Computation Conference, pp. 688–699. Springer, Seattle (2004)
13. Jin, Y., Wang, H., Chugh, T., Guo, D., Miettinen, K.: Data-driven evolutionary optimization: an overview and case studies. IEEE Trans. Evol. Comput. **23**(3), 442–458 (2018)
14. Jones, D.R., Schonlau, M., Welch, W.J.: Efficient global optimization of expensive black-box functions. J. Glob. Optim. **13**(4), 455–492 (1998)
15. Kourakos, G., Mantoglou, A.: Development of a multi-objective optimization algorithm using surrogate models for coastal aquifer management. J. Hydrol. **479**, 13–23 (2013)

16. Lim, D., Ong, Y.S., Jin, Y., Sendhoff, B.: A study on metamodeling techniques, ensembles, and multi-surrogates in evolutionary computation. In: Proceedings of the 9th Annual Conference on Genetic and Evolutionary Computation, pp. 1288–1295. Machinery, London (2007)
17. Loshchilov, I., Schoenauer, M., Sebag, M.: Comparison-based optimizers need comparison-based surrogates. In: International Conference on Parallel Problem Solving from Nature, pp. 364–373. Springer, Krakow (2010)
18. Parzen, E.: A new approach to the synthesis of optimal smoothing and prediction systems. In: Mathematical Optimization Techniques, pp. 75–108. Stanford University, California (1963)
19. Rahat, A.A., Everson, R.M., Fieldsend, J.E.: Alternative infill strategies for expensive multi-objective optimisation. In: Proceedings of the Genetic and Evolutionary Computation Conference, pp. 873–880. Association for Computing Machinery, New York (2017)
20. Ratle, A.: Accelerating the convergence of evolutionary algorithms by fitness landscape approximation. In: International Conference on Parallel Problem Solving from Nature, pp. 87–96. Springer, Amsterdam (1998)
21. Sacks, J., Welch, W.J., Mitchell, T.J., Wynn, H.P.: Design and analysis of computer experiments. Stat. Sci. 4(4), 409–423 (1989)
22. Schonlau, M., Welch, W.J., Jones, D.R.: Global versus local search in constrained optimization of computer models. Lecture Notes-Monogr. Ser. 34, 11–25 (1998)
23. Stronger, D., Stone, P.: Polynomial regression with automated degree: a function approximator for autonomous agents. Int. J. Artif. Intell. Tools 17(01), 159–174 (2008)
24. Tian, J., Tan, Y., Zeng, J., Sun, C., Jin, Y.: Multiobjective infill criterion driven gaussian process-assisted particle swarm optimization of high-dimensional expensive problems. IEEE Trans. Evol. Comput. 23(3), 459–472 (2018)
25. Virginia, T., Michael, T.: Using approximations to accelerate engineering design optimization. In: 7th AIAA/USAF/NASA/ISSMO Symposium on Multidisciplinary Analysis and Optimization, p. 4800. American Institute of Aeronautics and Astronautics, Inc., St. Louis (1998)
26. Wang, H., Jin, Y., Doherty, J.: Committee-based active learning for surrogate-assisted particle swarm optimization of expensive problems. IEEE Trans. Cybern. 47(9), 2664–2677 (2017)
27. Wang, H., Jin, Y., Sun, C., Doherty, J.: Offline data-driven evolutionary optimization using selective surrogate ensembles. IEEE Trans. Evol. Comput. 23(2), 203–216 (2019)

Chapter 15
Real-World Applications

Abstract EAs have been widely applied in various fields; in this chapter, results of some of real-world applications from our research group, e.g., the design of antenna, the vehicle routing problem, and the contamination source identification in water distribution systems will be presented.

15.1 Antenna Design

The wireless communication system has been widely used in various fields. As an important part of the communication system, antenna design shows high standard and intelligence characteristics. The traditional antenna design, which is based on the designer's experience and supplemented by simulation softwares, relies too much on the developer's experience, resulting in long design cycle, low efficiency, and difficult adjustment.

Evolutionary antenna is to design antenna by EAs. Evolutionary antenna [2] has been studied for about 30 years, and a large number of antennas have been successfully designed. For instance, evolutionary antennas have been designed based on GAs [16, 20, 21], PSO [1, 17, 19], and DE [10, 12]. The two most attractive cases in the field of antenna design are the design of X-band antennas for the NASA Space Technology 5 spacecraft [8] and the S-band antennas for the NASA lunar atmosphere and dust environment explorer [15].

15.1.1 Antenna Basics

This section introduces the fundamental parameters which have a significant impact on the characteristics of an antenna, such as frequency, impedance, radiation patterns, directivity, gain, and so on.

© China University of Geosciences Press 2024
C. Li et al., *Intelligent Optimization*,
https://doi.org/10.1007/978-981-97-3286-9_15

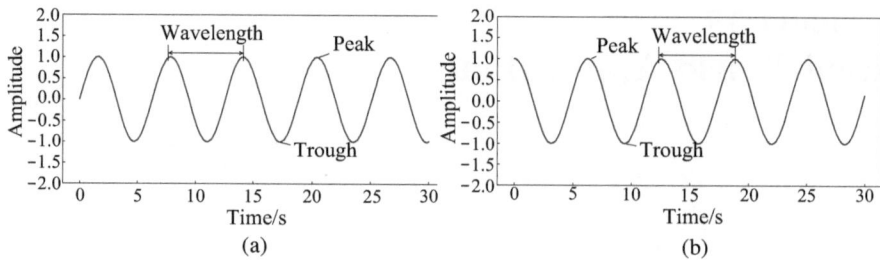

Fig. 15.1 The sinusoidal and cosine waves. (**a**) A sinusoidal wave (**b**) A cosine wave

15.1.1.1   Frequency

Frequency is one of the most important concepts in antenna theory. An antenna is used to transmit or receive electromagnetic waves. An electromagnetic wave is made up of a moving electric field and a moving magnetic field, both of which are usually associated. There are many kinds of waves, such as sinusoidal wave and cosine wave, which are plotted in Fig. 15.1a and b, respectively.

The wave is periodic, which means that it repeats itself every T seconds, and the unit for frequency is hertz (Hz). The frequency is the number of cycles the wave completes within 1 second. Mathematically, frequency ($freq$) is given as

$$freq = \frac{1}{T} \tag{15.1}$$

Wavelength λ is the distance after which the wave begins to repeat. The speed of wave propagation is determined by multiplying the frequency of oscillation by the size of each step taken during one complete period. The formula that relates the speed of light (c), frequency, and wavelength is written as

$$c = freq \times \lambda \tag{15.2}$$

The frequency is an indicator of how fast the wave oscillates. Since all electromagnetic waves transmit with the same speed, the slower it oscillates, the longer the wavelength is. And a longer wavelength means a slower frequency.

15.1.1.2   Frequency Bands

The signal bandwidth is the difference value between the high and low frequencies for the signal. For example, a signal that is transmitted between 40 and 60 MHz has a bandwidth of 20 MHz. This means that the energy is only contained between 40 and 60 MHz, and the energy for any other frequency range is almost ignored.

Table 15.1 Table of frequency bands

Frequency band	Range/Hz	Wavelength/m	Application
Extremely low frequency	3–30	10k–100k	Underwater communication
Super low frequency	30–300	1k–10k	AC power
Ultralow frequency	0.3k–3k	100–1k	/
Very low frequency	3k–30k	10–100	Navigational beacons
Low frequency	30k–300k	1–10	AM radio
Medium frequency	0.3M–3M	0.1–1	Aviation
High frequency	3M–30M	10^{-2}–10^{-1}	Shortwave radio
Very high frequency	30M–300M	10^{-3}–10^{-2}	FM radio
Ultrahigh frequency	0.3G–3G	10^{-4}–10^{-3}	Television, GPS
Superhigh frequency	3G–30G	10^{-5}–10^{-4}	Wireless communication
Extremely high frequency	30G–300G	10^{-6}–10^{-5}	Remote sensing
Visible spectrum	380T–750T	4.0×10^{-7}–7.8×10^{-7}	Human eye

Frequency bands along with the corresponding wavelengths are given in Table 15.1. Antenna designers can further divide the bands into various bands, such as "S-band," "X-band," and "Ku-band."

15.1.1.3 Impedance

The antenna can be equivalent to a lumped parameter impedance connected to the feeder terminal, which is called the input impedance of the antenna. The formula that relates voltage U_{in}, current I_{in}, and impedance Z_{in} (Ω) is given as

$$Z_{in} = \frac{U_{in}}{I_{in}} = R_{in} + jX_{in} \tag{15.3}$$

where R_{in} is input resistance; it can show the transmitting power of the antenna, including radiant power and thermal loss by the antenna itself; X_{in} is input reactance; it can show the energy stored by the antenna.

When designing an antenna, the antenna input impedance should be designed at 50 Ω as close as possible. The input impedance of antenna depends on its structure, working frequency, and the influence of surrounding environment. Approximate calculation or experimental methods are adopted to obtain input impedance in engineering.

15.1.1.4 Radiation Pattern

Radiation pattern defines the variation of the power radiated by antenna as a function of the direction away from the antenna. Figure 15.2 is an example of a three-dimensional radiation pattern.

Fig. 15.2 Radiation pattern
for an antenna

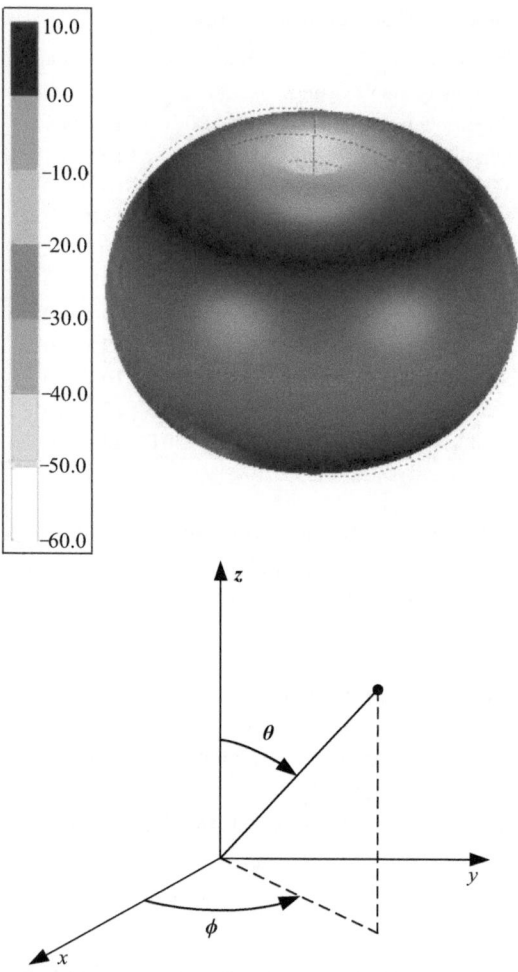

In this example, there is very little power transmitted in a vertical direction, which would correspond to the radiation directly overhead the antenna, where (θ, ϕ) represents the observation direction of the far field, in detail θ represents the elevation angle and φ is the azimuth angle, as shown in Fig. 15.2. The radiated power is maximum in the horizontal direction. The plot is useful for visualizing the directions that the antenna radiates.

In a word, radiation pattern is a figure which can allow us to visualize where the antenna transmits or receives power.

15.1.1.5 Directivity

Directivity (D_{ir}) is a fundamental parameter which measures an antenna's radiation pattern "directional." A normalized radiation pattern formula for D_{ir} is written as

$$D_{ir} = \frac{1}{\frac{1}{4\pi} \int_0^{2\pi} \int_0^{\pi} |F(\theta, \phi)|^2 \sin\theta \, d\theta \, d\phi} \qquad (15.4)$$

where the numerator for D_{ir} is the maximum value of F, and the denominator means the "average power radiated over all directions" for an antenna. This formula is just a parameter of the peak value of antenna radiated power divided by the average.

If a short dipole antenna is used as an example, $F(\theta, \phi) = \sin\theta$, and then $D_{ir} = 1.5$.

15.1.1.6 Efficiency

Due to various losses, e.g., conductor loss and medium loss, in antenna system, the practical electromagnetic wave power radiated into space is smaller than that transmitted to the antenna by the transmitter.

The antenna efficiency, which is a ratio of the radiated power P_Σ to the input power P_{in} of the antenna, is defined to describe the degree of antenna energy loss during transmission

$$\eta_a = \frac{P_\Sigma}{P_{in}} = \frac{P_\Sigma}{P_\Sigma + P_L} \qquad (15.5)$$

where P_L is loss power of antenna, such as dielectric losses of antenna structure.

Usually, antenna efficiency is a value between 0 and 1, and it is a percentage. Thus, it is necessary to increase the radiation resistance to improve the antenna efficiency.

15.1.1.7 Gain

Antenna gain is a parameter that comprehensively represents the efficiency and directional radiation capability of an antenna. Antenna gain represents the ability to radiate the input power in a specific direction. It is defined as the ratio of the power density of the antenna in the maximum radiation direction and that of the nondirectional antenna in that direction under the same input power and distance conditions. Antenna gain (G) is defined

$$G = \frac{4\pi}{P_{in}} \qquad (15.6)$$

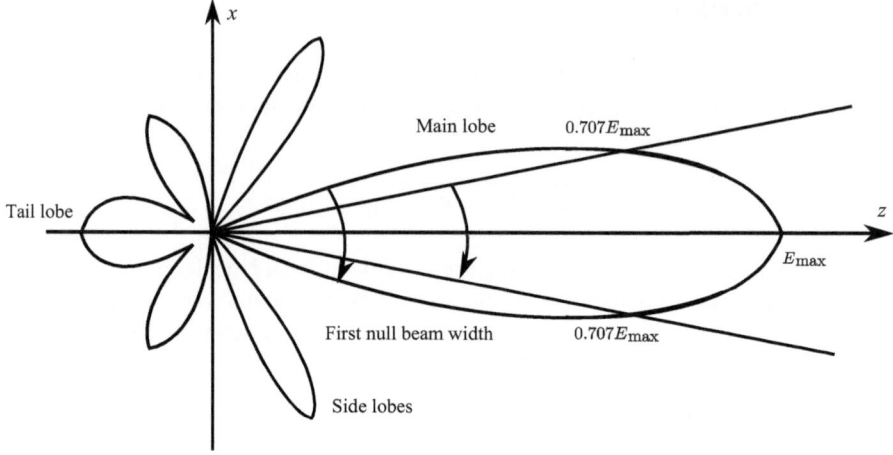

Fig. 15.3 Radiation pattern lobe

Antenna gain can be related to directivity (D_{ir}) and antenna efficiency (η_a) by

$$G = \eta_a D_{ir} \tag{15.7}$$

In a real-world problem, antenna gain is usually expressed in decibels

$$G_{dB} = 10 \lg(G) \tag{15.8}$$

15.1.1.8   Beamwidths and Sidelobes

The antenna radiation pattern is characterized by its beamwidth and sidelobes.

There are many lobes, which are shown in Fig. 15.3, in the direction diagram of an antenna. The main beam is the maximum beam within the radiation direction region. The tail lobe is the lobe opposite the main lobe. Other lobes are named sidelobes and are all smaller than the main lobes. These sidelobes are usually radiation in undesired radiation directions, and they can never be completely eliminated.

The width of the main lobe can measure whether the antenna power radiation is concentrated or not. It is the angle between the two vectors whose radiation power density is half of the maximum. It is also known as half power width or 3 dB width. The smaller the width of the main lobe, the more concentrated the radiation energy in the antenna. The null-to-null beamwidth is the angular separation from which the magnitude of the radiation pattern decreases to zero away from the main lobe, and the first null beamwidth shown in Fig. 15.3 is an example of null to null beamwidth.

The sidelobe level (SLL) is used to describe the strength of sidelobe relative to main lobe. The main lobe is surrounded by the first sidelobe, the second sidelobe, and so on in turn on both sides in the radiation pattern. The first sidelobe is usually the largest within the sidelobes. SLL is a ratio of the maximum sidelobe value to the maximum main beam value of the antenna. It is usually expressed in decibels

$$\xi_{dB} = 20 \lg \left| \frac{F(\theta_{SLL})}{F_{max}(\theta_0)} \right| \tag{15.9}$$

where $F(\theta_{SLL})$ is maximum sidelobe value; θ_{SLL} is the direction of maximum sidelobe value; θ_0 is the direction of maximum main lobe value; and $F_{max}(\theta_0)$ is maximum main lobe value.

15.1.1.9 Voltage Standing Wave Ratio

The impedance of the transmission line for the communication system must be well matched to the impedance of the antenna. The impedance of the antenna and the feeder are usually not matched, and the high-frequency energy will generate reflection and interfere with the part to produce standing wave. In order to characterize and measure the characteristics of standing wave of the antenna system, the concept of standing wave ratio is established.

VSWR is a function of the reflection coefficient, which describes the power reflected from the antenna. It is defined as

$$\text{VSWR} = \frac{1 + |\Gamma|}{1 - |\Gamma|} \tag{15.10}$$

where Γ is reflection coefficient. In mobile communication, the VSWR is often required to be less than 1.5.

The VSWR is a real and positive value for the antenna. The smaller the VSWR is, the better the transmission line is matched to the antenna, and the more power is transmitted to the antenna. The minimum of the VSWR is 1.0. That is to say, no power is reflected, and the antenna is ideal.

15.1.1.10 Polarization

Polarization is an important characteristic for any antenna. It is a rule that the direction of the E-field change over time Fig. 15.4. It represents the shape, orientation, and direction of rotation about the trajectory of the endpoint of the time-varying electric field vector. The polarization of antenna can be divided into linear polarization, circular polarization, and elliptical polarization according to the shape of electric field vector endpoint movement trajectory. According to the different directions of rotation, it can be divided into left polarization and right polarization in elliptica polarization and circular polarization.

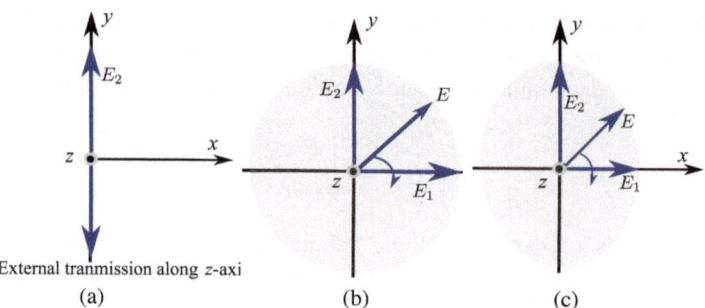

Fig. 15.4 Polarizations. (**a**) Linear polarization (**b**) Circular polarization (**c**) Elliptica polarization

1. Linear polarization

   Suppose a vector E-field that has only an x component or only a y component, and the propagation direction is in the positive z-axis. Instantaneous value is given by

$$E_x(z, t) = E_1 \sin(wt - kz) \qquad (15.11)$$

   where t represents the time; w represents angular frequency; z is the location in z-axis, while k is the wave number; kz denotes the initial phase of E_x; and E_1 is the amplitude along the x-axis in the linear polarization wave.

   Also, suppose that a vector E-field that has only a y component, and the propagation direction is in the positive z-axis. Instantaneous value is given by

$$E_y(z, t) = E_2 \sin(wt - kz) \qquad (15.12)$$

   where E_2 is the amplitude along the y-axis in the linear polarization wave.

   Obviously, at any fixed point in space, the traces of E-field vector endpoint with time are a straight line parallel to the x-axis. If the E-field stays along a single line, this field would be said to be linearly polarized as shown in Fig. 15.4a.

2. Circular polarization

   Assuming the phase difference is $\frac{\pi}{2}$ between two linearly polarized waves E_x and E_y, amplitudes are all E_m. The instantaneous values are given

$$E_x(z, t) = E_1 \sin(wt - kz)$$
$$E_y(z, t) = E_2 \sin\left(wt - kz + \frac{\pi}{2}\right) = E_2 \cos(wt - kz) \qquad (15.13)$$

   The instantaneous value of composite wave is given as

$$E(z, t) = \sqrt{(E_x^2(z, t) + E_y^2(z, t))} = E_m \qquad (15.14)$$

The angular separation α between the composite wave vector and x-axis is given

$$\tan \alpha = \frac{E_x(z, t)}{E_y(z, t)} = \tan \left[\frac{\pi}{2} - (wt - kz) \right] \tag{15.15}$$

And then

$$\alpha = \frac{\pi}{2} - (wt - kz) \tag{15.16}$$

The magnitude of the electric field remains unchanged, and the direction changes with time in Eq. (15.16). When the trajectory of the end of the electric field vector projects a circle on a plane perpendicular to the propagation direction, it is called circular polarization. If the polarization plane has a clockwise spiral relationship with the propagation direction of electromagnetic wave, it is right-handed circular polarization in Fig. 15.4b. On the contrary, if they have counterclockwise spiral relationship, it is said left-handed circular polarization.

3. Elliptical polarization
   Assuming two linearly polarized waves E_x and E_y that are orthogonal to each other have different amplitudes and phases

$$E_x(z, t) = E_1 \sin(wt - kz)$$
$$E_y(z, t) = E_2 \sin(wt - kz + \phi) \tag{15.17}$$

Component of composite wave E_x and E_y satisfy an equation

$$\left(\frac{E_x}{E_1} \right)^2 + \left(\frac{E_y}{E_1} \right)^2 - \frac{2E_x E_y}{E_1 E_2} \cos \phi = \sin^2 \phi \tag{15.18}$$

If the magnitude of the electric field also changes with time, and the horizontal component and the vertical component have different amplitudes, the projection trajectory of the end point of the resultant vector is an ellipse, and then the polarization wave formed is an ellipse polarization as shown in Fig. 15.4c.

The ratio of the major and minor axes of an ellipse is called the axial ratio (AR). Linear polarization is formed when the AR is infinite. The AR is equal to 1 for circularly polarized waves.

15.1.2   Antenna Design Problems

The antenna design problem is normally regarded as the optimization problem, which usually has many different objectives and constraints. The design problem can be modeled as a single objective optimization problem, constrained optimization problem, multi-objective optimization problem, or constrained multi-objective optimization problem. Common objectives and constraints are introduced below.

1. Objectives
   In the antenna design optimization, the objective can be divided into two categories. Designers can set optimization objective according to real-world design requirements.

   (1) Antenna performance parameters, such as G, AR, and VSWR. One or more performance parameters can be selected to construct the new objective.
   (2) Geometric parameters of the antenna, such as the reflector size, the director size, the active oscillator size, and the distance between the reflector and the active oscillator.

   Antenna designers can construct appropriate optimization objective according to their own preferences and design requirements.

2. Constraints
   When constraints are constructed, they can be divided into two categories. Designers can also set one or multiple constraints according to real-world design requirements.

   (1) The antenna performance parameters, such as G, AR, and VSWR, are limited to a certain range.
   (2) Restrictions between antenna geometry, such as the size of reflector length, are larger than the width.

   Designers can also add constraints among the structural parameters of the antenna.

   In addition, our research group proposed a new approach to enhance the robustness and stability in the antenna design [9, 11]. The design of antenna not only needs to satisfy multiple requirements but also is expected to be robust and stable. To enhance the robustness and stability of antenna design, the objective, which takes the variance of antenna performance over the frequency band and radiation space, is used. The smaller the variance, the more robust the antenna. The objective function is written as

$$f = \sum_{\theta} \sum_{\varphi} (var_G(\theta, \varphi) + var_{AR}(\theta, \varphi)) + var_{VSWR} \qquad (15.19)$$

where (θ, φ) is the single direction in spherical coordinates, θ represents the elevation angle, and φ is the azimuth angle; $var_G(\theta, \varphi)$, $var_{AR}(\theta, \varphi)$ and var_{VSWR} are variances of antenna gain, variances of axial ratio in the direction (θ, φ), and variances of VSWR, respectively, over the frequency band.

The objectives are shown as below

$$
\begin{aligned}
var_G(\theta, \varphi) &= \sum_{freq} (G(\theta, \varphi, freq) - \underset{freq}{mean}\, G(\theta, \varphi))^2 \\
var_{AR}(\theta, \varphi) &= \sum_{freq} (AR(\theta, \varphi, freq) - \underset{freq}{mean}\, AR(\theta, \varphi))^2 \\
var_{VSWR} &= \sum_{freq} (VSWR(freq) - \underset{freq}{mean}\, VSWR)^2
\end{aligned}
\tag{15.20}
$$

where $freq$ is a single frequency point, $G(\theta, \varphi, freq)$, $AR(\theta, \varphi, freq)$ and $VSWR(freq)$ mean the antenna gain, axial ratio, and VSWR, respectively, in the direction (θ, φ) and at the frequency point $freq$.

The constraints about VSWR, G, and AR are modeled with any vector of design parameters x

$$
\begin{aligned}
&\qquad\qquad VSWR(x|freq) - C_{VSWR} < 0 \\
&\qquad\qquad -G(x|\theta, \varphi, freq) - C_G < 0 \\
&\text{Linear Polarisation:}\quad C_{AR} - AR(x|\theta, \varphi, freq) < 0 \\
&\text{Circular Polarisation:}\ AR(x|\theta, \varphi, freq) - C_{AR} < 0
\end{aligned}
\tag{15.21}
$$

where C_{VSWR}, C_G, and C_{AR} are specified requirements for the VSWR, G, and AR, respectively.

15.1.3 Case Studies

In this chapter, we list the application of EAs for two types of antennas.

15.1.3.1 S-band Omnidirectional Antenna

First, the S-band omnidirectional antenna for the lunar satellite of the NASA mission, named the lunar atmosphere and dust environment explorer, was designed by DCNSGA-II [24].

Table 15.2 shows the requirements of the S-band omnidirectional antenna. One arm antenna was designed for LADEE. The wire of antenna arm has seven segments. The first wire segment, which is used as a transmission line to match a 50 impedance, is created from $(0, 0, 0)$ to $(0, 0, z1)$. The other six bent wire segments have the same amplitude and phase.

The structure of the antenna is shown in Table 15.3, and the geometry is shown in Fig. 15.5. There are 20 design variables $x = (r, z_1, x_2, y_2, z_2, x_3, y_3, z_3, x_4, y_4, z_4, x_5, y_5, z_5, x_6, y_6, z_6, x_7, y_7, z_7)$, where (x_i, y_i, z_i) is the position of the i-th junction of the antenna and r is the radius of the conduct wire. The decision space $\mathbb{X} = \{x | -25\ \text{mm} \le x_i, y_i \le 25\ \text{mm}, 0 \le z_i \le 80\ \text{mm}, 10\ \text{mm} \le r \le 20\ \text{mm}\}$. The gains

Table 15.2 Omnidirectional antenna requirements

Parameter	Requirements
Receive frequency	2025–2110 MHz
Transmit frequency	2200–2290 MHz
RF power	5 W
Antenna RF input	SMA
VSWR	<2.0:1
Gain pattern	> −2 dB for ±80° elevation from bore-sight for 0° ≤ ϕ ≤ 360° with 90% coverage
Polarization	RHCP, axial ratio ≤14 dB
Input impedance	50Ω
Antenna size	diameter, height <12.7 cm
Antenna mass	<200 g

Table 15.3 Representation of the structure of the omnidirectional antenna

	Start point	End point	Radius
Wire 1	(0,0,0)	(0,0,z_1)	r
Wire 2	(0,0,z_1)	(x_2, y_2, z_2)	r
Wire 3	(x_2, y_2, z_2)	(x_3, y_3, z_3)	r
Wire 4	(x_3, y_3, z_3)	(x_4, y_4, z_4)	r
Wire 5	(x_4, y_4, z_4)	(x_5, y_5, z_5)	r
Wire 6	(x_5, y_5, z_5)	(x_6, y_6, z_6)	r
Wire 7	(x_6, y_6, z_6)	(x_7, y_7, z_7)	r

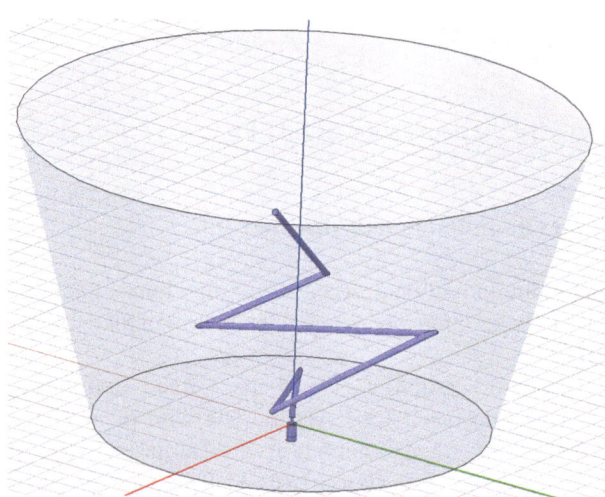

Fig. 15.5 Geometry of NASA's omnidirectional antenna

and AR are sampled with 30° and 5° increments over the angle range 0° ≤ ϕ ≤ 360° and −80° ≤ θ ≤ 80°, respectively, and the frequencies are sampled with 50 MHz increments over the band from 1.95 GHz to 2.35 GHz.

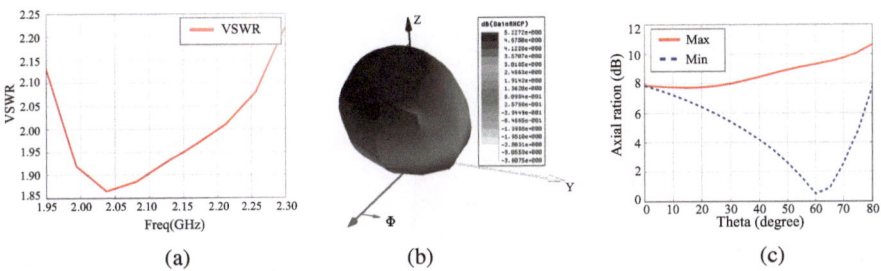

Fig. 15.6 Obtained VSWR, G, and AR of the NASA's omnidirectional antenna synthesized by DCNSGA-II. (**a**) VSWR (**b**) 3D RHCP G pattern (**c**) AR

Then, the omnidirectional antenna problem is modeled as

$$\min \quad f(\boldsymbol{x}) = \sum_{\theta} \sum_{\varphi} (var_G(\boldsymbol{x}|\theta, \varphi) + var_{AR}(\boldsymbol{x}|\theta, \varphi)) + var_{VSWR}(\boldsymbol{x})$$

$$\text{s.t.} \quad \text{VSWR}(\boldsymbol{x}|freq)) - 2 < 0$$

$$- (G(\boldsymbol{x}|\theta, \varphi, freq) - 2) < 0$$

$$\text{AR}(\boldsymbol{x}|\theta, \varphi, freq) - 14 < 0$$

$$\text{where} \quad \varphi = 0°, 30°, 60°, \cdots, 360°$$

$$\theta = -80°, -75°, -70°, \cdots, 80°$$

$$freq = 1.95\,\text{GHz}, 2.0\,\text{GHz}, \cdots, 2.35\,\text{GHz}. \tag{15.22}$$

During optimization process, the electromagnetic simulation software Ansys HFSS was used to simulate the design of the S-band omnidirectional antenna. Note that the simulation process is very time-consuming. It took about 2 months for the optimization with DCNSGA-II and the simulation software Ansys HFSS.

Figure 15.6 shows the results, which are as follows:

(1) **VSWR**: Fig. 15.6a shows the simulated VSWR at $2.0 - 2.3$ GHz. Whether for the receive frequency ($2.02-2.11$ GHz) or for the transmit frequency ($2.20-2.29$ GHz), the VSWR is less than 2 and satisfies the design requirements.

(2) **G**: The performance of G is required for $\pm 80°$ elevation from bore-sight for $0° \le \phi \le 360°$ with 90% coverage greater than -2 dB. Figure 15.6b shows the gain radiation pattern at the central frequency of 2.15 GHz, where about 90% coverage is greater than -2 dB.

(3) **AR**: Fig. 15.6c presents the AR of the central frequency of 2.15 GHz along with the change of the elevation angle. As shown in Fig. 15.6c, experimental results can satisfy the need of the AR, which is less than 14 dB.

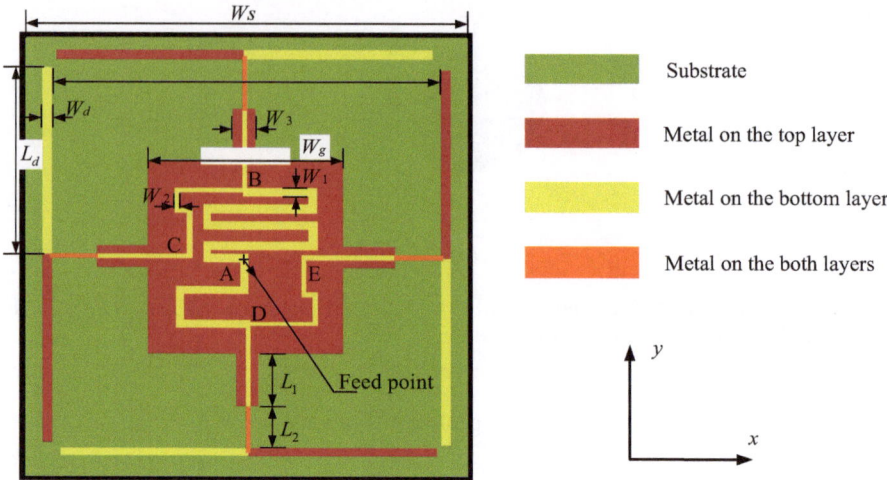

Fig. 15.7 Geometry of the CP antenna

From the data obtained, the evolved antenna can meet the need of NASA's LADEE mission.

15.1.3.2   Circularly Polarized Antenna

Another application is the modified low-profile wide-beamwidth circularly polarized antenna. The CP antenna is formed on a Rogers RO3010(tm) substrate, of which the relative permittivity $\varepsilon_r = 10.2$. Figure 15.7 exhibits its geometry, and it consists of three layers: top patch layer, lower patch layer, and substrate. The length and width (W_s) of the substrate are both 250 mm, and the height is L_s. Two pairs of parallel dipoles with distance of D_1 are orthogonally placed at the four sides of a square contour.

On the top layer, the length and width for the microstrip lines AB, BC, DE, and AD are $\lambda_g/4$ and W_1, $\lambda_g/4$ and W_2, $\lambda_g/4$ and W_2, $\lambda_g \times 3/4$ and W_1, respectively, where λ_g is the guided-wavelength at the central frequency of 425 MHz. The lower layer is made up of a square patch with a size of W_g, four rectangular patches with a width of W_3 and a length of L_1, and four L-shaped microstrip patches. The length of the L-shaped microstrip patch is about $\lambda_0/4$ ($\lambda_0/4 = L_2 + L_d$), where λ_0 is the vacuum wavelength at the central frequency of 425 MHz.

Table 15.4 presents the requirements of the CP antenna. Table 15.5 lists decision parameters and their search ranges. The G and AR are sampled in 5° increments over the angle region $0° \leq \phi \leq 360°$, and the frequencies are sampled in 10 MHz increments over the band from 400 MHz to 450 MHz.

Table 15.4 CP antenna requirements

Parameter	Requirements	Parameter	Requirements
Frequency	400–450 MHz	VSWR	<1.5:1
Input impedance	50Ω	Gain pattern	>3 dB
Polarization	RHCP	Antenna size	250 mm * 50 mm
Axial ratio	<4 dB		

Table 15.5 Decision variables and ranges of the CP antenna

Parameters	Ranges/mm	Parameters	Ranges/mm
L_s	[0.8, 50]	L_1	[0.1, 7.0]
L_2	[65, 75]	λ_0	[650, 750]
λ_g	[220, 290]	W_1	[0.4, 3.0]
W_2	[0.4, 3.0]	W_3	[1.2, 9.0]
W_d	[0.4, 3.0]		

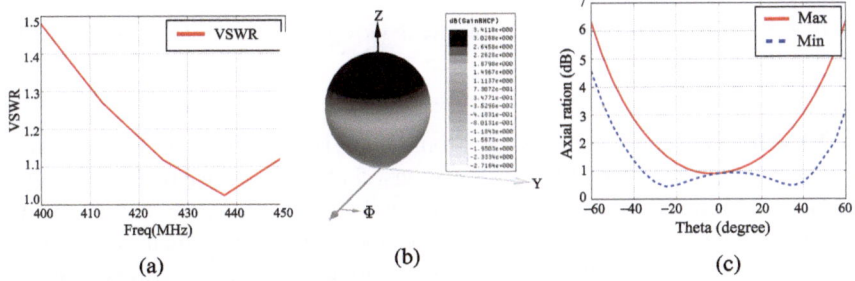

(a)   (b)   (c)

Fig. 15.8 Obtained VSWR, G, and AR of the CP antenna synthesized by DCNSGA-II. (**a**) VSWR (**b**) 3D RHCP G pattern (**c**) AR

Then, the COP is formulated as follows

$$\min\ f(x) = \sum_{\varphi}(var_G(0°, \varphi) + var_{AR}(0°, \varphi)) + var_{VSWR}$$

$$\text{s.t. } VSWR(x|freq) - 1.5 < 0$$

$$-(G(x|0°, \varphi, freq) - 3.0) < 0$$

$$AR(x|0°, \varphi, freq) - 4.0 < 0$$

$$\text{where } \varphi = 0°, 5°, 10°, \cdots, 360°$$

$$freq = 0.4\,\text{GHz}, 0.41\,\text{GHz}, \cdots, 0.45\,\text{GHz} \tag{15.23}$$

Similar to the design of the NASA's omnidirectional antenna, the electromagnetic simulation software Ansys HFSS was also used to simulate the CP antenna. Experimental results are shown in Fig. 15.8.

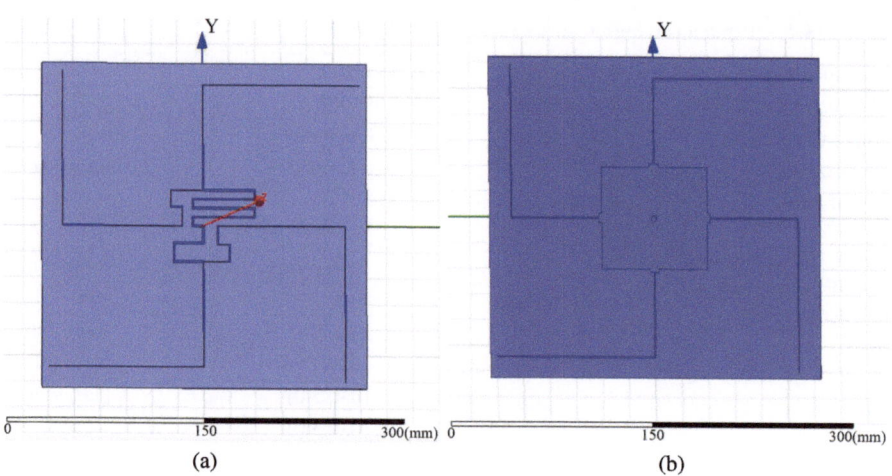

Fig. 15.9 Top-view and bottom-view geometries of the evolved CP antenna by DCNSGA-II. (**a**) Top-view (**b**) Bottom-view

(1) **VSWR**: The optimized VSWR from 400 MHz to 450 MHz is presented in Fig. 15.8a, where the VSWR is less than 1.5 at 400−450 MHz.
(2) **G**: Fig. 15.8b exhibits the 3D gain pattern. From the gain pattern, the maximum gain obtained is 3.41 dB, which satisfies the design requirement of the antenna gain. Meanwhile, the half-power beamwidth at 425 MHz has extended to about 92° in an angular range between −46° and 46°.
(3) **AR**: As shown in Fig. 15.8c, the axial ratio is less than 4 dB in an angular about range from −44° to 44° at central frequency of 425 MHz.

Figure 15.9a and b depict the top-view and bottom-view geometries of the optimized CP antenna. Based on the experimental data, we can conclude that the obtained CP antenna satisfies the design requirements well.

15.2   Vehicle Routing

With the development of online shopping, the logistics industry has attracted more and more attention. In 2018, the total expenditure on social logistics in China represented 14. 8% of GDP. Every 1% reduction in entire logistics expenses in GDP will reduce hundreds of billions of costs. In the same year, the total social logistics expense in China was 13.3 trillion yuan, among which transportation cost was 6.9 trillion yuan. Therefore, research on the vehicle routing problem (VRP) is of great significance in logistics transportation.

In simple terms, the VRP consists of a fleet of vehicles departing from a depot and a group of customers to be delivered by the fleet. Each customer is served only

Fig. 15.10 Schematic
diagram of the vehicle
routing problem

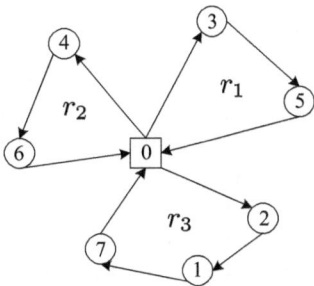

once by one vehicle. The task for this problem is to get a set of routes with minimum cost. VRP is an important topic in the field of operations research. At present, the primary method for VRP focuses on the exact algorithm and the heuristic algorithm. The exact algorithm can attain the global optimum of the VRP, but it is suitable for small-scale problems. Since exact algorithms could not obtain a solution in an acceptable time for a complex problem, more scholars have focused their attention on heuristic algorithms.

15.2.1   Basics of the Vehicle Routing Problem

In 1959, to simulate the problem to satisfy the demand for gas stations around the central hub, Dantzig and Ramser [6] proposed the truck dispatching problem for the first time. In 1964, Clarke and Wright [5] summarized the problem as an optimization problem frequently encountered in the field of logistics and transportation: how to use trucks to provide distribution services for customers distributed around the central depot, i.e., the vehicle routing problem, which is demonstrated in Fig. 15.10. In this figure, node 0 represents the depot, and nodes 1–7 represent the customers. The number of routes equals to the number of the vehicles.

Generally, depending on the focus of the decision-maker when performing the service and the requirements of customers, the goals of the vehicle routing problem usually are minimizing the driving distance, minimizing the driving time, minimizing the number of vehicles, and maximizing customer satisfaction. Constraint conditions characterize the problem.

Although different working scenarios may have different objectives and constraints, the specific characteristics of VRP are usually manifested in the relationship between the three entities: depot, vehicle, and customers. In general, constraints of the VRP exist in the description of these three entities.

15.2.1.1   Depot

There are three elements to describe the depot: the number of depots, the type of depot, and time constraint of depots. The number of depots can be multiple or single. The most studied case is the VRP of a single depot. The type of depot can be closed, open, and close-open. When the vehicle starts and ends in the same warehouse, it is a closed type; when the vehicle starts and ends differently, it is an open type. The time constraint is mainly the opening time of the depot, that is, the earliest time the vehicle can leave the depot and the latest time the vehicle returns.

15.2.1.2   Customers

The customers have four elements: the type of service, the demand, the time of service, and the time window constraint. The vehicle needs to deliver or pick up the demand of goods required by the customer, and delivery or pickup can be simultaneous. The demand of customers can be known in advance or unknown. The service time for every customer can be the same as or different from the demand of the customer. The time window constraint limits the time range in which the customer is allowed to be served.

15.2.1.3   Vehicles

This section generally describes vehicles in terms of quantity, type, capacity, time constraints, cost, and form of service. Quantity is the number of vehicles available in the depot. The type of these vehicles can be the same or different. Different types of vehicles have different capacities, cost, and time constraints. Constraints are often used to limit the driver's working hours. The cost is generally used to specify the vehicle usage fee required per unit distance and unit time.

Different service strategies are adopted according to the properties of cargos, e.g., whether the cargo is divisible and whether the cargo can bring backhaul. If it is indivisible, all needs of a customer must be completed by the same vehicle at one time; otherwise, the needs of a customer can be completed multiple times by different vehicles.

According to the above description of VRP, it can be seen that VRP has many factors, which simulate the scenario of real-world applications. The combination of different factors will form different types of VRP.

15.2.2   Variants of the Vehicle Routing Problem

So far, the VRP can be roughly divided into the following four categories: ① capacitated vehicle routing problem (CVRP); ② vehicle routing problem with time windows (VRPTW); ③ vehicle routing problem with simultaneous delivery and pickup (VRPDP); and ④ vehicle routing problem with uncertain characteristics. Of course, the VRP is not limited to these four categories. More details about variants of the VRP can refer to [3].

15.2.2.1   Capacitated VRPs

The CVRP, which is the first VRP version studied, is the basis of various VRP variants. In CVRP, the type of vehicles is the same, and the total demand of customers that serviced by the same vehicle must be less than the maximum capacity of the vehicle. Under this condition, it is required to minimize the cost of serving all customers, which is generally expressed in terms of the total length of the routes or the travel time. The formulation of CVRP can be defined as

$$\min \sum_{i=0}^{N} \sum_{j=0}^{N} \sum_{k=1}^{K} d_{ij} x_{ij}^{k} \tag{15.24}$$

$$\sum_{k=1}^{K} \sum_{i=0}^{N} x_{ij}^{k} = 1, j \in \{1, \cdots, N\} : i \neq j \tag{15.25}$$

$$\sum_{k=1}^{K} \sum_{j=0}^{N} x_{ij}^{k} = 1, i \in \{1, \cdots, N\} : i \neq j \tag{15.26}$$

$$\sum_{i=0}^{N} \sum_{j=0}^{N} x_{ij}^{k} \gamma_{i} \leq Q_{k}, \quad k \in \{1, \cdots, K\} \tag{15.27}$$

$$\sum_{j=1}^{N} x_{ij}^{k} = \sum_{j=1}^{N} x_{ji}^{k} \leq 1, i = 0, \mathrm{k} \in \{1, \cdots, K\} \tag{15.28}$$

$$\sum_{k=1}^{K} \sum_{j=1}^{N} x_{ij}^{k} \leq K, \quad i = 0 \tag{15.29}$$

where d_{ij} denotes the distance between customers i and j; K represents the number of vehicles used; x_{ij}^{k} denotes the number of times vehicle k accesses customer j from customer i; γ_{i} represents the demand of customer i; and Q_{k} is the maximum capacity of vehicle k.

Equation (15.24) represents the total distance traveled by all vehicles. Equations (15.25) and (15.26) are used to ensure that each customer can only be served by one specific vehicle and only once. Equation (15.27) is used to ensure that the total customer demand in each route does not exceed the maximum capacity of the vehicle. Equation (15.28) guarantees that each route starts at the depot and ends at the depot. Equation (15.29) guarantees that the number of routes does not exceed the total number of vehicles in the depot.

15.2.2.2   VRPs with Time Windows

The VRPTW arises, since some customers have a demand for delivery timeliness, such as takeaway. The VRPTW requires that all delivery services must be completed within a period preset by each customer. This period is called a time window $[b_i, e_i]$. Currently, there are two forms of time window constraints: soft time windows and hard time windows. In a hard time window constraint, vehicles are only allowed to deliver to customers before the specified latest time. In the soft time window constraint, vehicles are allowed to be delivered after the specified latest time but subject to corresponding penalties, i.e., delay time. In these two types of time windows, if the vehicle arrives at the customer before the earliest specified time, the vehicle must wait; therefore a waiting time occurs.

15.2.2.3   VRPs with Simultaneous Delivery and Pickup

This type of problem is characterized by the fact that the vehicles not only deliver the goods to the customer but also pick them up from the customer and then deliver the goods to the depot. Compared with traditional VRP, which only considers distribution, this type of problem is more complicated, because the vehicle must always meet the basic constraints of the maximum capacity of the vehicle at any time during the distribution process.

15.2.2.4   VRPs with Uncertainties

Most of the problems mentioned above assume that the information required for the problem is known, that is, in a deterministic environment. But in practical applications, some of the information of the problem may be uncertain. Uncertainties come from different sources, including expected changes and unexpected events. This situation is particularly common in the VRP. For example, the customer's demand is uncertain and can only be determined when the customer is visited. When a vehicle with a planned route reaches the customer node, it may find that the customer's demand exceeds the load capacity of the vehicle. In this case, in order to produce a feasible solution, some additional decisions are usually required, which also results in some additional costs. Therefore, to solve the above problems, it is necessary to develop a model that explicitly considers all the uncertainties of the problem.

In VRP with uncertain characteristics, there are three factors that are generally considered:

(1) Stochastic demand. The number and location of customer nodes are fixed, but the amount of goods to be delivered to customers is random and meets a certain probability distribution, for example, the post office's mail collection service.
(2) Uncertain customers. Customer demands are determined, but customers order or cancel service with a certain probability.
(3) Uncertain time. The service time of the customer or the driving time of the vehicle is uncertain.

At present, there are two major types of benchmark problem test suites: the Solomon benchmark [22] and the multi-objective test set proposed by Castro-Gutierrez et al. [4]. For a more detailed summary of the VRP benchmark, please refer to the website.[1]

15.2.3 Multi-objective VRP with Real-Time Traffic Conditions

This section will take the multi-objective vehicle routing problem with real-time traffic conditions (MOVRPRTC) as an example to introduce the application of EAs in VRP. The MOVRPRTC fills the gap of existing VRPs that do not consider real road network and road conditions. Its model is based on a real road network. The connectivity between nodes can only be based on this road network. The road network is shown in Fig. 15.11. Moreover, the travel time between nodes will change dynamically according to real-time road conditions. In general, the MOVRPRTC is a problem that includes real-time traffic conditions and considers soft time window constraints.

MOVRPRTC is described as follows: in a road network topology $G = (\mathbb{V}, \mathbb{E})$, $\mathbb{V} = \{v_0, \cdots, v_N\}$ represents all the N nodes in the road network and $\mathbb{E} = \{\langle i, j \rangle | z_{ij} = 1\}$ denotes all the edges formed by the connection in the road network, where z_{ij} are elements of the matrix $\mathbf{Z}_{N \times N}$. If $z_{ij} = 1$, the link from v_i to v_j is feasible. The fleet starts from the depot v_0. They need to complete the delivery task of n customers $c_1, c_2, \cdots, c_n$ selected from $\mathbb{V}$ and then return to depot v_0. Each customer needs to be served once by one vehicle. The length d_{ij} of the path from the node v_i to v_j is fixed, but the travel time $T_{ij}(t)$ of the path between the node v_i and v_j changes with the time t. At any time, the load Q_k of the vehicle k cannot exceed its maximum capacity C. The time window for each customer c_i to be served is $[b_i, e_i]$, and if the arrival time $a_i(t)$ of the vehicle arriving at the customer c_i exceeds e_i, the vehicle is allowed to unload, but there will be a delay penalty for consideration of customer satisfaction. The waiting time of the vehicle will accumulate if the arrival time $a_i(t)$ of the vehicle arriving at the customer c_i

---

[1] http://neo.lcc.uma.es/vrp/vrp-instances/

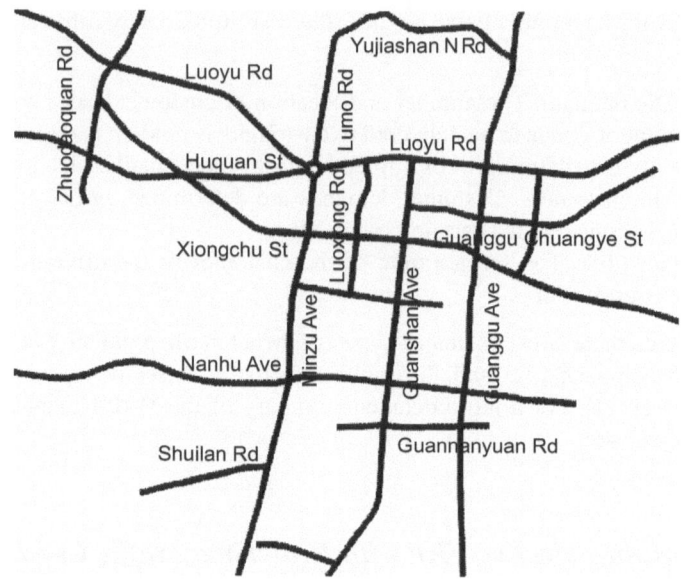

Fig. 15.11 Road network in Wuhan Guanggu Area

is earlier than b_i. The time for the vehicle to serve the customer c_i is s_i. γ_i and β_i denote the delivery demand and pickup demand of the customer c_i, respectively. After the vehicle leaves the depot, there will be new orders, requiring the vehicle to serve these new orders as much as possible based on the original plan. The generation of new orders follows a normal distribution. And the time when new orders appear is also randomly generated. But the generation time should be 60 minutes before its earliest time window b_i.

According to the above description, the multi-objective optimization problem can be defined as

$$\min\ f_1 = \sum_{k=1}^{K}\sum_{i=0}^{n}\sum_{j=0, j\neq i}^{n} d_{ij}x_j^k z_{ij} \tag{15.30}$$

$$\min\ f_2(t) = \sum_{k=1}^{K}\sum_{i=1}^{n}\max\{a_i(t)\cdot m_{ik} - e_i, 0\} \tag{15.31}$$

$$\min\ f_3(t) = \sum_{k=1}^{K}\sum_{i=1}^{n}\max\{b_i - a_i(t)\cdot m_{ik}, 0\} \tag{15.32}$$

$$\text{s. t. } g_1 = \sum_{k=1}^{K} m_{ik} = 1, i = 1, 2, 3, \cdots, n \tag{15.33}$$

$$g_2 = C_k^i - \sum_{i=1}^{n} (\gamma_i \cdot m_{ik}) + \sum_{i=1}^{n} (\beta_i \cdot m_{ik}) \leq C, k = 1, 2, 3, \cdots, K$$

$$(15.34)$$

$$g_3 = a_0 \leq e_0 \tag{15.35}$$

where K represents the number of vehicles used; x_{ij}^k denotes the number of times vehicle k accesses v_j from v_i; if the customer c_i has been served by vehicle k, then $m_{ik} = 1$, otherwise $m_{ik} = 0$; C_k^i is the capacity of vehicle k when the vehicle k reaches customer i; f_1 represents the total distance traveled by all vehicles; f_2 is the sum of the delay time of all vehicles; f_3 is the sum of the waiting time of all vehicles, which is used to measure the delivery efficiency; g_1 means that each customer is served once; g_2 means that the vehicle cannot be overload at any time; and g_3 means that the time when the vehicle returns to the depot cannot exceed the closure time of the depot.

In this model, the opening time of the depot, the maximum capacity of each vehicle, and the number of customers (smaller than 802) are free to set. Demands and time windows of customers are determined according to the real data of the online shopping after desensitization.

Its dynamic characteristics are reflected in two aspects: ① The traffic conditions will change in real-time, i.e., the vehicle needs to adjust the traveling route according to the real-time traffic conditions; ② New orders appear after the vehicle departs from the depot. The vehicle should serve these new orders as many as possible.

15.2.4 Algorithm Framework and Results

The VRP is a NP-hard combinatorial optimization problem. Besides, it can be a multi-objective optimization problem, a constraint optimization problem, and a dynamic optimization problem. To solve the multi-objective optimization problem in VRP, the multi-objective optimization algorithms such as NSGA-II [7] and MOEA/D [25] can be used. The constraint-handling techniques introduced in Chap. 9 can be used to handle the constraints in VRP.

In order to obtain a better solution of the above MOVRPRTC, we propose a two-stage optimization method: an offline and an online optimization mechanism to solve this dynamic multi-objective constrained optimization problem. In the offline optimization phase, the search operator, multi-objective optimization algorithm and constraint optimization techniques are used to simulate the delivery of orders that need to be delivered in the next working day. In this phase, we use historical traffic data to build the model to simulate the traffic condition of the next working day.

The algorithm for offline optimization is called dynamic constrained multi-objective evolutionary algorithm framework with adaptive local search (DCMOEA-ALS), which is based on the dynamic constraint multi-objective genetic algorithm framework (DCMOEA) [24]. During online optimization phase, that is, after the vehicle leaves the depot, the vehicle will adjust the distribution route in real time based on real-time traffic conditions and new order data.

DCMOEA is a framework for handling constraints, particularly good at dealing with equality constraints. Different from the original framework, MOVRPRTC is a combinatorial optimization problem and does not consider the niche-count objective in [24]. Inspired by DCMOEA, f_2 and f_3 in MOVRPRTC can be converted into constraints. Then the problem can be converted into VRPRTC

$$\min f_1 = \sum_{k=1}^{K} \sum_{i=0}^{n} \sum_{j=0, j=i}^{n} d_{ij} x_{ij}^k z_{ij} \tag{15.36}$$

$$\text{s.t. } \boldsymbol{g} = \{f_2, f_3, g_1, g_2, g_3\} \tag{15.37}$$

According to DCMOEA which transforms a COP to a dynamic constrained multi-objective problem (DCMOP), a VRPRTC is converted to a DCMOVRPTRC

$$\text{DCMOVRPRTC}^{(s)} = \begin{cases} \min(f(x), \text{cv}(x)) \\ \text{s.t. } \boldsymbol{g}(x) \leq \boldsymbol{\varepsilon}^{(s)} \end{cases} \tag{15.38}$$

$$\text{cv}(x) = \frac{1}{4} \sum_{i=1}^{4} \frac{\hat{g}_i(x)}{\max_{x \in \mathbb{P}(0)} \{\hat{g}_i(x)\}} \tag{15.39}$$

$$\hat{g}_i(x) = \max \{g_i(x), 0\}, i = 1, 2, \cdots, 4 \tag{15.40}$$

where x in Eq. (15.39) denotes a solution of DCMOVRPRTC, which consists of a set of routes; S is a given number of changes, $s = 0, 1, \cdots, S$; $\boldsymbol{\varepsilon}^{(s)}$ is the dynamic constraint boundary, and s is the state; A change means a reduction of the constraint boundary from state s to state $s + 1$.

The purpose of adopting DCMOEA is to solve the constraint optimization difficulties in MOVRPRTC with its constraint handling technology. In order to deal with the multi-objective optimization difficulty in MOVRPRTC, NSGA-II can be used. However, when the number of problems exceeds three, the possibility of non-domination will increase, which will lead to poor convergence of solutions. After converting MOVRPTC to DCMOVRPRTC, the number of objectives becomes two. This can alleviate the difficulty of multi-objective optimization.

15.2.4.1   Solution Initialization

Since the MOVRPRTC contains the road network, a solution should include not only the service sequence of customers but also the traveling sequence of nodes of the road network for each vehicle. Figure 15.12 shows an example of the solution, where graph A indicates the service sequence of the customers and graph B indicates

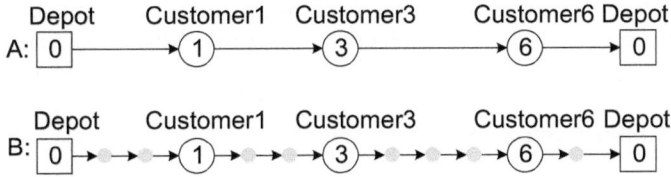

Fig. 15.12 Solution representation

the traveling sequence of the nodes. Gray dots between two adjacent customers indicate the nodes in the road network that the vehicle needs to travel from the current customer to the next customer. The optimization algorithm only operates sequence A to reallocate customers for vehicles, while sequence B is only used for objective evaluation, where a traffic condition prediction model is used during the evaluation.

Random initialization can be used to get the initial solution. The random initialization method randomly selects one of the customers and places it in the first vehicle. If the maximum capacity of the first vehicle is exceeded, then add it to another vehicle, and repeat the above steps until all customers are assigned. The random initialization is simple but is not an effective method for obtaining a high quality initial population. We use an agglomerative hierarchical clustering algorithm to group customers into clusters by taking both the distance between customers and their time windows. Each cluster represents the customers served by one vehicle.

15.2.4.2 Adaptive Local Search Strategy

The fundamental tasks of solving MOVRPRTC are the allocation of customers for vehicles and the sequencing of customer servings. Therefore, the search operator should be designed for these two tasks. The operator used to determine the order of customer service only adjusts the order of customers in a single route. The operator used to determine customer assignment can adjust the customer order in two routes. A priori knowledge of the problem can be used during the work of these operators, such as swapping customers with long delay time before customers with long waiting time. We designed eight heuristic search operators for these two types of tasks.

To adaptively select one operator, each operator has the same weight at the beginning of a run, and the weight of each operator is controlled during the search based on the performance of the offspring produced by the operator in each iteration. For example, if the offspring obtained by an operator dominates its parent, then its weight will increase by 1; otherwise the weight will not change. In each iteration, the search operator is selected according to its weight based on the roulette wheel rule. However, continuously positive feedback may cause the failure of learning the best operator due to biased search behavior. To solve this problem, if the generated offspring does not dominate their parents more than a predefined number of iterations, the weight of the search operator is reset to the initial value.

15.2.4.3   Online Optimization

The online optimization is to choose an optimal solution from the solution set obtained at the offline optimization stage. The best solution is defined as the solution with the minimum delay time, since this objective directly reflects customer satisfaction.

The task of online optimization is to serve as many new orders as possible and to adjust the route of each vehicle based on real-time traffic conditions. The adaptive local search mechanism can still be applied to this stage. It can adjust the customer service sequence in a single route in combination with the traffic conditions predicted by the traffic condition prediction model and estimate the impact of newly inserted customers on the overall plan.

15.2.4.4   Experimental Results

The algorithm was implemented in C++ in OFEC platform. The depot opening time is from 8:00 to 24:00. The vehicle is the same with a maximum capacity of 3 tons. The number of customers known before the working day is 100, and the number of new customers is 15.

In order to prove the effectiveness of DCMOEA-ALS, we compared the initial solution and the final solutions generated by DCMOEA-ALS in the objective space. As shown in Fig. 15.13, the initial solution set is completely dominated by the final solution set. Table 15.6 shows the comparison of the best solution of the initial population and the best one of the final set of solutions. All the three original objectives have been improved, especially for the delay time and waiting time with a drop ratio of 99.3% and 77.66%, respectively.

The results of the online optimization are shown in Table 15.7. The final solution used 11 vehicles, with an average delay of 79.10 minutes per vehicle. According to the results, each vehicle serves about 15 customers, so the average delay time for each customer is about 5 minutes.

This section just provides an example for solving VRP. If you are interested in VRP, you can try to solve the standard VRP test suites using the algorithm introduced in this chapter. In addition, readers can try to build their own VRP models, such as VRP models for drone or unmanned vehicle distribution.

Fig. 15.13 Dominance of initial solutions and final solutions

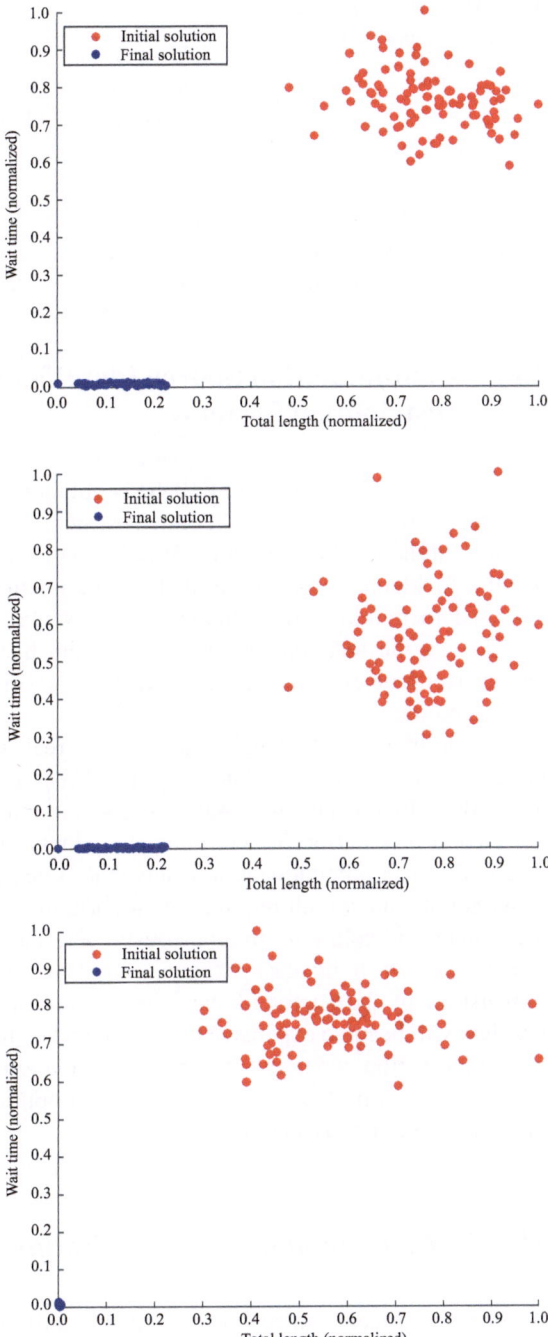

Table 15.6 The objectives of the initial solution and the final solution

	Total length	Wait time	Delay time
Initial solution	608.629 km	4884.91 min	2941.29 min
Final solution	**504.505** km	**1091.35** min	**20.5629** min
Reduce ratio	17.11%	77.66%	99.3%

Table 15.7 The results of online optimization

	Total length	Wait time	Delay time	Remaining orders
Begin	504.505 km	1091.35 min	20.5629 min	15
End	595.602 km	815.303 min	870.053 min	0

15.3   Contamination Source Identification in Water Distribution Systems

The water distribution system (WDS) is vulnerable to a range of threats, including uncertain natural disasters, deliberate destruction, and system failures. For example, when a pollutant is injected into a WDS, it will spread rapidly and will risk people's health. To identify and locate such contamination events, water quality sensors are used to monitor contaminants in the WDS. Based on the observations, we can deduce the location, initiation time, and historical injection rate by solving an inverse problem with an optimization algorithm under a water distribution simulation model.

The problem is challenging due to the real-time, nonuniqueness/multimodal, large-scale, and expensive characteristics. The real-time property requires the search to be data-driven, i.e., the search starts immediately after the contamination is detected at any sensor and continuously adapts to changes when new observation data come. The nonuniqueness means that more than one solutions conform to the observation data, which requires the algorithm to have the capability of searching more than one solution simultaneously. The large-scale property means that the search space will increase exponentially due to the increase of the number of dimensions of the vector of the injection rate as observation data increase. The problem will become expensive to simulate when the scale of the network increases.

In this section, we transform the contamination source identification problem to an optimization problem using a simulation-optimization model and then seek the optimal solution by EAs [14].

15.3.1   Contamination Source Identification Problem

An inverse problem can be constructed to identify the contamination source characteristics in the WDS, where the input is a set of concentration observations at sensors and the objective is to minimize the error between predicted concentration and actual observation at sensors on the network using a water distribution simulation model.

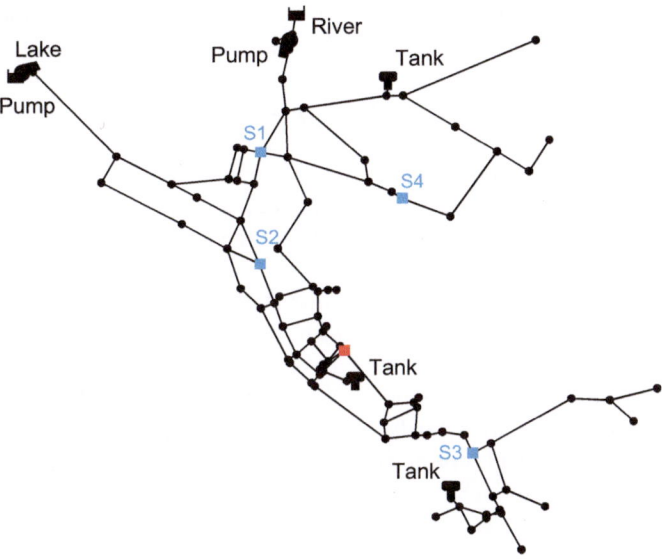

Fig. 15.14 An example of water distribution network

15.3.1.1   Water Distribution System

The water distribution simulation software EPANET 2.0 is used to establish a water network diagram, and a small-scale network is shown in Fig. 15.14. The small-scale network has one lake, one river, and two pump stations serving as the main water source. Three tanks are set in the city for auxiliary supplement. The solid black points are the customer nodes from which pollutants can be injected. Links between junctions are pipes. Suppose that the red square in the figure is the pollutant injection node, and the blue squares are the nodes where monitors are set. Due to the constant change of water quality and hydraulic power, the pollutant will begin to diffuse along with the water flow. If they spread to the location of monitoring nodes, contamination events will be found in the water network.

15.3.1.2   Simulation-Optimization Model

The simulation-optimization model in contamination source identification problem has two sub-models: a simulation model and an optimization model. The simulation model mainly describes the water movement in a WDS, including the water flow model and the solute transport model. The optimization model can transform the problem into an optimization problem of describing the location of the contamination source, the injection start time, and the injection history information.

The water flow model and water quality model in urban water supply network can be realized by the hydraulic simulation function of water quality in EPANET

2.0. The model can simulate the diffusion of solutes in contamination events and provide concentration data at each node. We define

$$y_j(t) = \mu(\boldsymbol{x}(t)), j = 1, \cdots, K \tag{15.41}$$

where $y_j(t)$ denotes the concentration data of contaminants detected by sensor j at time step t (each sensor is set at a different node); $\boldsymbol{x}(t) = (x_u, \boldsymbol{x}_l(t))$ denotes information of a contamination event at a single source at time step t, $x_u \in \mathbb{N}$ denotes the location of the contamination source; $\boldsymbol{x}_l(t) = (x_0, x_{t_0}, x_{t_0+1}, \cdots, x_t) \in \mathbb{R}^{t-t_0+2}$ denotes the injection profile, where t_0 is the starting time and $(x_{t_0}, x_{t_0+1}, \cdots, x_t)$ is a time series of injection rate from t_0 to t; μ is a simulation model of water distribution systems with the input $\boldsymbol{x}(t)$ and the observed output $\boldsymbol{y}(t)$ of K sensors.

Given the water quality hydraulic simulation model, the problem of locating and tracing contamination sources can be converted into a dynamic bi-level optimization problem. The upper-level optimization task is to find the location (x_u) of a contamination event, and the lower-level task is to find the injection history profile $(\boldsymbol{x}_l(t))$, which is defined as a dynamic optimization problem whose variables will increase as time goes on.

In this section, the square root of the mean squared error between the observed data of a possible solution and the actual observed data of contamination sources is used as the objective, which is shown below

$$\min_{x_u, \boldsymbol{x}_l(t)} f_u(x_u, \boldsymbol{x}_l(t)) = \sqrt{\frac{\sum_{j=1}^{K} \sum_{s=0}^{t} [y_j(s) - \check{y}_j(s)]^2}{K \cdot t}}$$
$$\text{s. t.} \quad \boldsymbol{x}_l(t) \in \arg\min_{\boldsymbol{x}_l(t)} \{f_l(x_u, \boldsymbol{x}_l(t))\} \tag{15.42}$$

where $\check{y}_j(s)$ denotes the observed data at sensor j at time step s of the actual contamination source.

Therefore, the objective is to minimize the cumulated error over all sensors so far, since the initial observed data are obtained, i.e., to find a solution that complies with the actual observation data. The two levels of problems all aim to minimize the prediction error but have different tasks.

15.3.2   Adaptive Multi-population Algorithm

According to the difficulties of the contamination source identification, we use an adaptive multi-population evolutionary optimization algorithm [13] to determine the real-time characteristics of contamination sources, where each population aims to locate and track a different global optimum. The algorithm adaptively adjusts the number of populations using a feedback learning mechanism. To effectively locate an optimal solution for a population, a coevolutionary strategy is used to identify the location and the injection profile separately.

The algorithm framework has three basic components: clustering, parallel search, and diversity increasing components. The framework starts with a set of randomly initialized individuals at $T = 0$ (a counter that keeps the number of times the diversity increasing component is triggered after clustering) and then clusters these new individuals to a set of populations. Populations simultaneously search for global optima and merge if their best individuals locate at the same node on the network. When the coverage ratio over all nodes does not change over two successive generations, it will trigger the diversity increasing procedure where a feedback learning method is used to control the number of populations to be added.

15.3.2.1   Generation of Multiple Populations

The k-means clustering method is adopted to generate multiple populations whose search areas do not overlap with each other. Firstly, a certain number of individuals are generated at node i with a probability p_i, which is the same ($p_i = 1/n$, n is the number of nodes of a network) for all nodes at the beginning of a run and is updated before the increase of populations

$$p_i(T) = 1 - \sum_{s=0}^{T} \kappa_i(s) / \max_{i \in n} \sum_{s=0}^{T} \kappa_i(s) \qquad (15.43)$$

where $\kappa_i(s)$ is the accumulated times of node i visited by individuals since the start of the run.

From Eq. (15.43), the more times a node is visited, the smaller the probability of being selected when generating new solutions. After that, the nodes covered by all new individuals are then clustered by the k-means clustering method.

15.3.2.2   Cooperative Coevolutionary Search

The representation of the solution to the problem is composed of discrete and continuous parts. The location variable is an integer, and the injection profile is represented by a vector of real values. To alleviate the difficulty in searching the solution of the hybrid representation, we use the cooperative coevolution (CC) strategy [18] to solve the upper-level and lower-level problems separately.

We use the genetic learning (GL) [23] to find the injection location of the upper-level problem, and the nodes with smaller errors have larger probability to be selected for GL. And we use DE with four classical difference operators DE/rand/1, DE/best/1, DE/target-to-best/1, and DE/best/2 to optimize the initial time and historical injection rate. During the optimization process, the number of improvements of individuals made by these four differential operators is counted, and then the selection probability of each operator is calculated based on these data to guide the selection of one of the four operators.

15.3.2.3 Population Management

Population removal and population increase are two critical components of the adaptive multi-population framework, especially in dynamic environments.

The population removal component aims to remove redundant populations and hence to save computational resources. After the clustering, each population will cover a unique search area containing one or serval geographically closed nodes on the network. As the search goes on, some of the populations may move toward the same area and, finally, converge at the same node of that area. This causes redundant search, which should be prevented. To prevent more than one population from searching in the same area, we check their best solutions. When their best solutions are located at the same node on the network, these populations are deemed to be overcrowded. The competition mechanism will be triggered. The surviving individuals form a new population, and the remaining individuals are removed.

The population increase component aims to increase the diversity in the areas which have not been searched or not sufficiently searched so far and hence to find more global optima solutions. Increasing populations mean increasing the diversity for exploring unexplored areas. We increase populations when all populations enter a stable state on the network indicated by the coverage ratio over all nodes. Then, inspired by [13], a feedback learning strategy is adopted to estimate the number of populations to increase.

Finally, we can generate random solutions to increase the population diversity. Different from the traditional random immigrant scheme where new solutions are uniformly randomly generated at all nodes without considering the distribution of the current and historical solutions on the network, we generate new solutions at node with the probability of the distribution of solutions. In this way, infrequently visited nodes have large probabilities of being explored, which is very helpful for finding new global optima.

Here, to respond to changes, the population increase is determined only based on the current evolutionary status of the whole populations but not necessarily at the moment of a change occurring, i.e., the distribution of all solutions on the network does not change means that the algorithm becomes converging and needs to be diversified to enhance the exploration capability.

15.3.3 Experimental Results

The water network mentioned in Fig. 15.14 was used to study the performance of the algorithm.

Table 15.8 Configurations for three test instances

Instance	Location	Start time	Duration/min	Injection rate
1-1	113	0:00	60	5, 10, 15, 20, 15, 10
1-2	157	2:00	120	30, 25, 20, 15, 10, 5, 5, 10, 15, 20, 25, 30
1-3	267	4:00	240	30, 5, 30, 5, $\cdots$, 30, 5, 30, 5

15.3.3.1   Experimental Setup

Three different test cases, which are listed in Table 15.8, were generated based on the network, where the injection rate changes every 10 minutes. Four sensors were set on nodes 113, 147, 211, and 120.

In the experiments, the simulation duration was set to 24 hours and the observation data was collected from the sensors every 10 minutes. Given the above configurations in Table 15.8, we assume that the range of [0h, 4h] for the start time and [5g, 30g] for the injection mass. Therefore, new solutions are randomly initialized within these ranges. The initial number of populations was set to 20 ($k = 20$), and a fixed single population size is 50 ($m = 50$). To reduce the complexity of the problem, we assume the injection duration is known for all algorithms. All the results in this section are averaged over 20 independent runs, and each test case was run for 200,000 objective evaluations.

We use a success rate and the prediction error to evaluate the performance of an algorithm. A successful run is a run where the true injection node is found, so the success rate is the number of successful runs over the total number of runs. Note that the prediction error is averaged over all successful runs, not over the total number of runs.

15.3.3.2   Results and Discussions

As shown in Fig. 15.15, the population diversity is increased for five times to maintain the diversity. Table 15.9 presents the success rate and prediction error obtained by the algorithm on each test instance. It can be seen that the success rates are very high, and the minimum success rate is 0.95. The average prediction error of the three test cases is small. The experimental results show that the simulation-optimization method has good performance for the contamination source identification problem.

Figure 15.16a shows the time-varying mean of contaminants concentrations observed at four detection points on instance 1-3, where the solid red curve refers to the actual observation data, the solid dark curve refers to the contaminants concentrations simulated by the best solution found by the algorithm, and the gray dashed curves are for the remaining solutions. It can be seen that the general trend of the dark solid curves is very close to the actual data, i.e., the optimal solutions found by the algorithm are sound. And the algorithm can find many optimal solutions.

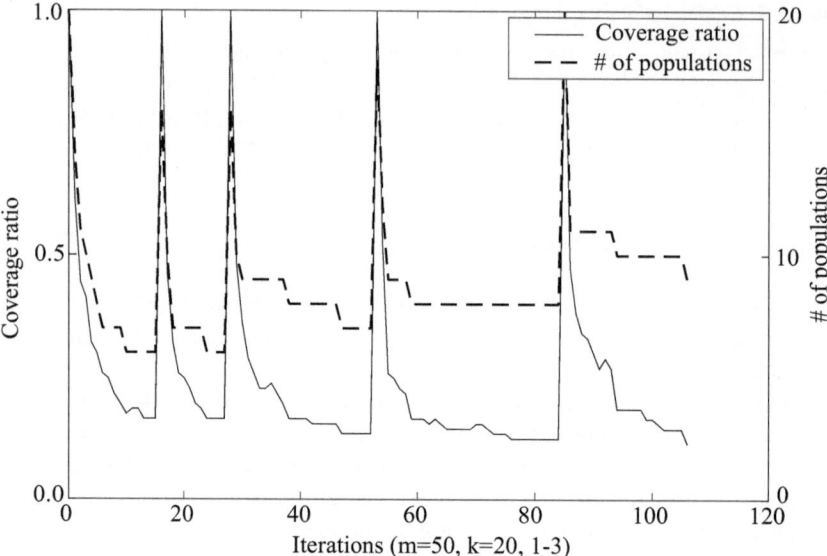

Fig. 15.15 The change in the number of populations and the coverage ratio on instance 1-3

Table 15.9 Success rate, prediction error, and standard deviation of three test cases		Instance 1-1	Instance 1-2	Instance 1-3
	Success rate	1	0.95	1
	Prediction error	0 ± 0	2.6 ± 1.9	13.0 ± 3.0

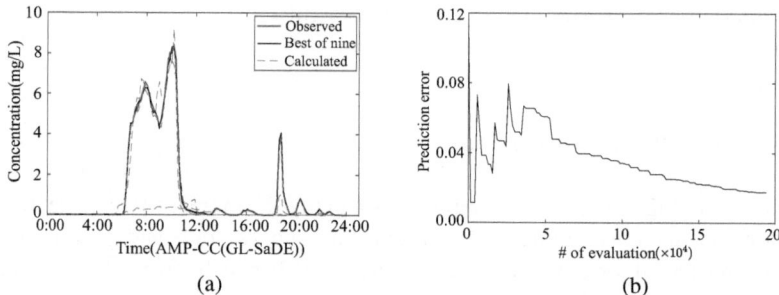

Fig. 15.16 Experimental results on instance 1-3. (**a**) Solutions found by the algorithm (**b**) The change in the prediction error

When we observe the change in the prediction error of the best solutions obtained by the algorithm in Fig. 15.16b, the multi-population-based algorithm can quickly locate the injection event in the beginning of the run.

This section provides an adaptive multi-population framework to solve the contamination source identification problem in the water distribution system. The problem is defined as a dynamic bi-level optimization problem. To handle the real-time and nonuniqueness characteristics of the problem, an adaptive mechanism is used to enhance the exploring ability of multi-population methods and a cooperative coevolution strategy for the bi-level optimization problem. The AMP framework can automatically remove redundant populations and add a proper number of populations at a proper moment, which makes it adaptable to different problems.

References

1. Afshinmanesh, F., Marandi, A., Shahabadi, M.: Design of a single-feed dual-band dual-polarized printed microstrip antenna using a boolean particle swarm optimization. IEEE Trans. Antennas Propag. **56**(7), 1845–1852 (2008)
2. Altshuler, E.E., Linden, D.S.: Wire-antenna designs using genetic algorithms. IEEE Antennas Propag. Mag. **39**(2), 33–43 (1997)
3. Braekers, K., Ramaekers, K., Van Nieuwenhuyse, I.: The vehicle routing problem: state of the art classification and review. Comput. Ind. Eng. **99**, 300–313 (2016)
4. Castro-Gutierrez, J., Landa-Silva, D., Pérez, J.M.: Multi-objective vehicle routing problem with time windows dataset generator. https://github.com/psxjpc/MOVRPTW-Generator (2011). Accessed 11 Dec 2019
5. Clarke, G., Wright, J.W.: Scheduling of vehicles from a central depot to a number of delivery points. Oper. Res. **12**(4), 568–581 (1964)
6. Dantzig, G.B., Ramser, J.H.: The truck dispatching problem. Manag. Sci. **6**(1), 80–91 (1959)
7. Deb, K., Pratap, A., Agarwal, S., Meyarivan, T.: A fast and elitist multiobjective genetic algorithm: Nsga-ii. IEEE Trans. Evol. Comput. **6**(2), 182–197 (2002)
8. Hornby, G.S., Lohn, J.D., Linden, D.S.: Computer-automated evolution of an x-band antenna for nasa's space technology 5 mission. Evol. Comput. **19**(1), 1–23 (2011)
9. Hu, C., Zeng, S., Jiang, Y., Sun, J., Sun, Y., Gao, S.: A robust technique without additional computational cost in evolutionary antenna optimization. IEEE Trans. Antennas Propag. **67**(4), 2252–2259 (2019)
10. Jaafar, H., Collardey, S., Sharaiha, A.: Optimized manipulation of the network characteristic modes for wideband small antenna matching. IEEE Trans. Antennas Propag. **65**(11), 5757–5767 (2017)
11. Jiao, R., Sun, Y., Sun, J., Jiang, Y., Zeng, S.: Antenna design using dynamic multi-objective evolutionary algorithm. IET Microw. Antennas Propag. **12**(13), 2065–2072 (2018)
12. Kurup, D.G., Himdi, M., Rydberg, A.: Synthesis of uniform amplitude unequally spaced antenna arrays using the differential evolution algorithm. IEEE Trans. Antennas Propag. **51**(9), 2210–2217 (2003)
13. Li, C., Nguyen, T.T., Yang, M., Mavrovouniotis, M., Yang, S.: An adaptive multipopulation framework for locating and tracking multiple optima. IEEE Trans. Evol. Comput. **20**(4), 590–605 (2015)
14. Li, C., Yang, R., Zhou, L., Zeng, S., Mavrovouniotis, M., Yang, M., Yang, S., Wu, M.: Adaptive multipopulation evolutionary algorithm for contamination source identification in water distribution systems. J. Water Resour. Plan. Manag. **147**(5), 04021014 (2021)
15. Lohn, J.D., Linden, D.S., Blevins, B., Greenling, T., Allard, M.R.: Automated synthesis of a lunar satellite antenna system. IEEE Trans. Antennas Propag. **63**(4), 1436–1444 (2015)
16. Pantoja, M.F., Ruiz, F.G., Bretones, A.R., Garcia, S.G., Martin, R.G., Arbesú, J.G., Romeu, J., Rius, J., Werner, P., Werner, D.H.: Ga design of small thin-wire antennas: comparison with sierpinsky-type prefractal antennas. IEEE Trans. Antennas Propag. **54**(6), 1879–1882 (2006)
17. Pavlenko, T., Reustle, C., Dobrev, Y., Gottinger, M., Jassoume, L., Vossiek, M.: Design and optimization of sparse planar antenna arrays for wireless 3-d local positioning systems. IEEE Trans. Antennas Propag. **65**(12), 7288–7297 (2017)

18. Potter, M.A., De Jong, K.A.: A cooperative coevolutionary approach to function optimization. In: Davidor, Y., Schwefel, H.P., Männer, R. (eds.) Parallel Problem Solving from Nature, pp. 249–257. Springer, Berlin (1994)
19. Robinson, J., Rahmat-Samii, Y.: Particle swarm optimization in electromagnetics. IEEE Trans. Antennas Propag. **52**(2), 397–407 (2004)
20. Rogers, S.D., Butler, C.M.: Wide-band sleeve-cage and sleeve-helical antennas. IEEE Trans. Antennas Propag. **50**(10), 1409–1414 (2002)
21. Smith, J.S., Baginski, M.E.: Thin-wire antenna design using a novel branching scheme and genetic algorithm optimization. IEEE Trans. Antennas Propag. **67**(5), 2934–2941 (2019)
22. Solomon, M.M.: Solomon benchmark. http://web.cba.neu.edu/~msolomon/problems.htm (1987). Accessed 11 Dec 2019
23. Xia, Y., Li, C.: Memory-based statistical learning for the travelling salesman problem. In: IEEE Congress on Evolutionary Computation, pp. 2935–2941. IEEE, Vancouver (2016)
24. Zeng, S., Jiao, R., Li, C., Li, X., Alkasassbeh, J.S.: A general framework of dynamic constrained multiobjective evolutionary algorithms for constrained optimization. IEEE Trans. Cybern. **47**(9), 2678–2688 (2017)
25. Zhang, Q., Li, H.: MOEA/D: A multiobjective evolutionary algorithm based on decomposition. IEEE Trans. Evol. Comput. **11**(6), 712–731 (2007)

Appendix A
Mathematical Background

A.1 Mathematical Analysis

A.1.1 Set

Definition A.1 (Set) Set is a collection of one or more deterministic element(s).

The uppercase letters such as $\mathbb{A}, \mathbb{B}, \mathbb{S}, \mathbb{T}, \ldots$ are usually used to represent the set, while the lowercase letters such as $a, b, x, y, \ldots$ are used to represent the elements. If x is an element of set $\mathbb{S}$, then x belongs to $\mathbb{S}$, denoted as $x \in \mathbb{S}$. If y is not an element of set $\mathbb{S}$, then y does not belong to $\mathbb{S}$, which is denoted by $y \notin \mathbb{S}$.

There are some peculiar sets.

Definition A.2 (Empty Set) An empty set has no elements, denoted by $\emptyset$ or by { }.

Definition A.3 (Subset) If every element of $\mathbb{A}$ is also in set $\mathbb{B}$, then $\mathbb{A}$ is a subset of $\mathbb{B}$ (denoted by $\mathbb{A} \subseteq \mathbb{B}$). For example, the set of even numbers is a subset of the set of real numbers.

A.1.1.1 Properties

Property A.1 (Certainty) Ambiguity is not allowed in sets.

Property A.2 (Uniqueness) Any two elements in one set are considered to be different, that is, no two members are identical.

Property A.3 (Disorder) There is no necessary order between elements in terms of the characteristics of the set. In a set, each element has the same status, and the elements are disordered. However, an ordering relation can be defined on a set, and the elements can be sorted according to the ordering relation.

A.1.1.2   Operations

(1) **Intersection**. The intersection of $\mathbb{A}$ and $\mathbb{B}$, written as $\mathbb{A} \cap \mathbb{B}$, is the set of all elements that are in both $\mathbb{A}$ and $\mathbb{B}$. If $\mathbb{A} \cup \mathbb{B} = \varnothing$, we say that $\mathbb{A}$ and $\mathbb{B}$ are disjoint.

(2) **Union**. The union of two sets $\mathbb{A}$ and $\mathbb{B}$, written as $\mathbb{A} \cup \mathbb{B}$, is the set of all elements that are in $\mathbb{A}$ or $\mathbb{B}$ (or both).

(3) **Complement**. If $\mathbb{S}$ is a universal set and $\mathbb{A}$ is any subset of $\mathbb{S}$, the complement of $\mathbb{A}$ is the set of all elements excluded from $\mathbb{A}$ in $\mathbb{U}$, denoted by

$$A' = \{x \ : \ x \in \mathbb{S} \text{ and } x \notin \mathbb{A}\}$$

(4) **Cartesian Product**. The Cartesian Product of sets $\mathbb{A} = \{a_1, \cdots, a_m\}$ and $\mathbb{B} = \{b_1, \cdots, b_n\}$ where m, n are positive numbers is set $\mathbb{C}$ denoted by

$$\mathbb{C} = \mathbb{A} \times \mathbb{B} = \{(a_1, b_1), \cdots, (a_1, b_n), (a_2, b_1), \cdots, (a_2, b_n), \cdots,$$
$$(a_m, b_1), \cdots, (a_m, b_n)\}$$

$\mathbb{A}^2$ indicates the Cartesian Product $\mathbb{A} \times \mathbb{A}$ and in general $\mathbb{A}^n = \underbrace{\mathbb{A} \times \mathbb{A} \times \cdots \times \mathbb{A}}_{n}$

if $\mathbb{A}$ is multiplied n times.

A.1.1.3   Properties of Set Operations

(1) **Associative property**. $\mathbb{A} \cup (\mathbb{B} \cup \mathbb{C}) = (\mathbb{A} \cup \mathbb{B}) \cup \mathbb{C}$, $\mathbb{A} \cap (\mathbb{B} \cap \mathbb{C}) = (\mathbb{A} \cap \mathbb{B}) \cap \mathbb{C}$

(2) **Commutative property**. $\mathbb{A} \cap \mathbb{B} = \mathbb{B} \cap \mathbb{A}$, $\mathbb{A} \cup \mathbb{B} = \mathbb{B} \cup \mathbb{A}$

(3) **Inversion Law (De Morgan's Laws)**. $(\mathbb{A} \cup \mathbb{B})' = \mathbb{A}' \cap \mathbb{B}'$, $(\mathbb{A} \cap \mathbb{B})' = \mathbb{A}' \cup \mathbb{B}'$

(4) **Distributive property**.
$\mathbb{A} \cap (\mathbb{B} \cup \mathbb{C}) = (\mathbb{A} \cap \mathbb{B}) \cup (\mathbb{A} \cap \mathbb{C})$, $\mathbb{A} \cup (\mathbb{B} \cap \mathbb{C}) = (\mathbb{A} \cup \mathbb{B}) \cap (\mathbb{A} \cup \mathbb{C})$

A.1.1.4   Display Methods

(1) **Enumeration method**. Enumeration is the way to enumerate the elements of a set one by one. For example, set $\mathbb{A}$ consisting of four letters a, b, c, d can be represented by $\mathbb{A} = \{a, b, c, d\}$, and so on.

(2) **Description method**. The description method represents the nature of elements in a set. Assuming that set $\mathbb{S}$ is composed of all elements with a certain property f, the set can be represented by describing the common attributes of elements in the set: $\mathbb{S} = \{x | f(x)\}$. For example, set $\mathbb{B} = \{x | x^2 = 2\}$ consists of the elements which is the square root of 2.

Fig. A.1 Relationship of set

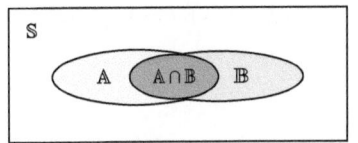

(3) **Image method**. Image method, also known as Venn diagram method, represents sets by a rectangle or circle on a two-dimensional plane. Figure A.1 is an intuitive graphical representation of sets.

(4) **Symbolic method**. There are some special symbols for common collections.

$\mathbb{N}$: the set of nonnegative integers or natural numbers, denoted by $\{0, 1, 2, 3, \ldots\}$.

$\mathbb{N}^*$ or $\mathbb{N}^+$: the set of positive integers, denoted by$\{1, 2, 3, \ldots\}$.

$\mathbb{Z}$: the set of integers, denoted by $\{\ldots, -1, 0, 1, \ldots\}$.

$\mathbb{Q}$: the set of rational numbers, denoted by $\left\{\ldots, -\frac{2}{3}, \frac{1}{7}, -\frac{4}{5}, \ldots\right\}$.

$\mathbb{Q}^+$: the set of positive rational numbers, denoted by $\left\{\ldots, \frac{2}{3}, \frac{1}{7}, \frac{4}{5}, \ldots\right\}$.

$\mathbb{Q}^-$: the set of negative rational numbers, denoted by $\left\{\ldots, -\frac{2}{3}, -\frac{1}{7}, -\frac{4}{5}, \ldots\right\}$.

$\mathbb{R}$: the set of real numbers, denoted by $\left\{\ldots, -\frac{2}{3}, e, \pi, \ldots\right\}$.

$\mathbb{R}^+$: the set of positive real numbers, denoted by $\left\{\ldots, 1, \frac{1}{7}, \pi, \ldots\right\}$.

$\mathbb{R}^-$: the set of negative real numbers, denoted by $\left\{\ldots, -e, -\frac{1}{7}, -3, \ldots\right\}$.

$\mathbb{C}$: the set of complex numbers, denoted by $\{\ldots, 3 + 2i, \ldots\}$. i is imaginary.

$\varnothing$: an empty set (a collection that does not contain any element).

A.1.2   Function

Definition A.4 (Map) Let $\mathbb{X}$ and $\mathbb{Y}$ be any non-empty sets, where x varies over elements of $\mathbb{X}$ and y varies of $\mathbb{Y}$. A **function** from $\mathbb{X}$ to $\mathbb{Y}$ is a subset f of $\mathbb{X} \times \mathbb{Y}$, the set of all pairs (x, y), with the property that for each $x \in \mathbb{X}$, there is a unique $y \in \mathbb{Y}$ such that $(x, y) \in f$, written as

$$f : \mathbb{X} \to \mathbb{Y}$$

to indicate that f **maps** $\mathbb{X}$ to $\mathbb{Y}$ and

$$y = f(x)$$

to call $f(x)$ the **value** of f at x or the **range** of f. $\mathbb{X}$ is called the **source** or **domain** of f, written in $\mathbb{D}_f = \mathbb{X}$, and $\mathbb{Y}$ the **target** or **codomain** of f.

It can also be written as

$$\mathbb{R}_f = f(\mathbb{X}) = \{f(x)|x \in \mathbb{X}\}$$

It should be noted that $\mathbb{R}_f \subset \mathbb{Y}$ and there is only a unique $y = f(x)$ for each $x \in \mathbb{X}$.

Definition A.5 (Function) The term function is synonymous with mapping for map.

Property A.4 (Bounded) Let $f(x)$ be defined in interval $\mathbb{I}$. If $\exists M > 0, \forall x \in [a, b]$, there is always $|f(x)| \leq M$, then $f(x)$ is bounded on interval $\mathbb{I}$. Otherwise, $f(x)$ is unbounded on the interval $\mathbb{I}$.

Property A.5 (Monotonic) Let the domain of the function $f(x)$ be $\mathbb{D}$ and the interval $\mathbb{I}$ be included in $\mathbb{D}$. If for any two points x_1 and x_2 on the interval, when $x_1 < x_2$, there is always $f(x_1) < f(x_2)$, then the function $f(x)$ is monotonically increasing in interval $\mathbb{I}$; on the contrary, if there is always $f(x_1) > f(x_2)$ when $x_1 < x_2$, then the function $f(x)$ is said to be monotonically decreasing in the interval $\mathbb{I}$.

Monotonically increasing or decreasing functions are collectively called monotonic functions.

Property A.6 (Parity) Let $f(x)$ be a real-valued function, if $f(-x) = -f(x)$, then $f(x)$ is an odd function. From a geometric point of view, an odd function is symmetric about the origin, because its image does not change after a 180 degree rotation around the origin. $f(x) = x$, $g(x) = \sin(x)$ and $p(x) = \sin h(x)$ are all odd functions.

Similarly, Let $f(x)$ be a real-valued function, if $f(-x) = f(x)$, then $f(x)$ is an even function. From a geometric point of view, an even function is symmetric about the y-axis because its graph does not change after mapping the y-axis. For example, $f(x) = |x|$, $g(x) = x^2$ and $p(x) = \cos(x)$ are all even functions.

Property A.7 (Periodic) Let the domain of the function $f(x)$ be $\mathbb{D}$. If there is a positive number t such that $f(x + t) = f(x)$ is constant for any x, then $f(x)$ is called a periodic function. t is called the period of $f(x)$, and usually we say that the period of the periodic function refers to the smallest positive period. The domain $\mathbb{D}$ of the periodic function is an unbounded interval of at least one side. If $\mathbb{D}$ is bounded, the function is not periodic.

Property A.8 (Concavity) Let the function $f(x)$ is continuous in the domain $\mathbb{D}$. If for any two points x, y ($x \neq y$) in $\mathbb{D}$ and $\alpha \in [0, 1]$, there is always $f(\alpha x + (1 - \alpha)y) \leq \alpha f(x) + (1 - \alpha)f(y)$, $f(\alpha x + (1 - \alpha)y) < \alpha f(x) + (1 - \alpha)f(y)$, then the first inequality is called the **convex** function on the interval; the second inequality is called the **strict convex** function, which is shown in Fig. A.2a.

Similarly, if there is always $f(\alpha x + (1 - \alpha)y) \geq \alpha f(x) + (1 - \alpha)f(y)$, $f(\alpha x + (1 - \alpha)y) > \alpha f(x) + (1 - \alpha)f(y)$, then the first inequality is called the **concave** function on the interval; the second inequality is called the **strict concave** function, which is shown in Fig. A.2b.

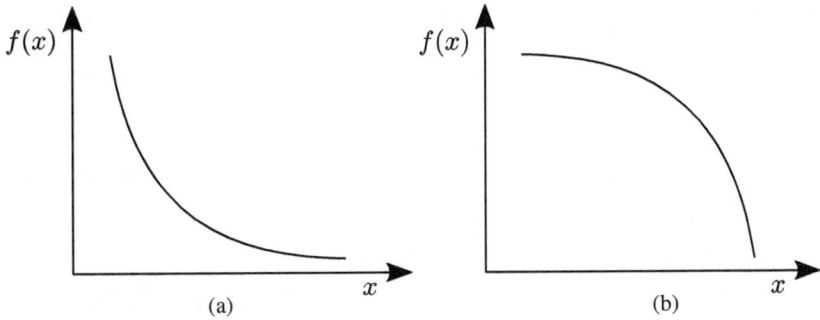

Fig. A.2 Concavity. (**a**) Convex function (**b**) Concave function

A.1.2.1   Common Functions

(1) **Polynomial functions**. A n degree polynomial function is written as

$$f(x) = a_n x^n + a_{n-1} x^{n-1} + a_{n-2} x^{n-2} + \ldots + a_0$$

Common polynomial functions are constant function, linear function, quadratic function.

(2) **Basic elementary functions**. Basic elementary functions include power functions, exponential functions, logarithmic functions, trigonometric functions, and inverse trigonometric functions.

(3) **Other common functions**.

-   Implicit functions. If the function $y = f(x)$ is determined by the equation $F(x, y) = 0$, then y is called the implicit function of x.
-   Hyperbolic functions. Hyperbolic sine: $\sinh(x) = \frac{e^x - e^{-x}}{2}$, hyperbolic cosine: $\cosh(x) = \frac{e^x + e^{-x}}{2}$, hyperbolic tangent: $\tanh(x) = \frac{\sin h(x)}{\cos h(x)}$, hyperbolic cotangent: $\cot h(x) = \frac{1}{\tan h(x)}$.
-   Gaussian function. $f(x) = a e^{-(x-b)^2 / 2c^2}$.

A.1.2.2   Sequences

Sequence describes a succession of certain things. A sequence of real numbers looks like

$$a_1, a_2, a_3, \cdots$$

and $1, \frac{1}{2}, \frac{1}{3}, \frac{1}{4}, \cdots$ is a real-number sequence, and it seems to approach 0. A sequence can be denoted by (a_n), (b_n), and the value of (a_n) at $n \in \mathbb{N}$ is given by a_n, which is called the n-th **term**.

Definition A.6 (Convergence of Sequences) Let $a \in \mathbb{R}$, $n_0 \in \mathbb{N}$. If for every $\varepsilon > 0$, the sequence (a_n) satisfies the condition

$$\forall n \geq n_0, |a_n - a| < \varepsilon$$

Then the (a_n) **converges** to a or that a is a **limit** of (a_n). This can be written as $a_n \to a$ while $n \to \infty$.

A sequence which is not convergent is **divergent**.

Property A.9 (Unique) A convergent sequence (a_n) has a unique limit.

Property A.10 (Bounded) A convergent sequence (a_n) is bounded.

If $n_1, n_2, \cdots$ are positive integers that $n_k < n_{k+1}$ ($k \in \mathbb{N}$), then the sequence (a_{n_k}), a succession of $a_{n1}, a_{n2}, \cdots$, is called a **subsequence** of (a_n). It should be noted that $n_1 < n_2 < \cdots$ implies that $n_k \to \infty$ as $k \to \infty$.

If the sequence $a_n \to a$, then all of its subsequences are convergent and have the same limit a.

Definition A.7 (Series) Given a sequence $(a_n)_{n=m}^{\infty}$ in $\mathbb{R}$, consider the associated sequence $(s_n)_{n=m}^{\infty}$ defined by

$$s_n = a_m + a_{m+1} + \cdots + a_n = \sum_{k=m}^{n} a_k$$

The number s_n is called the n-**th partial sum** of the infinite series $\sum_{n=m}^{\infty} a_n$, and the series converges if the sequence (s_n) converges.

A.1.2.3   The Limit of Function

Since a convergent sequence $a_n \to a$ can be regarded as a function: $a_n = f(n)$, where n is the variable and $n \in \mathbb{N}_+$, then the function value $f(n)$ is limit to a when n is approaching ∞. The certain value a is called the **limit** of function f.

Definition A.8 (The Limit of Function) Suppose that $\mathbb{D} \subseteq \mathbb{R}$, $a \in \mathbb{R}$ and $\mathbb{D}$ contains $(a - r, a)$ and $(a, a + r)$ for some $r > 0$, and considering a function $f : \mathbb{D} \to \mathbb{R}$. Then we say a limit of f tends to α exists, if there is a real number α satisfying: (x_n) any sequence in $\mathbb{D} \backslash \{a\}$ and $x_n \to a \Rightarrow f(x_n) \to \alpha$.

The above is the definition about limit of function when variable x is approaching a certain real number a and the domain contains an open interval about a except possibly the point a itself.

When x tends to be ∞ and the corresponding function value $f(x)$ is infinitely to a certain real value a, a is called the limit of $f(x)$ when $x \to \infty$. It can be written as

$$\lim_{x \to \infty} f(x) = a \text{ or } f(x) \to a(x \to \infty).$$

A.1.2.4   The Continuity of Function

Suppose that $\mathbb{D} \subseteq \mathbb{R}$, $f : \mathbb{D} \to \mathbb{R}$ and a point $p \in \mathbb{D}$, f is **continuous** at p if any sequence (x_n) in $\mathbb{D}$ and $x_n \to p \Rightarrow f(x_n) \to f(p)$. If f is continuous at every $p \in \mathbb{D}$, then f is **continuous** on $\mathbb{D}$.

Definition A.9 (Continuity) The function $f(x)$ is continuous at x_0 can be described as $\forall \varepsilon > 0, \exists \delta > 0$, when $|x - x_0| < \delta$, there is $|f(x) - f(x_0)| < \varepsilon$.

Property A.11 If function f is continuous at each point of an interval, then f continuous in an interval.

Property A.12 If function $f(x)$ and $g(x)$ is continuous at point x_0, then $f \pm g$, $f \times g$, $\frac{f}{g}(g(x_0) \neq 0)$ are continuous at point x_0.

Property A.13 Basic elementary functions are continuous in their domain.

The following theorems are about $f(x)$ when continuous in a closed interval $[a, b]$ where a, b are both normal numbers.

Theorem A.1 (Bounded and Maximum or Minimum) $f(x)$ *is bounded, and its maximum or minimum value can be acquired in its closed interval.*

Theorem A.2 (Zero Point Theorem) $f(x)$ *is continuous in closed interval $[a, b]$, and their signs are opposite each other ($f(a) \times f(b) < 0$). Then there is a point p enabled $f(p) = 0$ in (a, b).*

Theorem A.3 (Intermediate Value Theorem) $f(x)$ *is continuous in closed interval $[a, b]$ and $f(a) = m, f(b) = n$, there is at least one point $p \in (a, b)$ enabled $f(p) = c$ when $a < p < b, m < c < n$.*

Corollary A.1 *The range of $f(x)$ which continuous in $[a, b]$ is $[A, B]$, and A is the minimum of $f(x)$ while B is the maximum of $f(x)$.*

A function has "disconnections" or "breaks" means following three situations:

(1) There is no definition in $x = x_0$.
(2)] There is a definition but $\lim\limits_{x \to x_0} f(x)$ does not exits .
(3) $\lim\limits_{x \to x_0} f(x) \neq f(x_0)$ though there is definition and $\lim\limits_{x \to x_0} f(x)$ exits. f is **discontinuous** at x_0 and x_0 called the **discontinuous point**.

Dirichlet function is a typical discontinuous function.

Example A.1 (Dirichlet Function) Let $\mathbb{Q}$ be rational set and $f : \mathbb{R} \to \mathbb{R}$, then f is discontinuous at every $p \in \mathbb{R}$ if f defined by

$$f(x) = \begin{cases} 1 & x \in \mathbb{Q} \\ 0 & x \in \mathbb{R} \backslash \mathbb{Q} \end{cases}$$

A.1.2.5 Derivative and Differentiation

Differentiation is a process that relates to a real-valued function f and f' which is the derivative of f.

Definition A.10 (Differentiable) Let $\mathbb{D}$ be a subset of $\mathbb{R}$ and an element $p \in \mathbb{D}$ is an interior point of $\mathbb{D}$, if there is $r > 0$ such that $(p - r, p + r) \in \mathbb{D}$. If the following limit exists

$$\lim_{h \to 0} \frac{f(p + h) - f(p)}{h}$$

then the function $f : \mathbb{D} \to \mathbb{R}$ is said to be **differentiable** at the interior point p.

The value of the limit is called the **derivative** of f at p, which denoted by $f'(p)$. f is **differentiable** on $\mathbb{D}$ if it is differentiable at every point of $\mathbb{D}$ where each point is an interior point. In this case, a new function from $\mathbb{D}$ to $\mathbb{R}$ is obtained whose value at $p \in \mathbb{D}$ is $f'(p)$ and is called the **derivative (function)** of f, denoted by f'.

Derivative f' has several alternative notations like

$$\frac{df}{dx}, \text{ or } \frac{dy}{dx} \text{ when writing } y = f(x)$$

Sometimes, $f'(p)$ is denoted like

$$\frac{df}{dx}\bigg|_{x=p}, \text{ or } \frac{dy}{dx}\bigg|_{x=p}$$

Property A.14 (Carathẽodory's Lemma) $\mathbb{D} \subseteq \mathbb{R}$ and p is an interior point of $\mathbb{D}$, then $f : \mathbb{D} \to \mathbb{R}$ is differentiable at p only if there exists a function $g : \mathbb{D} \to \mathbb{R}$ such that $\forall x \in \mathbb{D}, f(x) - f(p) = (x - p)g(x)$, and g is continuous at p. Furthermore, if the above conditions hold, then $f'(p) = g(p)$.

Corollary A.2 *$\mathbb{D} \subseteq \mathbb{R}$ and p is an interior point of $\mathbb{D}$, if function $f : \mathbb{D} \to \mathbb{R}$ is differentiable at p, then f is continuous at p.*

If f' is differentiable on $[a, b]$, then f'' is the derivative of f' on $[a, b]$, which means that f is **twice differentiable** on $[a, b]$. Generally, f's n-th derivative on $[a, b]$, $f^{(n)}$ is similarly defined for any $n \in \mathbb{N}$.

Theorem A.4 (Rolle's Theorem) *If function $f : [a, b]$ is continuous on $[a, b] \in \mathbb{R}$, differentiable on (a, b), and $f(a) = f(b)$, then there is at least a point $p \in (a, b)$ such that $f'(p) = 0$.*

Theorem A.5 (Lagrange's Mean Value Theorem) *If function $f : [a, b]$ is continuous on $[a, b] \in \mathbb{R}$ and differentiable on (a, b), then there is at least a point $p \in (a, b)$ such that*

$$f(b) - f(a) = f'(p)(b - a)$$

Theorem A.6 (Cauchy's Mean Value Theorem) *Let function $f, g : [a, b] \to \mathbb{R}$ be continuous on $[a, b] \in \mathbb{R}$ and differentiable on (a, b), and $g'(x) \neq 0$ for all x in (a, b), then there is at least a point $p \in (a, b)$ such that*

$$g'(p)[f(b) - f(a)] = f'(p)[g(b) - g(a)]$$

Theorem A.7 (Taylor's Theorem) *Let $\mathbb{I}$ be an interval, and $f : \mathbb{I} \to \mathbb{R}$ satisfy the condition that $f^{(n+1)}(t)$ exists at each $t \in \mathbb{I}$. Fix $c \in \mathbb{I}$, and define $P_n(t) = \sum_{k=0}^{n} \frac{f^{(k)}(c)}{k!}(t - c)^k$, then for each $x \in \mathbb{I}$, there exists ξ between c and x such that*

$$f(x) = P_n(x) + \frac{f^{(n+1)}(\xi)}{(n + 1)!}(x - c)^{n+1}$$

Let $p \in \mathbb{R}$, and $f(x) = (1 + x)^p$ for $-1 < x < +\infty$. Then Taylor series $\sum_{n=0}^{\infty} \frac{f^{(k)}(0)x^k}{k!} a_n$ converges to $f(x)$ in the interval $-1 < x < 1$.

A.1.2.6   Absolute Minimum and Maximum

Absolute Extrema of the function $f : [a, b] \to \mathbb{R}$ in its entire domain (see Fig. A.3) can be classified into **Absolute Minimum and Maximum**. For example, the **absolute maximum** of f is at x_1, while the **absolute minimum** at x_2.

Generally speaking, local maximum and minimum are the wave crests and troughs of a curve if they are not located in the interval's end points. It is very hard to tell whether a local minimum is truly less than a local maximum (see points x_3 and b in Fig. A.3). Sometimes, an **absolute minimum or maximum** is also called a **global minimum or maximum**.

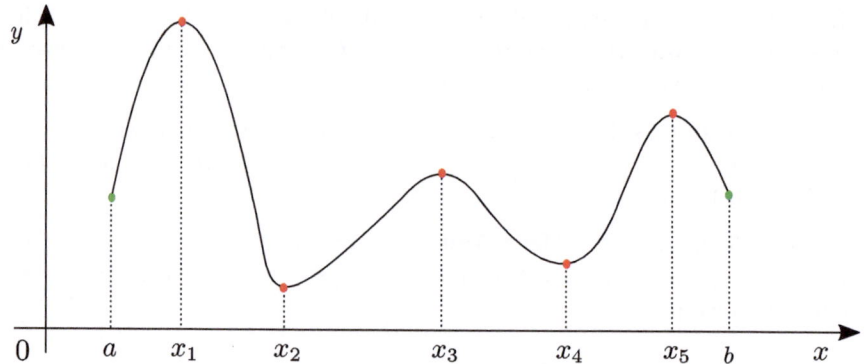

Fig. A.3 Graph of function defined on the interval $[a, b] \in \mathbb{R}$ with local minima at $x = a, x_2, x_4, b$ and with local maxima at $x = x_1, x_3, x_5$

A.2   Mathematical Algebra

A.2.1   Vector

Definition A.11 (Vector) $\mathbb{R}^n$, the Cartesian product of $\mathbb{R}$ by n times, is a set of ordered n-tuples of real numbers. For example, the n-tuple

$$a = \begin{bmatrix} 1.1 \\ e \\ \sqrt{2} \\ 0 \end{bmatrix}$$

is a vector of $\mathbb{R}^4$. A numeric vector $\lambda \in \mathbb{R}^1$ is called **scalar**.

Definition A.12 (Length) For any vector a in $\mathbb{R}^n$, the length of $a = \begin{bmatrix} a_1 \\ \vdots \\ a_n \end{bmatrix}$ is

defined as

$$\|a\| = \sqrt{a_1^2 + \ldots + a_n^2}$$

Suppose that numeric vectors $a = \begin{bmatrix} a_1 \\ \vdots \\ a_n \end{bmatrix}, b = \begin{bmatrix} b_1 \\ \vdots \\ b_n \end{bmatrix}, c = \begin{bmatrix} c_1 \\ \vdots \\ c_n \end{bmatrix} \in \mathbb{R}^n$ and scalars λ, r, s, then the following operations are defined by

Definition A.13 (Sum) $a + b = \begin{bmatrix} a_1 + b_1 \\ \vdots \\ a_n + b_n \end{bmatrix}$.

Definition A.14 (Product of a Vector by a Scalar) $\lambda a = \begin{bmatrix} \lambda a_1 \\ \vdots \\ \lambda a_n \end{bmatrix}$. More general,

$$r(a + b) = ra + rb, \ (r + s)a = ra + sa.$$

Definition A.15 (Scalar Product (also Dot Product or Inner Product))

$$a \cdot b = \langle a, b \rangle = ab = a_1 b_1 + \cdots + a_n b_n.$$

What's more, the following properties are valid:

Property A.15 (Commutative Law) $a + b = b + a$.

Property A.16 (Symmetry) $ab = ba$.

Property A.17 (Associativity) $\lambda(ba) = (\lambda a)b = a(\lambda b)$.

Property A.18 (Distributivity) $a(b + c) = ab + ac$.

Property A.19 (Triangle Inequality) $||a + b|| \leq ||a|| + ||b||$.

Property A.20 (Cauchy-Schwartz Inequality) $|a \cdot b| \leq ||a|| \, ||b||$.

Definition A.16 (Angle in $\mathbb{R}^n$) The angle between two nonzero vectors a and b in $\mathbb{R}^n$, $n \geq 2$, is defined as

$$\theta = \arccos \frac{a \cdot b}{||a|| \, ||b||}.$$

A.2.1.1   Vector Space

Definition A.17 (Vector Space (or Linear Space)) A vector space consists of a field $\mathbb{F}$ of scalars, a set $\mathbb{V}$ of vectors and two rules called **vector addition** and **scalar multiplication**, respectively.

Definition A.18 (Euclidean Space (or Cartesian Space)) Let k be a positive integer and the Euclidean k-space is the set of all sequences $(a_1, a_2, \cdots, a_k)$ of k real numbers with addition and multiplication operations. k-space or k-dimension Euclidean space is denoted by $\mathbb{R}^k$ and the ith component of $(a_1, a_2, \cdots, a_k)$ is the number a_i.

Fig. A.4 Graph of $\mathbb{R}^3$
Euclidean space

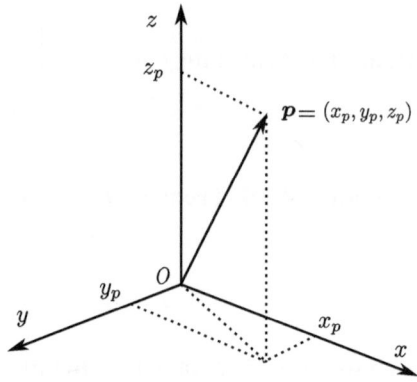

Theorem A.8 *For each positive integer k, $\mathbb{R}^k$ is a vector space and $\mathbb{R}^1 = \mathbb{R}$ in particular.*

Figure A.4 shows the $\mathbb{R}^3$ Euclidean space, and each point in the space has its coordinates (x, y, z); for example, point p has coordinates (or vector) (x_p, y_p, z_p).

Definition A.19 (Basis) If a finite subset $\mathbb{B}$ of a vector space $\mathbb{X}$ satisfies

(1) $\mathbb{B}$ can span $\mathbb{X}$,
(2) $\mathbb{B}$ is a set of independent vectors, then $\mathbb{B}$ is called a basis for $\mathbb{X}$.

A.2.1.2   Dimension

The **Dimension Theorem** defines that all bases of a vector space must have the same number of vectors. If a vector space $\mathbb{X}$ has two bases $\mathbb{A} = \{a_1, \cdots, a_m\}$ and $\mathbb{B} = \{b_1, \cdots, b_n\}$ with positive integers m and n, then there must hold $m = n$. The number of vectors in the basis is the dimension of the vector space.

Definition A.20 (Dimension) The dimension of vector space $\mathbb{X}$ is a positive integer m can be described as $\mathbb{X}$ has a basis of m vectors, denoted $dim(\mathbb{X}) = m$. The dimension of the vector space $\mathbb{O}$ is defined to be 0, and it has an empty set for the basis.

Definition A.21 (Norm) For $u \in \mathbb{F}^n$, its norm $\|u\|$ is defined as the nonnegative square root of $\langle u, u \rangle$ which is $\sqrt{\langle u, u \rangle}$.

Definition A.22 (Orthogonality in $\mathbb{R}^n$) Two vectors a, b in $\mathbb{R}^n$ are defined to be **orthogonal** to each other if $a \cdot b = 0$. Generally, a basis $u_1, \cdots, u_n$ of an n-dimensional inner product space $\mathbb{F}$ is **orthogonal** if $\langle u_i, u_j \rangle = 0$ for all $i \neq j$, and each vector has unit length: $\|u_i\| = 1, i = 1, \cdots, n$.

The basis is **orthonormal** and the simplest orthonormal basis in the Euclidean space $\mathbb{R}^n$ is the **standard basis** which are

$$
e_1 = \begin{bmatrix} 1 \\ 0 \\ \vdots \\ 0 \\ 0 \end{bmatrix} \quad
e_2 = \begin{bmatrix} 0 \\ 1 \\ 0 \\ \vdots \\ 0 \end{bmatrix} \quad \cdots \quad
e_n = \begin{bmatrix} 0 \\ 0 \\ \vdots \\ 0 \\ 1 \end{bmatrix}.
$$

A.2.1.3   Systems of Linear Equations

Suppose a field $\mathbb{F}$ and given elements in $\mathbb{F}$ which are a_{ij}, $1 \leq i \leq m$, $1 \leq j \leq n$, the system of m linear equations in n unknowns can be represented as finding n scalars $x_1, \cdots, x_n$ which are the elements of $\mathbb{F}$ satisfying Eqs. (A.1).

$$
a_{11}x_1 + a_{12}x_2 + \cdots + a_{1n}x_n = b_1
$$
$$
a_{21}x_1 + a_{22}x_2 + \cdots + a_{2n}x_n = b_2
$$
$$
\cdots
$$
$$
a_{m1}x_1 + a_{m2}x_2 + \cdots + a_{mn}x_n = b_m \tag{A.1}
$$

The solution of the system can be any n-tuple $(x_1, \cdots, x_n)$ of elements of $\mathbb{F}$ as long as satisfying each of Eqs. (A.1). The system is **homogeneous** if $b_1 = b_2 = \cdots = b_m = 0$.

Selecting m scalars $c_1, \cdots, c_m$ and multiplying the jth equation by c_j, and then adding all the equations together, thus a **linear combination** is acquired

$$
(c_1 a_{11} + \cdots + c_m a_{m1})x_1 + \cdots + (c_1 a_{1n} + \cdots + c_m a_{mn})x_n = c_1 b_1 + \cdots + c_m b_m \tag{A.2}
$$

Definition A.23 (Linearly Dependent) If there are distinct vectors $a_1, \ldots, a_k$ in set $\mathbb{A}$ and real numbers $r_1, \ldots, r_k$ not all 0, then a **linear relation** between $a_1, \ldots, a_k$ described as

$$
r_1 a_1 + \ldots + r_k a_k = 0.
$$

A.2.2 *Matrix*

Definition A.24 (Matrix) A matrix is a rectangular array of numbers, such as $\begin{bmatrix} \pi & 0.64 \\ e & \sqrt{5} \end{bmatrix}$. And a general matrix of size $r \times c$ represents

$$A = \begin{bmatrix} a_{11} & a_{12} & \cdots & a_{1c} \\ a_{21} & a_{22} & \cdots & a_{2c} \\ \vdots & \vdots & \ddots & \vdots \\ a_{r1} & a_{r2} & \cdots & a_{rc} \end{bmatrix}$$

where r denotes the number of rows in A while c the number of columns.

A **column vector** can be regarded as an $r \times 1$ matrix and **row vector** a $1 \times c$ matrix.

Definition A.25 (Square) The square is the matrix where $r = c$. For example, matrix A is a square.

$$A = \begin{bmatrix} 1 & 3 \\ 1 & 4 \end{bmatrix}$$

Only square matrices can be used to calculate **determinants**, and the determinant of a square matrix $A \in \mathbb{F}^{n \times n}$ is denoted as $\det(A)$ or $|A|$.

$$\det(A) = \sum_{j=1}^{n} a_{1j} (-1)^{1+j} \det(A_{1j}) \tag{A.3}$$

where $A_{1j} \in \mathbb{F}^{(n-1) \times (n-1)}$ is obtained from A by deleting its first row and the j-th column.

Definition A.26 (Null Matrix) Null matrix's elements are zeros. The null matrix of $\mathbb{R}^2$ is

$$O = \begin{bmatrix} 0 & 0 \\ 0 & 0 \end{bmatrix}.$$

Definition A.27 (Symmetric Matrix) A symmetric matrix A is a **square** where the elements follow $\forall i, j : a_{ij} = a_{ji}$. An example for symmetric matrix $A \in \mathbb{R}^3$ is

$$A = \begin{bmatrix} 5 & 3 & 0 \\ 3 & 2 & 4 \\ 0 & 4 & 1 \end{bmatrix}.$$

Definition A.28 (Identity Matrix) Identity matrix is a diagonal matrix with each diagonal entry as 1 denoted by

$$
\mathbf{I} = \begin{bmatrix} 1 & 0 & \cdots & \cdots & 0 \\ 0 & 1 & 0 & \cdots & 0 \\ \vdots & 0 & \ddots & \cdots & 0 \\ \vdots & \vdots & \vdots & \ddots & 0 \\ 0 & 0 & \cdots & 0 & 1 \end{bmatrix} = \mathrm{diag}(1, \cdots, 1)
$$

and an identity matrix of order n can be written as $\mathbf{I}_n$.

Definition A.29 (Inverse) A square matrix A of order n is **invertible** when there is a matrix B of the same order having

$$
AB = \mathbf{I} = BA
$$

and B is the **inverse** of A written as A^{-1}. Not all square matrices are invertible; for instance, $\mathbf{I}$ is invertible but O is not.

Definition A.30 (Transpose) The transpose of matrix $A \in \mathbb{F}^{m \times n}$ is $A^{\mathrm{T}} \in \mathbb{F}^{n \times m}$ and the (i, j)th entry of A^{T} = the (j, i)th entry of A.

Some properties of transpose operation are listed:

(1) $(A + B)^{\mathrm{T}} = A^{\mathrm{T}} + B^{\mathrm{T}}$.
(2) $(\lambda A)^{\mathrm{T}} = \lambda A^{\mathrm{T}}$.
(3) $(AB)^{\mathrm{T}} = B^{\mathrm{T}} A^{\mathrm{T}}$.
(4) If A is invertible, then A^{T} is invertible either, and $(A^{-1})^{\mathrm{T}} = (A^{\mathrm{T}})^{-1}$.

Definition A.31 (Orthogonal Matrix) If the product between a square matrix $A \in \mathbb{R}^n$ and its transpose is the identity matrix, then A is an orthogonal matrix.

$$
AA^{\mathrm{T}} = \mathbf{I} = A^{\mathrm{T}} A.
$$

A.2.2.1   Operations

(1) **Matrix Addition (Sum).** The premise for matrix addition is that the dimensions of the matrix are the same. The matrix addition principle is adding the elements of the corresponding position. Let A, B and their sum $C \in \mathbb{R}^{r,c}$, then their elements follow $\forall i, j : c_{ij} = a_{ij} + b_{ij}$. For example,

$$
\begin{bmatrix} 1 & 5 & 3 \\ 3 & 2 & 0 \end{bmatrix} + \begin{bmatrix} 1 & 0 & 0 \\ 1 & 2 & 2 \end{bmatrix} = \begin{bmatrix} 2 & 5 & 3 \\ 4 & 4 & 2 \end{bmatrix}.
$$

(2) **Scalar Product**. The product of a scalar by a matrix is multiplying the scalar by each element of the matrix. Let $A \in \mathbb{R}^{r,c}$ and $\lambda \in \mathbb{R}$, then $B = \lambda A$ equals to $\forall i, j : b_{ij} = \lambda a_{ij}$. For example, the product of the scalar $\lambda = 2$ by the matrix $A = \begin{bmatrix} 3 & 0 & -2 \\ -1 & \pi & 1 \end{bmatrix}$ is

$$\lambda A = \begin{bmatrix} 2 \times 3 & 2 \times 0 & 2 \times (-2) \\ 2 \times (-1) & 2 \times \pi & 2 \times 1 \end{bmatrix} = \begin{bmatrix} 6 & 0 & -4 \\ -2 & 2\pi & 2 \end{bmatrix}.$$

(3) **Vector Product (also Cross Product)**. The vector product is an operator associating a vector with two vectors ($R^3 \times R^3 \rightarrow R^3$). Let vectors $a, b \in R^3$ and their angle θ, the vector product $a \times b$ (sometimes denoted as $a \otimes b$) is the same to the (symbolic) determinant of the matrix A where

$$A = \begin{bmatrix} e_1 & e_2 & e_3 \\ a_1 & a_2 & a_3 \\ b_1 & b_2 & b_3 \end{bmatrix}$$

and $\{e_1, e_2, e_3\}$ is orthonormal.

(4) **Matrix Multiplication**. The premise of matrix multiplication is that the number of columns of the first matrix is equal to the number of rows of the second matrix. If an $m \times n$ matrix A multiply an $n \times r$ matrix B (the number of columns of A equals rows of B), then an $m \times r$ matrix AB obtained.

The matrix multiplication principle is summing the row i, column j entry of AB, which is $\sum_{k=1}^{n} a_{ik}b_{kj}$, where a_{ij} and b_{ij} are the row i, column j entries of A and B respectively. For example,

$$\begin{bmatrix} 1 & 2 & 3 \\ 4 & 5 & 6 \end{bmatrix} \begin{bmatrix} 2 \\ 3 \\ 4 \end{bmatrix} = \begin{bmatrix} 1 \times 2 + 2 \times 3 + 3 \times 4 \\ 4 \times 2 + 5 \times 3 + 6 \times 4 \end{bmatrix} = \begin{bmatrix} 20 \\ 47 \end{bmatrix}.$$

Definition A.32 (Eigenvalues and Eigenvectors) An eigenvalue of an $n \times n$ matrix A is a number λ such that $Av = \lambda v$ for some $n \times 1$ nonzero column vector v, then v is an **eigenvector** of A (also called right eigenvector). A left eigenvector of A would be a row vector w satisfying $wA = \lambda w$. When A and v are multiplied, v just gets stretched by the constant λ.

A.2.2.2   Matrix of Linear System

Linear system can be written in the single matrix equation form: $\mathbf{AX} = \mathbf{B}$ (see Eq. (A.4)), where matrix $\mathbf{A}$ is the **coefficient matrix** of the linear system and $\mathbf{B}$ the **constant matrix**.

$$A = \begin{bmatrix} a_{11} & a_{12} & \cdots & a_{1n} \\ a_{21} & a_{22} & \cdots & a_{2n} \\ \vdots & \vdots & \cdots & \vdots \\ a_{m1} & a_{m2} & \cdots & a_{mn} \end{bmatrix}, X = \begin{bmatrix} x_1 \\ x_2 \\ \vdots \\ x_m \end{bmatrix}, B = \begin{bmatrix} b_1 \\ b_2 \\ \vdots \\ b_m \end{bmatrix} \tag{A.4}$$

Definition A.33 (Quadratic Forms) The square of linear form $ax + by$ is

$$(ax + by)^2 = a^2 x^2 + 2abxy + b^2 y^2.$$

It is a typical quadratic form and can be written as

$$(ax + by)^2 = a^2 x^2 + 2abxy + b^2 y^2 = \begin{bmatrix} x & y \end{bmatrix} \begin{bmatrix} a^2 & ab \\ ab & b^2 \end{bmatrix} \begin{bmatrix} x \\ y \end{bmatrix}.$$

The quadratic form of n variables $x_1, \cdots, x_n$, also known as **the matrix of the quadratic form**, can be written as

$$Q = q_{11}x_1{}^2 + 2q_{12}x_1 x_2 + \cdots + q_{nn}x_n{}^2$$

$$= \begin{bmatrix} x_1 & x_2 & \cdots & x_n \end{bmatrix} \begin{bmatrix} q_{11} & q_{12} & \cdots & q_{1n} \\ q_{12} & q_{22} & \cdots & q_{2n} \\ \vdots & \vdots & \ddots & \vdots \\ q_{1n} & q_{2n} & \cdots & q_{nn} \end{bmatrix} \begin{bmatrix} x_1 \\ x_2 \\ \vdots \\ x_n \end{bmatrix}.$$

If an $n \times n$ symmetric matrix K satisfies $X^{\mathrm{T}} K^{\mathrm{T}} X > 0, \mathbf{0} \neq X \in \mathbb{R}^n$, then K is a positive definite matrix denoted by $K > 0$.

A.3   Probability and Stochastics Processes

Probability is associated with the various possible outcomes that might be obtained and the possible events that might occur when an experiment is carried out. An **experiment** is any real or hypothetical process where the possible outcomes can be identified in advanced and an **event** is a defined set of possible outcomes of the experiment. The collection of all possible results of an experiment is the **sample space** of the experiment, and the **sample point** or **element** refers to a result in the collection.

For example, all possible outcomes of rolling a die consist of a sample space $\mathbb{S} = \{1, 2, 3, 4, 5, 6\}$, and the event A describing "the number is odd" is the subset $\mathbb{A} = \{1, 3, 5\}$.

A.3.1   Probability

Definition A.34 (Probability) Probability gives a measure of what is likely to happen, and it is associated with the observable frequency of occurrence.

Let $N_n(A)$ denote the number of times the event A is observed in n throws, and then the **relative frequency** of A is

$$P_n(A) = \frac{N_n(A)}{n}.$$

Let n go to infinity, leading $P_n(A)$ to settle down to a steady value, and the empirical **limiting frequency** of the event A is

$$P(A) = \lim_{n \to \infty} P_n(A).$$

Assuming that P is the same for all similar sequences of trials and according to Law of Large Numbers, the probability of A that occurs is

$$P(A) = \frac{|\mathbb{A}|}{|\mathbb{S}|} = \frac{3}{6} = \frac{1}{2}.$$

Definition A.35 (Conditional Probability) The conditional probability of the event B is the probability that B occurs under the condition that event A has occurred, denoted as

$$P(B|A) = \frac{P(AB)}{P(A)}.$$

Property A.21 (Multiplication Formula for Conditional Probability)

$$P(AB) = P(A)P(B|A) = P(B)P(A|B).$$

Property A.22 (Full (or Total) Probability Formula) Let the events $B_1, \cdots, B_k$ form a partition of the space $\mathbb{S}$ and $P(B_j) > 0$ for $j = 1, \cdots, k$. Then for every event A in $\mathbb{S}$,

$$P(A) = \sum_{j=1}^{k} P(B_j)P(A|B_j).$$

Definition A.36 (Independent Events) If the occurrence of B has changed without affecting the probability of A, then A and B are independent. Described as

$$P(A \cap B) = P(A)P(B).$$

Theorem A.9 (Bayesian Formula) *Let the events $B_1, \cdots, B_k$ form a partition of the space $\mathbb{S}$ and $P(B_j) > 0$ for $j = 1, \cdots, k$ and event A which $P(A) > 0$, then for $i = 1, \cdots, k$,*

$$P(B_i \,|\, A) = \frac{P(B_i)P(A \,|\, B_i)}{\sum_{j=1}^{k} P(B_j)P(A \,|\, B_j)}.$$

Definition A.37 (Random Variable) A real-valued function defined on an experiment's sample space $\mathbb{S}$ is a random variable.

Considering an experiment in which a person is randomly selected from some population, his or her height measured in centimeters is a random variable.

Definition A.38 (Distribution) Let $\mathbb{C}$ be a subset of the real number and random variable $X \in \mathbb{C}$ event, then $P(X \in \mathbb{C})$ denote the probability that the value of X will belong to $\mathbb{C}$. The distribution of X is the collection of all probabilities of $P(X \in \mathbb{C})$ for all sets $\mathbb{C}$ of real numbers.

Definition A.39 (Discrete Random Variable) X is a discrete random variable if X can take only a finite number k of different values $x_1, \cdots, x_k$ or an infinite sequence of different values $x_1, x_2, \cdots$.

Definition A.40 (Probability Function) If a random variable X has a discrete distribution, the probability function of X is defined as the function f such that for every real number x,

$$f(x) = P(X = x).$$

The closure of the set $x : f(x) > 0$ is the **support(or distribution)** of X.
Here are some common discrete distributions:

(1) **Bernoulli Distribution.** A random variable X that takes only two values 0 and 1 with $P(X = 1) = p$ has the Bernoulli distribution with parameter p. It can also be described as X is a Bernoulli random variable with parameter p. The probability function of X is

$$f(x) = p^x(1 - p)^{1-x}, x \in \{0, 1\}.$$

(2) **Uniform Distribution on Integers.** A random variable X which is equally likely to be each of the integers $a, \cdots, b(a \leq b)$ or has probability $\frac{1}{n}$ on each integer on a set of k integers has the uniform distribution on the integers. The probability function of X on the integers $a, \cdots, b(a \leq b)$ is

$$f(x) = \begin{cases} \dfrac{1}{b - a + 1} & x = a, \cdots, b \\ 0 & \text{otherwise} \end{cases}$$

Definition A.41 (Continuous Random Variable) X is a continuous random variable if X can take only every value in an interval of real numbers (bounded or unbounded).

The **probability density function** can characterize the distribution of a continuous random variable.

Definition A.42 (Probability Density Function) If X has a continuous distribution, then the probability that X takes a value in the interval is the integral of nonnegative function f over the interval. Suppose two integers $a, b, (a < b)$, there are

$$P(a \le X \le b) = \int_a^b f(x)dx$$

$$P(a < X) = \int_a^{+\infty} f(x)dx$$

$$P(X < b) = \int_{-\infty}^b f(x)dx$$

For all x, any probability density functions satisfy $f(x) \ge 0$ and $\int_{-\infty}^{+\infty} f(x)dx = 1$. Here are some common continuous distributions:

(1) **Uniform Distribution on an Interval.** Given two real numbers $a, b(a < b)$, and a random variable X which is known $a \le X \le b$, and for every subinterval of $[a, b]$, the probability that X will belong to that subinterval is proportional to the length of it. The probability density function of X on the integers $a, \cdots, b(a \le b)$ is

$$f(x) = \begin{cases} \dfrac{1}{b-a} & a \le x \le b, a \ne b \\ 0 & \text{otherwise} \end{cases}$$

(2) **Exponential Distribution.** If a random variable X has the exponential distribution with parameter $\lambda > 0$, then the probability density function of X is

$$f(x) = \begin{cases} \lambda e^{-\lambda x} & x \ge 0 \\ 0 & x < 0 \end{cases}$$

Definition A.43 (Mean of Discrete Random Variable) Let X be a discrete random variable with probability function f, and at least one of the sums $\sum_{x>0} xf(x)$, $\sum_{x<0} xf(x)$ is finite, then the **mean(also expectation or expected value)** of X is defined to be

$$E(X) = \sum_{x} x f(x)$$

However, if both of the sums are infinite, then $E(X)$ does not exist.

Definition A.44 (Mean of Continuous Random Variable) Let X be a continuous random variable with probability density function f, and at least one of the integrals $\int_{0}^{+\infty} x f(x) dx$, $\int_{-\infty}^{0} x f(x) dx$ is finite, then the **mean** of X is defined to be

$$E(X) = \int_{-\infty}^{+\infty} x f(x) dx.$$

However, if both of the integrals are infinite, then $E(X)$ does not exist.

If X is a random variable whose expectation $E(X)$ exists and a, b are finite constants, here are some basic properties.

Theorem A.10 (Linear Function) *If the random variable $Y = aX + b$, then*

$$E(Y) = a E(X) + b.$$

Corollary A.3 *If $P(X \geq a) = 1$, then $E(X) \geq a$. If $X = a$ with probability 1, then $E(X) = a$. If $P(X \leq a) = 1$, then $E(X) \leq a$.*

Theorem A.11 *For n random variables $X_1, \cdots, X_n$ with each finite expectation $E(X_i), i = 1, \cdots, n$, there is*

$$E(X_1 + \cdots + X_n) = E(X_1) + \cdots + E(X_n).$$

Definition A.45 (Variance) Let X be a random variable with finite mean $E(X) = \mu$, then the variance of X, denoted by $var(X)$, is defined as

$$var(X) = E[(X - \mu)^2].$$

However, if $E(X)$ does not exist or is infinite, the variance of X does not exist. In addition, the nonnegative square root of $var(X)$ is called **Standard Deviation** of X.

Just as $E(X)$ depends only on the distribution, the variance and standard deviation of X depend only on the distribution of random variable X.

Theorem A.12 (Alternative Method for Calculating the Variance) *For every random variable X, there is*

$$var(X) = E(X^2) - [E(X)]^2.$$

Theorem A.13 *If $X_1, \cdots, X_n$ are independent random variables with finite means and $a_1, \cdots, a_n$ and b are arbitrary constants, there are*

$$var(X_1 + \cdots + X_n) = var(X_1) + \cdots + var(X_n),$$

$$var(a_1 X_1 + \cdots + a_n X_n + b) = a_1{}^2 var(X_1) + \cdots + a_n{}^2 var(X_n).$$

Covariance of random variables indicates the relationship between them or their tendency to vary together rather than independently.

Definition A.46 (Covariance) Suppose that random variables X, Y have expectation $E(X) = \mu_X$, $E(Y) = \mu_Y$, respectively. The covariance of X and Y, denoted by $cov(X, Y)$, is defined as

$$cov(X, Y) = E[(X - \mu_X)(Y - \mu_Y)].$$

Theorem A.14 *For all random variables X, Y with $\sigma_X{}^2 < +\infty$, $\sigma_Y{}^2 < +\infty$, there is*

$$cov(X, Y) = E(XY) - E(X)E(Y).$$

The magnitude of covariance is influenced by the overall magnitudes of random variables and thus defined **Correlation** to measure the association between random variables which is not driven by arbitrary changes in the scales of one or the other random variable.

Definition A.47 (Correlation) Let X, Y be random variables with finite variances $\sigma_X{}^2, \sigma_Y{}^2$; then the correlation of X, Y denoted by $\rho(X, Y)$ is defined as

$$\rho(X, Y) = \frac{cov(X, Y)}{\sigma_X \sigma_Y}.$$

Theorem A.15 *Let $var(X) < +\infty$, $var(Y) < +\infty$, then*

$$var(X + Y) = Var(X) + var(Y) + 2cov(X, Y).$$

Extending to the variance of the sum of n random variables, there is

Theorem A.16 *Let $X_1, \cdots, X_n$ are random variables such that $Var(X_i) < +\infty$ for $i = 1, \cdots, n$, there is*

$$Var\left(\sum_{i=1}^{n} X_i\right) = \sum_{i=1}^{n} Var(X_i) + 2\sum_{i<j}\sum Cov(X_i, X_j).$$

A set of common discrete distributions and continuous distributions are listed in Tables A.1 and A.2.

Table A.1 Discrete distributions

	Bernoulli with parameter p	Binomial with parameters n, p
Probability function	$f(x) = p^x(1-p)^{1-x}$ for $x = 0, 1$.	$f(x) = \binom{n}{x} p^x(1-p)^{n-x}$ for $x = 0, \cdots, n$
Mean	p	np
Variance	$p(1-p)$	$np(1-p)$
	Uniform on the integers $a, \cdots, b$	Poisson with mean λ
Probability function	$f(x) = \frac{1}{b-a+1}$ for $x = a, \cdots, b$	$f(x) = e^{-\lambda}\frac{\lambda^x}{x!}$ for $x = 0, 1, \cdots$
Mean	$\frac{b+a}{2}$	λ
Variance	$\frac{(b-a)(b-a+2)}{12}$	λ

Table A.2 Continuous distributions

	Beta with parameters α, β	Exponential with parameters λ
Probability density function	$f(x) = \frac{\Gamma(\alpha+\beta)}{\Gamma(\alpha)\Gamma(\beta)}x^{\alpha-1}(1-x)^{\beta-1}$ for $0 < x < 1$	$f(x) = \begin{cases} \lambda e^{-\lambda x}, & x \geq 0 \\ 0, & x < 0 \end{cases}$ for $\lambda > 0$
Mean	$\frac{\alpha}{\alpha+\beta}$	$\frac{1}{\lambda}$
Variance	$\frac{\alpha\beta}{(\alpha+\beta)^2(\alpha+\beta+1)}$	$\frac{1}{\lambda^2}$
	Normal	Uniform on the interval $[a, b]$
Probability density function	$f(x) = \frac{1}{\sqrt{2\pi}\sigma}e^{\frac{-(x-\mu)^2}{2\sigma^2}}$	$f(x) = \frac{1}{b-a}$ for $a < x < b$
Mean	μ	$\frac{(a+b)}{2}$
Variance	σ^2	$\frac{(b-a)^2}{12}$
	Gamma with parameters α, β	Pareto with parameters x_0, α_0
Probability density function	$f(x) = \frac{\beta^\alpha}{\Gamma(\alpha)}x^{\alpha-1}e^{-\beta x}$ for $x > 0$	$f(x) = \frac{\alpha x_0^\alpha}{x^{\alpha+1}}$ for $x > x_0$
Mean	$\frac{\alpha}{\beta}$	$\frac{\alpha x_0}{\alpha-1}$ if $\alpha > 1$
Variance	$\frac{\alpha}{\beta^2}$	$\frac{\alpha x_0^2}{(\alpha-1)^2(\alpha-2)}$ if $\alpha > 2$

Definition A.48 (Stochastic Process) A stochastic process describes a sequence of random variables $X_1, X_2, \cdots$ with discrete time parameter, and the initial state of the process is the first random variable X_1; and for $n = 2, 3, \cdots$, the random variable X_n is the state of the process at time n.

A.3.1.1 Markov Chains

A Markov chain is defined regarding the conditional distributions of future states given the present and past states. This type of stochastic process with discrete time parameter n is a Markov chain, if the conditional distributions of all X_{n+j} for $j > 1$ given $X_1, \cdots, X_n$ depend only on X_n and not on the earlier states $X_1, \cdots, X_{n-1}$ for each time n.

For each b and possible sequence of states $x_1, x_2, \cdots, x_n, n = 1, 2, \cdots$, there is

$$P(X_{n+1} \leq b \,|\, X_1 = x_1, X_2 = x_2, \cdots, X_n = x_n) = P(X_{n+1} \leq b \,|\, X_n = x_n).$$

Appendix B
Open Framework of Evolutionary Computation

Since researchers from different teams in the field of EC have their own preferences in programming languages and research topics, most teams publish source code on their own experimental platforms.

However, when people try to use algorithms, benchmark problems, and measurements published by other teams, the work of translating code always takes a lot of time. On the other hand, people outside the EC field have difficulty using these source codes without any manuals or documentation.

For the above reasons, some teams proposed open platforms and libraries. Table B.1 shows some representative ones.

Among this platforms or libraries:

(1) DEAP, Jenetics, and GAUL both supply a bunch of EA operators (e.g., initialization, crossover, mutation, selection, migration). GAUL additionally includes some multi-population schemes and some traditional stochastic optimization algorithm. However, both of them have only a small amount of complete EAs and very few benchmark functions.
(2) PlatEMO contains more than 50 multi-objective EAs and more than 100 multi-objective test problems. However, EAs other than MOEAs cannot be developed in this platform.

So we propose the Open Framework of Evolutionary Computation (OFEC), which is an open-source C++ project that aims to develop a common platform for designing and testing almost all kinds of EAs.

Table B.1 Some open
platforms and libraries of EA

Name	Language	Reference
DEAP	python	https://deap.readthedocs.io/en/master/
Jenetics	java	http://jenetics.io/
GAUL	C++	http://gaul.sourceforge.net/
PlatEMO	matlab	http://bimk.ahu.edu.cn/12957/list.htm

B.1   Features of OFEC

B.1.1   Comprehensiveness

OFEC encompasses most of the representative EAs, including early proposed classical EAs (e.g., GA, GP, EP, and ES), recent EAs for continuous optimization (e.g., DE and PSO), EAs for combinatorial optimization (e.g., EDA and ACO), representative variants of above EAs, and traditional single-solution heuristics (e.g., hill climbing and simulated annealing).

The platform also covers benchmark functions of a variety of mainstream optimization problems, including global optimization problems, multimodal optimization problems, multi-objective optimization problems, constrained optimization problems, dynamic optimization problems, large-scale optimization problems, expensive optimization problems, combinatorial optimizations (e.g., TSP, QAP), and some real-world optimization problems.

B.1.2   Easy of Use

In OFEC, we have implemented most of the classical EAs, and users can easily develop variants of these EAs. They only need to write parts that are different from the classical version. If a user wants to run EAs on a problem other than the existing ones, then he merely needs to write the added problem itself and totally leave alone the details of how the algorithm call his problem. In addition, we also provide with many commonly used utilities and metrics. All these works contribute to reducing the amount of code that users need to write.

B.1.3   Scalability

In OFEC, solution, individual, population, multi-population, and some EA framework such as NSGAII are both defined as class template. In this way, Solution class and Individual class can support different encodings of decision variables and objectives. And population classes and EA framework classes can support different kinds of individuals.

B.2   Procedures of using OFEC

B.2.1   Running OFEC

A typical running process of OFEC is as follow.

```
ParamMap params;

params["problem_name"] = string("Sphere");
params["number_of_variables"] = 2;

params["algorithm_name"] = string("Canonical-DE");
params["population_size"] = 100;
params["scaling_factor"] = 0.5;
params["crossover_rate"] = 0.6;

params["maximum_evaluations"] = 20000;
params["number_of_runs"] = 50;

int id_param = ADD_PARAM(params);
int id_pro = ADD_PRO(id_param, 0.5);
int id_alg = ADD_ALG(id_param, id_pro, 0.5, -1);
GET_PRO(id_pro).initialize();
GET_ALG(id_alg).initialize();
GET_ALG(id_alg).run();
```

Which does the following steps:

(1) Generate a `ParamMap` instance and write necessary experimental parameters (which determine the algorithm, the problem, the number of runs, the measurement interval and so on) to it.
(2) Use `ADD_PARAM()` macro to add the `ParamMap` instance to the global instance manager and obtain its global ID.
(3) Use `ADD_PRO()` macro and the ID of `ParamMap` instance to generate a `Problem` instance inside the global instance manager and also obtain its global ID.
(4) Use `ADD_ALG()` macro, the ID of `ParamMap` instance and the ID of the `Problem` instance to generate a `Algorithm` instance inside the global instance manager and also obtain its global ID.
(5) Use `GET_PRO()` macro and the ID of the `Problem` instance to access the problem and initialize it.
(6) Use `GET_ALG()` macro and the ID of the `Algorithm` instance to access the algorithm and initialize it.
(7) Use `GET_ALG()` macro and the ID of the `Algorithm` instance to access the algorithm and run it.

B.2.2   Adding a Problem

Suppose we want to add a continuous optimization problem `TestFunction`, we need to define it as a class derived from class `Continuous`.

```
class TestFunction : public Continuous {
protected:
void initializ_() override;
void evaluateObjective(Real *var, vector<Real> &obj) override;
}
```

Here we skip the definition of its member functions. We just need to know that `initialize_()` determines the optima and the search domain of the problem, and `evaluateObjective` defines how to evaluate the objective values of the given decision variables.

Next we need to register `TestFunction` by two steps:

(1) In the file */run/include_problem.h* include the header file.

```
#include "../instance/problem/\ldots../test_function.h"
```

(2) In the file */run/register_problem.cpp*, use the `REGISTER(...)` macro to register.

```
REGISTER(Problem, TestFunction, "Test_funtion", set<ProTag>({
    ProTag::GOP, ProTag::ConOP }));
```

B.2.2.1   Adding an Algorithm

Suppose we want to add CrDE (a variant of DE for MMOP), we first need to define a CrDE population class inherited from DE population.

```
class PopCrDE final : public PopDE<IndDE> {
public:
explicit PopCrDE(size_t size_pop, int id_pro);
int evolve(int id_pro, int id_alg, int id_rnd) override;
};
```

And override its member function `evolve()` as follow.

```
int PopCrDE::evolve(int id_pro, int id_alg, int id_rnd) {
int tag = NormalEval;
for (size_t i = 0; i < size(); ++i) {
mutate(i, id_rnd, id_pro);
m_inds[i]->recombine(m_CR, m_recombine, id_rnd, id_pro);
tag = m_inds[i]->trial().evaluate(id_pro, id_alg);
if (tag == Terminate)
return tag;
int idx = nearestNeighbour(i).begin()->second;
if (m_inds[i]->trial().dominate(*m_inds[idx]))
m_inds[idx]->solut() = m_inds[i]->trial();
}
m_iter++;
return tag;
}
```

Then define a CrDE class inherited from class `algorithm`, and make `PopCrDE` as its data member.

```
class CrDE final : public Algorithm {
private:
PopCrDE m_pop;
protected:
void initialize_() override;
void run_() override;
};
```

And write the initialization and running process.

```
void CrDE::initialize_() {
m_pop_size = 100;
m_F = 0.5;
m_CR = 0.9;
}

void CrDE::run_() {
m_pop.resize(m_pop_size, m_id_pro);
m_pop.CR() = m_CR;
m_pop.F() = m_F;
m_pop.initialize(m_id_pro, m_id_rnd);
```

(continued)

```
m_pop.evaluate(m_id_pro, m_id_alg);
while (!terminating()) {
m_pop.evolve(m_id_pro, m_id_alg, m_id_rnd);
}
}
```

After these steps, the implementation of CrDE is done.

B.3   Components of OFEC

OFEC consists of four components:

(1) *Utility.* data structures, machine learning algorithms, random number genera-
    tors, vector and matrix for math calculation, sorting algorithms, and so on.
(2) *Problem.* domain, encoding, solution, optima, virtual base problems, and
    problem instances.
(3) *Algorithm.* termination criterion, individual, population, multi-population, vir-
    tual base algorithms, and algorithm instances.
(4) *Measurement.* methods to record and measure all kinds of algorithms.

B.3.1   Problem Component

In OFEC, as shown in Fig. B.1, combinatorial problems are directly inherited
from the base class Problem, while most continuous problems in the platform
are inherited from Problem's derived class Continuous. Benchmark functions

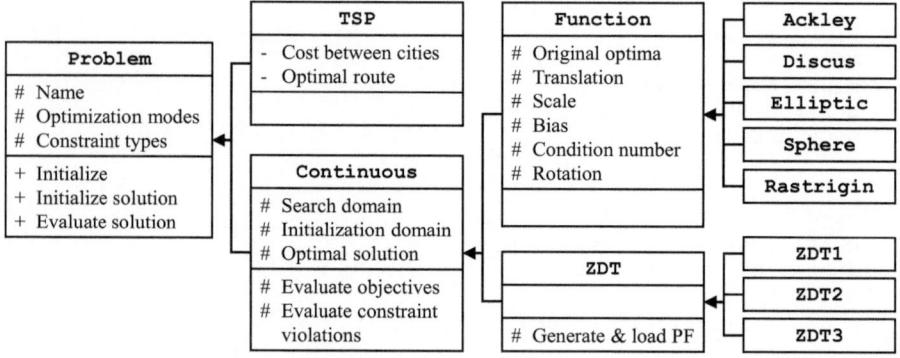

Fig. B.1 Class diagram of problems

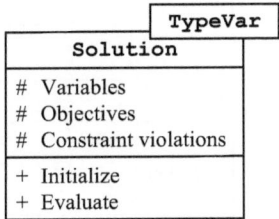

Fig. B.2 Class diagram of solution

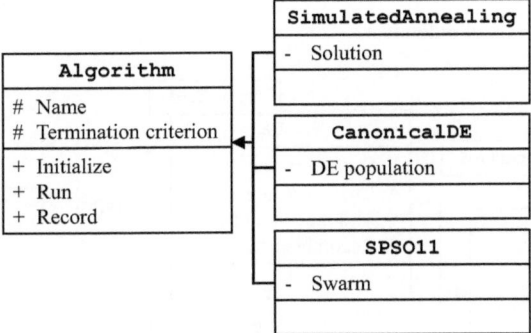

Fig. B.3 Class diagram of algorithm

frequently transformed and used by composition problem have a common base class Function, which is also derived from class Continuous.

As shown in Fig. B.2, the template class Solution consists of decision variables, objective values, and constraint violation. The encoding of the decision variables is determined by the template parameter. It can call the bound problem instance to initialize its decision variables or evaluate its objectives and constraint violation.

B.3.2   Algorithm Component

As shown in Fig. B.3, all algorithm objects in OFEC must be defined as classes inherited from the base class Algorithm. A single solution Algorithm such as simulated annealing should own a solution class as data member, while population-based EAs should own the corresponding population class as data member.

Derived from the template class Solution, the template class Individual add some properties given in the population, which is shown in Fig. B.4.

As shown in Fig. B.5, OFEC also defines the population as a template class, with the template parameters being the type of the individual, which must be classes inherited from the template class Individual.

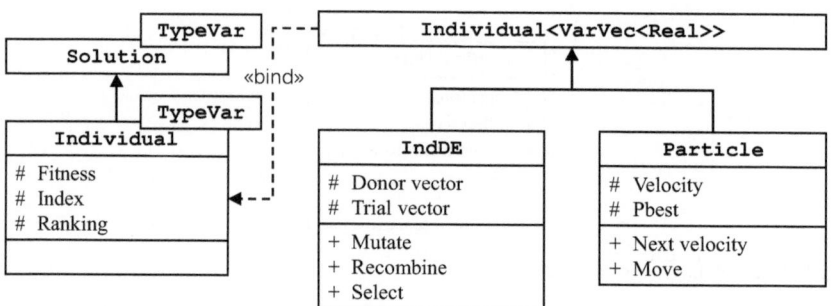

Fig. B.4 Class diagram of individual

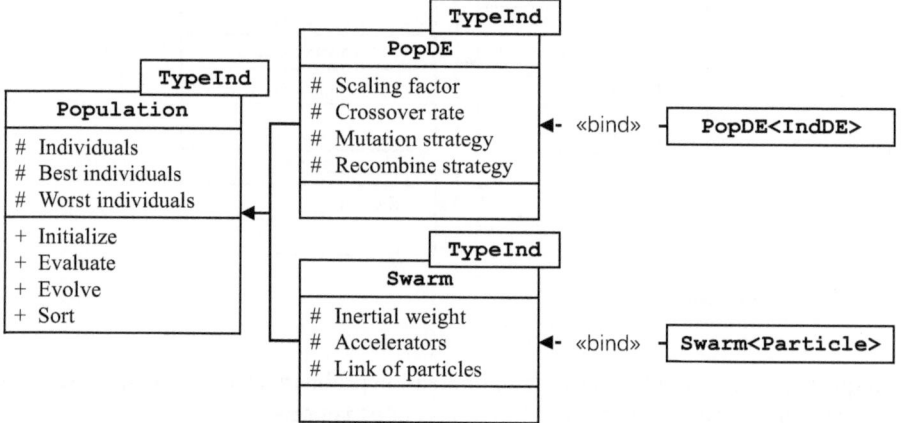

Fig. B.5 Class diagram of population in OFEC

B.4 Implemented Algorithms and Problems

Algorithms reimplemented in OFEC are as follows:

(1) Global optimization algorithms.

- Self-adaptive genetic algorithms with simulated binary crossover (GA-SBX) [5].
- Differential evolution (DE) [27].
- Paticle swarm optimiaztion (PSO) [2].
- Covariance matrix adaptation evolution strategy (CMA-ES) [16].

(2) Multimodal optimization algorithms.

- Differential evolution with neighborhood mutation (NCDE, NSDE) [26].
- Locally informed particle swarm model (LIPS) [25].
- Automatic niching differential evolution (ANDE) [29].

 - Crowding-based differential evolution (CrDE), defined in the 2004 IEEE congress on evolutionary computation.
 - Niching the CMA-ES via nearest-better clustering (NEA2), defined in the proceedings of the 12th annual conference companion on Genetic and evolutionary computation.

(3) Multi-objective optimization algorithms.

 - A fast and elitist multiobjective genetic algorithm (NSGA-II) [7].
 - An evolutionary many-objective optimization algorithm using reference-point-based nondominated sorting approach (NSGA-III) [6].
 - A multi-objective evolutionary algorithm based on decomposition (MOEA/D) [21].
 - A grid-based evolutionary algorithm for many objective optimization (GrEA) [31].
 - A constrained decomposition approach with grids for evolutionary multiobjective optimization (CDG-MOEA) [3].

(4) Dynamic optimization algorithms.

 - Self-adapting multi-swarm [1].
 - A particle swarm model using speciation (SPSO) [24].
 - A clustering particle swarm optimizer (CPSO) [30].
 - A differential evolution for dynamic optimization problems (DynDE), defined in the proceedings of the IEEE congress on evolutionary computation.

(5) Dynamic multi-objective optimization algorithms.

 - Dynamic multi-objective optimization and decision-making using modified NSGA-II-a case study on hydro-themal power scheduling (DNSGA-II).
 - Population prediction strategy (PPS) [33].
 - Kalman-filter prediction (KFP) [23].
 - Steady-state and Generational Evolutionary Algorithm (SGEA) [20].

(6) Constrained optimization algorithms.

 - A general framework of dynamic constrained multi-objective evolutionary algorithms for constrained optimization (DCNSGA-II-DE) [32].

(7) Discrete optimization algorithms.

 - Ant system (AS) [9].
 - Ant colony system (ACS) [10].
 - Max–min ant system (MMAS) [28].
 - A novel set-based particle swarm optimization method for discrete optimization problems (B-PSO) [4].

The platform also covers benchmark functions of a variety of mainstream optimization problems, including:

(1) Global optimization problems.

- Classical functions.
- Black-box opitimization benchmark (BBOB) [15].
- Problem definitions and evaluation criteria for the CEC 2005 special session on real-parameter optimization (CEC2005).
- Problem definitions and evaluation criteria for the CEC2015 competition on learning-based real-parameter single objective optimization (CEC2015).

(2) Multimodal optimization problems.

- Classical functions.
- Benchmark functions for CEC2013 special session and competition on niching methods for multimodal function optimization (CEC2013).
- Novel benchmark functions for continuous multimodal optimization with comparative results (CEC2015).

(3) Multi-objective optimization problems.

- Scalable test problems for evolutionary multiobjective optimization (DTLZ) [8].
- Comparison of multi-objective evolutionary algorithms: Empirical results (ZDT) [34].
- A scalable multi-objective test problem toolkit (WFG) [18].
- A multi-objective evolutionary algorithm using dynamic weight design method (GLT) [14].

(4) Dynamic optimization problems.

- Moving peak benchmark (MPB), defined in the proceedings of the 1999 congress on evolutionary computation.
- Benchmark Generator for CEC'2012 Competition on Evolutionary Computation for Dynamic Optimization Problems (DCBG and DRPBG).

(5) Dynamic multi-objective optimization problems.

- Dynamic multi-objective optimization problems: test cases, approximations, and applications (FDA) [12].
- A competitive-cooperative coevolutionary paradigm for dynamic multiobjective optimization (dMOPs) [13].
- Benchmarks for dynamic multi-objective optimization algorithms (HE) [17].
- Evolutionary dynamic multiobjective optimization: benchmarks and algorithm comparisons (JY) [19].
- Benchmark problems for CEC2018 competition on dynamic multi-objective optimization (CEC2018).

(6) Constrained optimization problems.

   – Problem definitions and evaluation criteria for the CEC 2017 competition
     and special session on constrained single-objective real-parameter optimiza-
     tion (CEC2017).

(7) Large-scale optimization problems.

   – Benchmark functions for the CEC'2013 special session and competition on
     large-scale global optimization (CEC2013).

(8) Expensive optimization problems.

   – Problem definitions and evaluation criteria for the CEC 2014 special session
     and competition on single-objective real-parameter numerical optimization
     (CEC2014).
   – Problem definitions and evaluation criteria for CEC 2015 special session
     on bound constrained single-objective computationally expensive numerical
     optimization (CEC2015).

(9) Discrete optimization problems.

   – One max problem, defined in the proceedings of the fourth international
     conference on genetic algorithms.
   – Multimensional Knapsack Problem (MKP), defined in the proceedings of the
     IEEE CEC1999.
   – Quadratic Assignment Problem (QAP) [22].
   – Traveling Salesman Problem (TSP) [11].

References

1. Blackwell, T.: Particle swarm optimization in dynamic environments. In: Yang, S., Ong, Y.S., Jin, Y. (eds.) Evolutionary Computation in Dynamic and Uncertain Environments, pp. 29–49. Springer, Berlin (2007)
2. Bratton, D., Kennedy, J.: Defining a standard for particle swarm optimization. In: 2007 IEEE Swarm Intelligence Symposium, pp. 120–127. IEEE Computational Intelligence Society. IEEE, Honolulu (2007)
3. Cai, X., Mei, Z., Fan, Z., Zhang, Q.: A constrained decomposition approach with grids for evolutionary multiobjective optimization. IEEE Trans. Evol. Comput. 22(4), 564–577 (2018)
4. Chen, W.N., Zhang, J., Chung, H.S.H., Zhong, W.L., Wu, W.G., Shi, Y.h.: A novel set-based particle swarm optimization method for discrete optimization problems. IEEE Trans. Evol. Comput. 14(2), 278–300 (2010)
5. Deb, K., Beyer, H.: Self-adaptive genetic algorithms with simulated binary crossover. Electron. Commerce 9(2), 197–221 (2001)
6. Deb, K., Jain, H.: An evolutionary many-objective optimization algorithm using reference-point-based nondominated sorting approach, part I: solving problems with box constraints. IEEE Trans. Evol. Comput. 18(4), 577–601 (2014)
7. Deb, K., Pratap, A., Agarwal, S., Meyarivan, T.: A fast and elitist multiobjective genetic algorithm: NSGA-II. IEEE Trans. Evol. Comput. 6(2), 182–197 (2002)

8. Deb, K., Thiele, L., Laumanns, M., Zitzler, E.: Scalable test problems for evolutionary multiobjective optimization. In: Abraham, A., Jain, L., Goldberg, R. (eds.) Evolutionary Multiobjective Optimization, pp. 105–145. Springer, London (2005)
9. Dorigo, M.: Optimization, learning and natural algorithms. Ph.D. Thesis, Politecnico di Milano, Milan (1992)
10. Dorigo, M., Gambardella, L.M.: Ant colony system: a cooperative learning approach to the traveling salesman problem. IEEE Trans. Evol. Comput. 1(1), 53–66 (1997)
11. Dorigo, M., Maniezzo, V., Colorni, A.: Positive feedback as a search strategy. Tech. rep., Dipartimento di Elettronica, Politecnico di Milano, Italy (1991)
12. Farina, M., Deb, K., Amato, P.: Dynamic multiobjective optimization problems: test cases, approximations, and applications. IEEE Trans. Evol. Comput. 8(5), 425–442 (2004)
13. Goh, C.K., Tan, K.C.: A competitive-cooperative coevolutionary paradigm for dynamic multiobjective optimization. IEEE Trans. Evol. Comput. 13(1), 103–127 (2008)
14. Gu, F., Liu, H.L., Tan, K.C.: A multiobjective evolutionary algorithm using dynamic weight design method. Int. J. Innov. Comput. Inform. Control 8(5(B)), 3677–3688 (2012)
15. Hansen, N., Finck, S., Ros, R., Auger, A.: Real-parameter black-box optimization benchmarking 2009: noiseless functions definitions. Tech. rep., INRIA, Paris (2009)
16. Hansen, N., Ostermeier, A.: Completely derandomized self-adaptation in evolution strategies. Evol. Comput. 9(2), 159–195 (2001)
17. Helbig, M., Engelbrecht, A.P.: Benchmarks for dynamic multi-objective optimisation algorithms. ACM Comput. Surv. 46(3), 1–39 (2014)
18. Huband, S., Barone, L., While, L., Hingston, P.: A scalable multi-objective test problem toolkit. In: International Conference on Evolutionary Multi-Criterion Optimization, pp. 280–295. Springer, Berlin (2005)
19. Jiang, S., Yang, S.: Evolutionary dynamic multiobjective optimization: Benchmarks and algorithm comparisons. IEEE Trans. Cybern. 47(1), 198–211 (2016)
20. Jiang, S., Yang, S.: A steady-state and generational evolutionary algorithm for dynamic multiobjective optimization. IEEE Trans. Evol. Comput. 21(1), 65–82 (2017)
21. Li, H., Zhang, Q.: Multiobjective optimization problems with complicated pareto sets, MOEA/D and NSGA-II. IEEE Trans. Evol. Comput. 13(2), 284–302 (2009)
22. Maniezzo, V.: Exact and approximate nondeterministic tree-search procedures for the quadratic assignment problem. Informs J. Comput. 11(4), 358–369 (1999)
23. Muruganantham, A., Tan, K.C., Vadakkepat, P.: Evolutionary dynamic multiobjective optimization via kalman filter prediction. IEEE Trans. Cybern.46(12), 2862–2873 (2016)
24. Parrott, D., Li, X.: Locating and tracking multiple dynamic optima by a particle swarm model using speciation. IEEE Trans. Evol. Comput. 10(4), 440–458 (2006)
25. Qu, B.Y., Suganthan, P.N., Das, S.: A distance-based locally informed particle swarm model for multimodal optimization. IEEE Trans. Evol. Comput. 17(3), 387–402 (2013)
26. Qu, B.Y., Suganthan, P.N., Liang, J.J.: Differential evolution with neighborhood mutation for multimodal optimization. IEEE Trans. Evol. Comput. 16(5), 601–614 (2012)
27. Storn, R., Price, K.: Differential evolution–a simple and efficient heuristic for global optimization over continuous spaces. J. Global Optim. 11(4), 341–359 (1997)
28. Stützle, T., Hoos, H.: Improving the ant system: A detailed report on the max-min ant system. Tech. rep., FG Intellektik, FB Informatik, TU Darmstadt, Germany (1996)
29. Wang, Z.J., Zhan, Z.H., Lin, Y., Yu, W.J., Wang, H., Kwong, S., Zhang, J.: Automatic niching differential evolution with contour prediction approach for multimodal optimization problems. IEEE Trans. Evol. Comput. 24(1), 114–128 (2020)
30. Yang, S., Li, C.: A clustering particle swarm optimizer for locating and tracking multiple optima in dynamic environments. IEEE Trans. Evol. Comput. 14(6), 959–974 (2010)
31. Yang, S., Li, M., Liu, X., Zheng, J.: A grid based evolutionary algorithm for many objective optimization. IEEE Trans. Evol. Comput. 17(5), 721–736 (2013)

32. Zeng, S., Jiao, R., Li, C., Li, X., Alkasassbeh, J.S.: A general framework of dynamic constrained multiobjective evolutionary algorithms for constrained optimization. IEEE Trans. Cybern. **47**(9), 2678–2688 (2017)
33. Zhou, A., Jin, Y., Zhang, Q.: A population prediction strategy for evolutionary dynamic multiobjective optimization. IEEE Trans. Cybern. **44**(1), 40–53 (2014)
34. Zitzler, E., Deb, K., Thiele, L.: Comparison of multiobjective evolutionary algorithms: Empirical results. Evol. Comput. **8**(2), 173–195 (2000)

Index

© China University of Geosciences Press 2024
C. Li et al., *Intelligent Optimization*,
https://doi.org/10.1007/978-981-97-3286-9